T0214076

# Lecture Notes in Computer Science 12692

More information about this subseries at http://www.springer.com/series/7407

Christine Zarges · Sébastien Verel (Eds.)

# Evolutionary Computation in Combinatorial Optimization

21st European Conference, EvoCOP 2021
Held as Part of EvoStar 2021
Virtual Event, April 7–9, 2021
Proceedings

 Springer

*Editors*
Christine Zarges
Aberystwyth University
Aberystwyth, UK

Sébastien Verel
Université du Littoral Côte d'Opale
Calais, France

ISSN 0302-9743 ISSN 1611-3349 (electronic)
Lecture Notes in Computer Science
ISBN 978-3-030-72903-5 ISBN 978-3-030-72904-2 (eBook)
https://doi.org/10.1007/978-3-030-72904-2

LNCS Sublibrary: SL1 – Theoretical Computer Science and General Issues

This Springer imprint is published by the registered company Springer Nature Switzerland AG
The registered company address is: Gewerbestrasse 11, 6330 Cham, Switzerland

# Preface

Combinatorial optimization is one of the main research areas in the context of algorithmic problem solving. A combinatorial optimization problem is solved when one or several solutions optimizing one or several objective functions are found from a large finite set of candidate solutions. This ubiquitous approach is successful in many real-world applications such as applications in the domains of transportation, energy, planning, resource management, scheduling, system design, and many more. With new challenges from these domains, the complexity and size of combinatorial optimization problems increase, and therefore, new research questions arise in order to understand and design more efficient search algorithms. Evolutionary algorithms and related methods, which can be more or less bio-inspired, are intuitive, flexible, and powerful approaches able to solve such problems. They are also capable of dealing with the trade-off between solution quality and running time, which makes such approaches very attractive. Thus, the task for the scientific community and practitioners is to develop efficient evolutionary algorithms able to reach high-quality solutions within the available computation resources according to the properties of the combinatorial problem under consideration. The following papers show the most recent theoretical and experimental research in this area.

This volume contains the proceedings of EvoCOP 2021, the 21th European Conference on Evolutionary Computation in Combinatorial Optimisation. Originally planned as a hybrid conference with on-site participation in Seville, Spain, the conference was later turned into an online-only event due to the COVID-19 pandemic.

The EvoCOP conference series started in 2001, with the first workshop specifically devoted to evolutionary computation in combinatorial optimization. It became an annual conference in 2004. EvoCOP 2021 was organized together with EuroGP (the 24th European Conference on Genetic Programming), EvoMUSART (the 10th International Conference on Artificial Intelligence in Music, Sound, Art and Design), and EvoApplications (the 24th International Conference on the Applications of Evolutionary Computation, formerly known as EvoWorkshops), in a joint event collectively known as EvoStar 2021. Previous EvoCOP proceedings were published by Springer in the *Lecture Notes in Computer Science* series (LNCS volumes 2037, 2279, 2611, 3004, 3448, 3906, 4446, 4972, 5482, 6022, 6622, 7245, 7832, 8600, 9026, 9595, 10197, 10782, 11452, and 12102). The table on the next page reports the statistics for each of the previous conferences.

This year, 14 out of 42 papers were accepted after a rigorous double-blind reviewing process, resulting in a 33% acceptance rate. We would like to acknowledge the high-quality and timely work of our diverse Program Committee members, who each year donate their time and expertise to maintain the high standards of EvoCOP and provide constructive feedback to help authors improve their papers. Decisions considered both the reviewers' report and the evaluation of the program chairs. The 14 accepted papers cover a wide spectrum of topics, ranging from the foundations of evolutionary algorithms and other search heuristics to their accurate design and application to combinatorial optimization problems. Fundamental and methodological aspects deal with runtime analysis, the structural properties of fitness landscapes, the study of core components of metaheuristics, the clever design of their search principles, and their careful selection and configuration. Applications cover problem domains such as scheduling, routing, search-based software engineering, and general graph problems. We believe that the range of topics covered in this volume of EvoCOP proceedings reflects the current state of research in the fields of evolutionary computation and combinatorial optimization.

| EvoCOP | LNCS vol. | Submitted | Accepted | Acceptance (%) |
|---|---|---|---|---|
| 2021 | 12692 | 42 | 14 | 33.3 |
| 2020 | 12102 | 37 | 14 | 37.8 |
| 2019 | 11452 | 37 | 14 | 37.8 |
| 2018 | 10782 | 37 | 12 | 32.4 |
| 2017 | 10197 | 39 | 16 | 41.0 |
| 2016 | 9595 | 44 | 17 | 38.6 |
| 2015 | 9026 | 46 | 19 | 41.3 |
| 2014 | 8600 | 42 | 20 | 47.6 |
| 2013 | 7832 | 50 | 23 | 46.0 |
| 2012 | 7245 | 48 | 22 | 45.8 |
| 2011 | 6622 | 42 | 22 | 52.4 |
| 2010 | 6022 | 69 | 24 | 34.8 |
| 2009 | 5482 | 53 | 21 | 39.6 |
| 2008 | 4972 | 69 | 24 | 34.8 |
| 2007 | 4446 | 81 | 21 | 25.9 |
| 2006 | 3906 | 77 | 24 | 31.2 |
| 2005 | 3448 | 66 | 24 | 36.4 |
| 2004 | 3004 | 86 | 23 | 26.7 |
| 2003 | 2611 | 39 | 19 | 48.7 |
| 2002 | 2279 | 32 | 18 | 56.3 |
| 2001 | 2037 | 31 | 23 | 74.2 |

We would like to express our appreciation to the various people and institutions making EvoCOP 2021 a successful event. First, we thank SPECIES, the Society for the Promotion of Evolutionary Computation in Europe and its Surroundings, which aims to promote evolutionary algorithmic thinking within Europe and wider, and more generally to promote inspiration of parallel algorithms derived from natural processes. We extend our acknowledgments to Nuno Lourenço from the University of Coimbra, Portugal, for his dedicated work with the submission and registration system, to João Correia from the University of Coimbra, Portugal, and Francisco Chicano from the University of Málaga, Spain, for EvoStar publicity, social media, and website, and to Sérgio Rebelo from the University of Coimbra, Portugal, for his important graphic design work. We wish to thank our prominent keynote speakers, Darrell Whitley from Colorado State University, USA, and Susanna Manrubia from the Spanish National

Centre for Biotechnology (CSIC), Madrid, Spain. Finally, we express our continued appreciation to Anna I. Esparcia-Alcázar from SPECIES, Europe, whose considerable efforts in managing and coordinating EvoStar helped towards building a unique, vibrant, and friendly atmosphere.

Special thanks go to the members of the EvoCOP Steering Committee (Christian Blum, Francisco Chicano, Carlos Cotta, Peter Cowling, Jens Gottlieb, Jin-Kao Hao, Jano van Hemert, Bin Hu, Arnaud Liefooghe, Manuel Lopéz-Ibáñez, Peter Merz, Martin Middendorf, Gabriela Ochoa, Luís Paquete, and Günther Raidl) for their hard work at and dedication to past editions of EvoCOP, making it one of the reference international events in evolutionary computation and metaheuristics for combinatorial optimization.

April 2021

Christine Zarges
Sébastien Verel

# Organization

EvoCOP 2021 was organized as a part of EvoStar 2021, jointly with EuroGP 2021, EvoMUSART 2021, and EvoApplications 2021.

## Conference Chairs

Christine Zarges      Aberystwyth University, UK
Sébastien Verel      University of the Littoral Opal Coast, France

## Publicity and e-Media Chair

João Correia      University of Coimbra, Portugal

## Web Chair

Francisco Chicano      University of Málaga, Spain

## Submission and Registration Manager

Nuno Lourenço      University of Coimbra, Portugal

## EvoCOP Steering Committee

Christian Blum      Artificial Intelligence Research Institute, Spain
Francisco Chicano      University of Málaga, Spain
Carlos Cotta      University of Málaga, Spain
Peter Cowling      Queen Mary University of London, UK
Jens Gottlieb      SAP SE, Germany
Jin-Kao Hao      University of Angers, France
Jano van Hemert      Optos, UK
Bin Hu      Austrian Institute of Technology, Austria
Arnaud Liefooghe      University of Lille, France
Manuel Lopéz-Ibáñez      University of Málaga, Spain
Peter Merz      Hannover University of Applied Sciences and Arts, Germany
Martin Middendorf      University of Leipzig, Germany
Gabriela Ochoa      University of Stirling, UK
Luís Paquete      University of Coimbra, Portugal
Günther Raidl      Vienna University of Technology, Austria

## Society for the Promotion of Evolutionary Computation in Europe and its Surroundings (SPECIES)

| | |
|---|---|
| Marc Schoenauer (President) | Inria Saclay - Île-de-France, France |
| Anna I. Esparcia-Alcázar (Vice-president, Secretary & EvoStar Coordinator) | SPECIES, Europe |
| Wolfgang Banzhaf (Treasurer) | Michigan State University, USA |

## Program Committee

| | |
|---|---|
| Khulood Alyahya | University of Exeter, UK |
| Soumen Atta | Masaryk University, Czech Republic |
| Matthieu Basseur | University of Angers, France |
| Alexander Brownlee | University of Stirling, UK |
| Maxim Buzdalov | ITMO University, Russia |
| Pedro Castillo | University of Granada, Spain |
| Josu Ceberio | University of the Basque Country, Spain |
| Francisco Chicano | University of Málaga, Spain |
| Carlos Coello Coello | CINVESTAV-IPN, Mexico |
| Carlos Cotta | University of Málaga, Spain |
| Bilel Derbel | University of Lille, France |
| Benjamin Doerr | École Polytechnique, France |
| Carola Doerr | CNRS and Sorbonne University, France |
| Paola Festa | University of Napoli Federico II, Italy |
| Bernd Freisleben | University of Marburg, Germany |
| Carlos García-Martínez | University of Córdoba, Spain |
| Adrien Goëffon | University of Angers, France |
| Andreia Guerreiro | University of Coimbra, Portugal |
| Jin-Kao Hao | University of Angers, France |
| Bin Hu | Austrian Institute of Technology, Austria |
| Thomas Jansen | Aberystwyth University, UK |
| Andrzej Jaszkiewicz | Poznań University of Technology, Poland |
| Ayush Joshi | University of Bath, UK |
| Ahmed Kheiri | Lancaster University, UK |
| Mario Köppen | Kyushu Institute of Technology, Japan |
| Frédéric Lardeux | University of Angers, France |
| Rhydian Lewis | Cardiff University, UK |
| Arnaud Liefooghe | University of Lille, France |
| Manuel López-Ibáñez | University of Málaga, Spain |
| Jose A. Lozano | University of the Basque Country, Spain |
| Gabriel Luque | University of Málaga, Spain |

# Contents

# A Novel Ant Colony Optimization Strategy for the Quantum Circuit Compilation Problem

Marco Baioletti[1]([✉]) [ID], Riccardo Rasconi[2] [ID], and Angelo Oddi[2] [ID]

[1] University of Perugia, Perugia, Italy
marco.baioletti@unipg.it
[2] Institute of Cognitive Sciences and Technologies (ISTC-CNR), Rome, Italy
{riccardo.rasconi,angelo.oddi}@istc.cnr.it

**Abstract.** Quantum Computing represents the most promising technology towards speed boost in computation, opening the possibility of major breakthroughs in several disciplines including Artificial Intelligence. This paper investigates the performance of a novel Ant Colony Optimization (ACO) algorithm for the realization (compilation) of nearest-neighbor compliant quantum circuits of minimum duration. In fact, current technological limitations (e.g., decoherence effect) impose that the overall duration (makespan) of the quantum circuit realization be minimized, and therefore the production of minimum-makespan compiled circuits for present and future quantum machines is of paramount importance. In our ACO algorithm (QCC-ACO), we introduce a novel pheromone model, and we leverage a heuristic-based Priority Rule to control the iterative selection of the quantum gates to be inserted in the solution.

The proposed QCC-ACO algorithm has been tested on a set of quantum circuit benchmark instances of increasing sizes available from the recent literature. We demonstrate that the QCC-ACO obtains results that outperform the current best solutions in the literature against the same benchmark, succeeding in significantly improving the makespan values for a great number of instances and demonstrating the scalability of the approach.

**Keywords:** Ant Colony Optimization · Quantum circuit compilation · Planning · Scheduling

## 1 Introduction

Quantum Computing explores the implications of using quantum mechanics to model information and its processing. The impact of quantum computing technology on theoretical/applicative aspects of computation as well as on the society in the next decades is considered to be immensely beneficial [16]. While classical computing revolves around the execution of logical gates based on two-valued *bits*, quantum computing uses *quantum gates* that manipulate multi-valued bits

C. Zarges and S. Verel (Eds.): EvoCOP 2021, LNCS 12692, pp. 1–16, 2021.
https://doi.org/10.1007/978-3-030-72904-2_1

(*qubits*) that can represent as many logical states (*qstates*) as are the obtainable linear combinations of a set of basis states (state *superpositions*). A quantum circuit is composed of a number of qubits and by a series of quantum gates that operate on those qubits, and whose execution realizes a specific quantum algorithm.

Executing a quantum circuit entails the chronological evaluation of each gate and the modification of the involved qstates according to the gate logic. Current quantum computing technologies like ion-traps, quantum dots, super-conducting qubits, etc. limit the qubit interaction distance to the extent of allowing the execution of gates between adjacent (i.e., *nearest-neighbor*) qubits only [6,13,23]. This has opened the way to the exploration of possible techniques and/or heuristics aimed at guaranteeing nearest-neighbor (NN) compliance in any quantum circuit through the addition of a number of so-called *swap* gates between adjacent qubits. The effect of a swap gate is to mutually exchange the qstates of the involved qubits, thus allowing the execution of the gates that require those qstates to rest on adjacent qubits. However, adding swap gates also introduces a time overhead in the circuit execution [4] that generally depends on the quantum hardware's topology; on the other hand, it is highly desirable to minimize the circuit's execution time (i.e., *makespan*), in order to mitigate the negative effects of *decoherence* and guarantee more stability to the quantum computation. The Quantum Circuit Compilation Problem (QCCP) can be described as the synthesis of a real quantum circuit to be executed on a specific quantum hardware.

The QCCP benchmarks used in this work[1] have been initially introduced and solved in [22] as a temporal planning problem. Subsequently, the same benchmark was tackled in [3,17], respectively through a hybrid approach that integrates Temporal Planning with Constraint Programming, and a heuristically-based Greedy Randomized Search (GRS) technique [12,19]. The results obtained in [17] have been further improved in [18] by means of a genetic algorithm. More recently, a similar version of the priority rule introduced in [17] has been used in [5] within a *rollout* procedure to further improve the best results, thus representing the benchmark's current bests.

In this work, we use a slightly modified version of the priority rule proposed in [5] within an Ant Colony Optimization (ACO) algorithm [9], and compare our results with those obtained in the same paper, significantly improving the makespan values for a considerable number of instances, and demonstrating the scalability of the approach.

ACO is a powerful metaheuristic, inspired by the foraging behaviour of colonies of ants that has been applied to many combinatorial optimization problems [9], and particularly to scheduling and permutation problems [2,14]; indeed, the QCC problem tackled here has in fact a strong scheduling component.

---

[1] A set of benchmark instances of different size belonging to the *Quantum Approximate Optimization Algorithm* (QAOA) class [10,11] tailored for the MaxCut problem and devised to be executed on top of a hardware architecture proposed by Rigetti Computing Inc. [20].

Among the evolutionary and swarm intelligence based algorithms, ACO seems to be well suited for the QCC problem because of its constructive nature. In fact, feasible solutions in ACO are built by means of an iterative process that starts from an empty solution and adds a component at a time consistently with the problem constraints, hence maintaining the feasibility of the solution at all times during the building process.

The paper is organized as follows. The next section proposes a formal statement of the tackled problem, whereas the subsequent section describes the novel ant colony optimization strategy proposed for its resolution. Finally, an empirical evaluation based on the benchmark proposed in [22] is performed, and some conclusions end the paper.

## 2   The QCC Problem

Formally, the Quantum Circuit Compilation Problem (QCCP) [17] is a tuple $P = \langle C_0, L_0, QM \rangle$, where: (i) $C_0$ is the input quantum circuit representing the execution of the algorithm of interest, (ii) $L_0$ is the initial assignment of qubits to qstates, and (iii) $QM$ is a representation of the quantum hardware.

- $C_0$ is the input quantum circuit expressed as a tuple $\langle Q, P\text{-}S, MIX, P, TC_0 \rangle$, where: (i) $Q = \{q_1, q_2, \ldots, q_N\}$ is the set of qstates which, from a planning & scheduling perspective represent the *resources* necessary for each gate's execution (see for example [15], Chap. 15); (ii) $P\text{-}S$ and $MIX$ respectively represent the set of *p-s* and *mix* gates that have to be scheduled, such that every $p\text{-}s(q_i, q_j)$ gate requires two qstates for execution, and every $mix(q_i)$ gate requires one qstate only; (iii) $P = \{1, ..., p\}$ is the number of times (i.e., *passes*) the gates in the $P\text{-}S$ and $MIX$ must be executed; (iv) $TC_0$ is a set of simple precedence constraints imposed on the $P\text{-}S$ and $MIX$ such that: (1) all the *p-s* gates belonging to the $k$-th pass ($P\text{-}S_k$) that involve a specific qstate $q_i$ must be executed before all the *mix* gates belonging to the same pass ($MIX_k$) that involve the same qstate $q_i$, for $k = 1, 2, \ldots, p$, and (2) all the *mix* gates belonging to the $k$-th pass ($MIX_k$) that involve a specific qstate $q_i$ must be executed before all the *p-s* gates belonging to the $(k+1)$-th pass ($P\text{-}S_{k+1}$) that involve the same qstate $q_i$, for $k = 1, 2, \ldots, (p-1)$. Lastly, the execution of every quantum gate requires the uninterrupted use of the involved qstates during its processing time, and each qstate $q_i$ can process at most one quantum gate at a time.
- $L_0$ is the initial assignment at the time origin $t = 0$ of qstates $q_i$ to qubits $n_i$. In this work, the $i$-th qstate $q_i$ is assigned to the $i$-th qubit $n_i$.
- $QM$ is a representation of the quantum hardware as an undirected multi-graph $QM = \langle V_N, E_{p\text{-}s}, E_{swap}, \tau_{mix}, \tau_{p\text{-}s}, \tau_{swap} \rangle$, where $V_N = \{n_1, n_2, \ldots, n_N\}$ is the set of qubits (nodes), $E_{p\text{-}s}$ ($E_{swap}$) is a set of undirected edges $(n_i, n_j)$ representing the set of *adjacent* locations the qstates $q_i$ and $q_j$ of the gates $p\text{-}s(q_i, q_j)$ ($swap(q_i, q_j)$) can potentially be allocated to. In addition, the *labelling* functions $\tau_{p\text{-}s} : E_{p\text{-}s} \to \mathbb{Z}^+$ and $\tau_{swap} : E_{swap} \to \mathbb{Z}^+$ respectively represent the durations of the gate operations $p\text{-}s(q_i, q_j)$ and

$swap(q_i, q_j)$ when the qstates $q_i$ and $q_j$ are assigned to the corresponding adjacent locations. Similarly, the *labelling* function $\tau_{mix} : V \to \mathbb{Z}^+$ represents the durations of the *mix* gate (which can be executed at any node $n_i$). Figure 1 shows an example of quantum hardware designs, with gate durations.

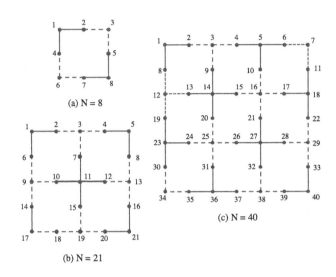

**Fig. 1.** Three quantum chip designs characterized by an increasing number of qubits ($N = 8, 21, 40$) inspired by Rigetti Computing Inc. Every qubit is located at a different location (node), and the integers at each node represent the qubit's identifier. Two qubits connected by an edge are adjacent, and each edge represents a 2-qubit gate (*p-s* or *swap*) that can be executed between those qubits. *p-s* gates executed on continuous edges have duration $\tau_{p\text{-}s} = 3$, while *p-s* gates executed on dashed edges have duration $\tau_{p\text{-}s} = 4$. Swap gates have duration $\tau_{swap} = 2$.

A feasible solution is a tuple $S = \langle SWAP, TC \rangle$, which extends the initial circuit $C_0$ to a circuit $C_S = \langle Q, P\text{-}S, MIX, SWAP, P, TC_S \rangle$, such that $TC_S = TC_0 \cup TC$, where $SWAP$ is a set of additional $swap(q_i, q_j)$ gates added to guarantee the adjacency constraints for the set of $P\text{-}S$ gates, and $TC$ is a set of additional simple precedence constraints such that: (i) for each qstate $q_i$, a total order $\preceq_i$ is imposed among the set $Q_i$ of operations requiring $q_i$, with $Q_i = \{op \in P\text{-}S \cup MIX \cup SWAP : op\ requires\ q_i\}$; (ii) all the $p\text{-}s(q_i, q_j)$ and $swap(q_i, q_j)$ gate operations are allocated on adjacent qubits in $QM$, and (iii) the graph $\langle \{P\text{-}S \cup MIX \cup SWAP\}, TC_S \rangle$ does not contain cycles.

Given a solution $S$, the makespan $mk(S)$ corresponds to the maximum completion time of the gate operations in $S$. An optimal solution $S^*$ is a feasible solution characterized by the minimum makespan.

# 3   A Novel Ant Colony Optimization Strategy

The solution pursued in this work to solve the QCCP exploits the Ant Colony Optimization (ACO) paradigm [9]. The proposed algorithm is called QCC-ACO and its main schema is depicted in Algorithm 1. The algorithm handles a colony of $n_a$ artificial ants, which indirectly communicate through the *pheromone* model and create solutions of the QCC problem. The main loop is repeated for a certain number of times, according to a given stop criterion, for instance, a computation time budget. At each iteration, every artificial ant builds a solution by means of a constructive procedure (BUILDSOLUTION()); then, pheromone values are updated (UPDATEPHEROMONEVALUES()) in order to select the best solutions.

---

**Algorithm 1.** QCC-ACO

---

**Require:** A problem $P = \langle C_0, L_0, QM \rangle$
  INITIALIZEPHEROMONEVALUES()
  **while not** termination criterion **do**
    **for** $i \leftarrow 1$ **to** $n_a$ **do**
      BUILDSOLUTION(P)
    **end for**
    UPDATEPHEROMONEVALUES()
  **end while**
  **return** BestSolution

---

---

**Algorithm 2.** BUILDSOLUTION

---

**Require:** A problem $P = \langle C_0, L_0, QM \rangle$
  $S \leftarrow$ INITSOLUTION$(P)$
  $t \leftarrow 0$
  **while** not all the $P\text{-}S$ and $MIX$ operations are inserted in $S$ **do**
    $op \leftarrow$ SELECTEXECUTABLEOPERATION$(P, S, t)$
    **if** $op \neq NULL$ **then**
      $S \leftarrow$ INSERTOPERATION$(op, S, t)$
    **else**
      $t \leftarrow t + 1$
    **end if**
  **end while**
  **return** $S$

---

In the next two subsections some details about the solution constructive procedure and the pheromone update strategy will be respectively provided.

## 3.1   Solution Construction Algorithm

The solution construction procedure used in QCC-ACO is described in Algorithm 2. The algorithm produces a complete solution $S$ for the given QCC problem $P$ by starting from an empty solution and by iteratively selecting (SELECTEXECUTABLEOPERATION) and adding (INSERTOPERATION) one operation (i.e., a quantum gate) at a time among those that can scheduled at the current instant $t$.

An operation $op$ can be scheduled at $t$ if the qstates required by $op$ are not used in $t$. If no operation can be scheduled at $t$, the time is increased to $t + 1$.

Clearly, one essential step of the BUILDSOLUTION() procedure is the SELECTEXECUTABLEOPERATION() method (shown in Algorithm 3) through which a new operation is selected for insertion in the partial solution, at each iteration. This method is based on a heuristic-based priority rule originally introduced in [17] and slightly modified in [5], whose technical details will be described in the following sections.

## 3.2   Gate Selection Procedure Based on Priority Rules

The priority rule used in our QCC-ACO exploits the distance between qstate pairs measured on the quantum hardware; in the following, we describe in detail the criteria upon which such distance is assessed.

Let $op \in Q_i$ be a general gate operation that involves qstate $q_i$, we define a *chain* $ch_i = \{op \in Q_i : op \in S\}$ as the set of gates involving $q_i$ and currently present in the partial solution $S$, among which a total order is imposed (see Fig. 2 for a graphical representation of a complete solution composed of a set of chains, one for each qstate $q_i$).

Let us also define $last(ch_i)$ as the last operation in the chain $ch_i$ according to the imposed total order and $n(last(ch_i))$ as the $QM$ node at which the last operation in the chain $ch_i$ terminates its execution. Given a partial solution $S$, the *state* $L_S$ is the tuple $L_S = \langle n(last(ch_1)), n(last(ch_2)), \ldots, n(last(ch_N)) \rangle$ of $QM$ locations (nodes) where each last chain operation $last(ch_i)$ terminates its execution.

Given the multi-graph $QM$ introduced in the QCC Problem section, we consider the distance graph $G_d(V, E_{p\text{-}s})$, so as to contain an undirected edge $(n_i, n_j) \in E_{p\text{-}s}$ when $QM$ can execute a $p\text{-}s$ gate on the pair $(n_i, n_j)$. In the graph $G_d$, an undirected path $p_{ij}$ between a node $n_i$ and a node $n_j$ is the list of edges $p_{ij} = ((n_i, n_{j1}), (n_{j1}, n_{j2}), \ldots, (n_{jk}, n_j))$ connecting the two nodes $n_i$ and $n_j$ and its length $l_{ij}$ is the number of edges in the path $p_{ij}$. Let $d_{ij}$ represent the minimal length among the set of all the paths between $n_i$ and $n_j$. The distance $d_{ij}$ between all nodes is computed only once at the beginning, by means of all-pairs shortest path algorithm, whose complexity is $O(|V|^3)$ in the worst case [7]. The distance $d^{L_S}$ associated to a given $p\text{-}s(q_i, q_j)$ gate that requires two qstates $q_i$ and $q_j$ w.r.t. the state $L_S$ of the partial solution $S$ is defined as:

$$d^{L_S}(p\text{-}s(q_i, q_j)) = d(nlast(ch_i), nlast(ch_j)) \tag{1}$$

Two qstates $q_i$ and $q_j$ are in adjacent locations in the state $L^S$ if $d^{L_S}(p\text{-}s(q_i, q_j)) = 1$. Intuitively, given a $p\text{-}s(q_i, q_j)$ gate and a partial solution $S$, the value $d^{L_S}(p\text{-}s(q_i, q_j))$ yields the minimal number of swaps for moving the two qstates $q_i$ and $q_j$ to adjacent locations on the machine $QM$.

The concept of distance defined on a single gate operation $p\text{-}s(q_i, q_j)$ can be extended to a set of gate operations, as follows. Let $S$ be a partial solution, $\overline{P\text{-}S}^S$ is the set of $p\text{-}s(q_i, q_j)$ gates that are not yet scheduled in $S$ and such that all predecessors gates according to the temporal order imposed by the set $TC_0$ (the set of simple precedences in the input circuit $C_0$) have already been scheduled in $S$. The authors in [17] proposed two different functions to measure the distance separating the set $\overline{P\text{-}S}^S$ from the *adjacent state*. The first sums the set of the distances $d^{L_S}(p\text{-}s(q_i, q_j))$:

$$D^S_{sum}(\overline{P\text{-}S}^S) = \sum_{p\text{-}s \in \overline{P\text{-}S}^S} d^{L_S}(p\text{-}s(q_i, q_j)) \tag{2}$$

The second returns the minimal value of the distance $d^{L_S}(p\text{-}s(q_i, q_j))$ in the set $\overline{P\text{-}S}^S$:

$$D^S_{min}(\overline{P\text{-}S}^S) = MIN_{p\text{-}s \in \overline{P\text{-}S}^S} d^{L_S}(p\text{-}s(q_i, q_j)) \tag{3}$$

Given the functions (2) and (3), it is now possible to assess the priority of each gate operation $op$ to possibly insert in the partial solution as follows:

$$f(S, op, \overline{P\text{-}S}^S) = \begin{cases} (D^{S'}_{sum}(\overline{P\text{-}S}^S \setminus \{op\}), 1) & \text{(p-s)} \\ (D^{S'}_{sum}(\overline{P\text{-}S}^{S'}), 1) & \text{(mix)} \\ (D^{S'}_{sum}(\overline{P\text{-}S}^{S'}), D^{S'}_{min}(\overline{P\text{-}S}^{S'})) & \text{(swap)} \end{cases} \tag{4}$$

where $S'$ is the new partial solution after the addition of the selected gate operation $op$. We are now in the position to describe in detail the SELECTEXECUTABLEOPERATION() procedure, which operates over three phases (see Algorithm 3).

In the first phase (*Phase1*), a set $\Omega'$ of operations (PS, MIX or SWAP) that can be time and resource feasibly scheduled at the current instant $t$ is selected through the ELIGIBLESET() procedure (*eligible* operations at time $t$). Subsequently, the values of $D^S_{sum}$ and $D^S_{min}$ of the solution $S$ (baseline values), are computed using formulas 2 and 3, respectively. From this point, a new set $\Omega$ of operations is built by further restricting $\Omega'$, in a fashion inspired to the Priority Rule described in [5]. Namely, $\Omega$ is built by immediately inserting all the eligible P-S and MIX operations previously stored in $\Omega'$ (if any), while the SWAP operations will be considered only if their execution produces a partial solution which is better than the current solution with respect to the $D_{sum}$ heuristic value or, being equal with respect to $D_{sum}$, it is better with respect to $D_{min}$.

The second phase (*Phase2*) is executed only if the first phase returns an empty $\Omega$ set, in which case the algorithm collects all the SWAP operations previously contained in $\Omega'$ whose execution produces a partial solution which is better than the current solution with respect to $D_{min}$ only.

The third phase (*Phase3*) chooses the operation to be returned by the procedure. If $\Omega$ is still empty, SELECTEXECUTABLEOPERATION returns NULL and

---

**Algorithm 3.** SELECTEXECUTABLEOPERATION

---

**Require:** a problem $P$, a partial solution $S$, a time $t$

    *//Phase1:*

    $\Omega' \leftarrow$ ELIGIBLESET$(S, t)$

    $\Omega \leftarrow \emptyset$

    Init $D^S_{sum}$ and $D^S_{min}$) resp. according to (2) and (3)

    **for all** $op \in \Omega'$ **do**

        $(op.D^S_{sum}, op.D^S_{min}) \leftarrow f(S, op, \overline{P\text{-}S^S})$

        **if** ($op$ is a MIX or a PS) **or** $(op.D^S_{sum} < D^S_{sum})$ **or** $(op.D^S_{sum} = D^S_{sum}$ **and**

        $op.D^S_{min} < D^S_{min})$ **then**

            $\Omega \leftarrow \Omega \cup \{op\}$

        **end if**

    **end for**

    *//Phase2:*

    **if** $\Omega = \emptyset$ **then**

        **for all** $op \in \Omega'$ **do**

            **if** ($op$ is a SWAP) **and** $(op.D^S_{min} < D^S_{min})$ **then**

                $\Omega \leftarrow \Omega \cup \{op\}$

            **end if**

        **end for**

    **end if**

    *//Phase3: op selection*

    **if** $\Omega = \emptyset$ **then**

        **return** NULL

    **end if**

    Evaluate $prob(op)$ for all $op \in \Omega$ according to (5)

    **return** $op$ chosen at random with probability $prob(op)$

---

the solution construction scheme continues by increasing the current value of $t$. Otherwise, a selection probability value $prob(op)$ is firstly computed for each operation contained in $\Omega$, according to the following formula:

$$prob(op) = \frac{\tau(op)^\alpha \eta(op)^\beta}{\sum_{op' \in \Omega} \tau(op')^\alpha \eta(op')^\beta} \tag{5}$$

where $\tau(op)$ is the pheromone value associated to the choice of $op$, $\eta(op)$ is the *desirability* value of $op$, and the parameters $\alpha$ and $\beta$ regulate the contribution of the pheromone and the desirability values to the probability of component selection.

The pheromone values will be described in the Pheromone Models section, while the desirability value for an operation $op$ is computed as:

$$\eta(op) = 1 - \frac{W \times op.\hat{D}^S_{sum} + op.\hat{D}^S_{min}}{W + 1}$$

where $op.\hat{D}^S_{sum}$ and $op.\hat{D}^S_{min}$ are the normalized values of $op.D^S_{sum}$ of $op.D^S_{min}$, respectively, and $W$ is a constant which enhances the contribution of $D^S_{sum}$ with respect to $D^S_{min}$.

Finally, the SELECTEXECUTABLEOPERATION() procedure selects an operation according to the probability distribution $prob(op)$.

### 3.3   Pheromone Models

Pheromone values could be associated directly to the operations, however this organization does not work well because P-S and MIX operations are present in all the feasible solutions (hence their pheromone value would be not significant), while SWAP operations can be used more times in the same solution. Therefore, it is necessary to associate a pheromone value to contextualized operations, i.e. to pairs $(op, c)$, where $c$ is a piece of information, denoting the context where $op$ is executed.

Using a technique similar to what is done in ACO approaches to scheduling [14] and to planning [1], pheromone values can be associated to the pairs $(op, t)$, where $op$ is the operation and $t$ is the start time of $op$. Pheromone values are organized as a matrix and the corresponding model is called Time Operation $(TO)$ model. Hence, in this model the (possible) start time of the operations is taken into account both in the operation selection phase (Eq. 5), where $\tau(op)$ is replaced with $\tau(op, t)$.

The pheromone values are updated with the following two steps procedure.

Firstly, an evaporation phase is performed, which lowers the value associated to each operation $op$ and each time step $t$ with the formula

$$\tau(op, t) \leftarrow (1 - \rho)\tau(op, t) \tag{6}$$

where $\rho \in (0, 1]$ is a parameter called *evaporation rate*.

Secondly, the values associated to the best solutions are increased. In QCC-ACO we decided to reward the best solution found so far $(S_{bs})$ or the best solution found in the current iteration $(S_{ib})$; the choice of which solution should be rewarded is made at random, the probability of rewarding $S_{ib}$ is 0.8, while $S_{bs}$ is rewarded with probability 0.2.

Let $S^*$ the solution to reward, the pheromone value of all the operations $op \in S^*$ is increased with the formula

$$\tau(op, t) \leftarrow \tau(op, t) + \frac{L}{mk(S^*)} \tag{7}$$

where $t$ is the time where $op$ is executed in $S^*$ and $L$ is constant (in the experiments its value is fixed to 10).

Another pheromone model, based on the TO model, is the Fuzzy Time Operation $(FTO)$ model. In this new model the value of pheromone of a given operation $op$ is spread around the time steps in the operation selection phase by fuzzyfing the time step $t$ when the pheromone model is queried. Internally, the pheromone values are stored and updated in a matrix $\tau(op, t)$ as in the $TO$ model. However, in the the operation selection phase, instead of employing $\tau(op, t)$, the value $\tau_w(op, t)$ is used in (5). This value is computed as the weighted average of the pheromone values of $op$ at times close to $t$: $t - w, t - w + 1, \ldots, t + w - 1, t + w$.

More in detail, $\tau_w(op, t)$ is computed with the following formula:

$$\tau_w(op, t) = \sum_{h=-w}^{w} \sigma(h)\tau(op, t + h)$$

where $w$ is the width of the window and

$$\sigma(h) = \frac{\exp(-\frac{h^2}{2})}{\sum_{k=-w}^{w} \exp(-\frac{k^2}{2})}$$

is the weight for the displacement $h = -w, \dots, w$. The width $w$ is a parameter of the model $FTO$, hence we will denote by $FTO_w$ the $FTO$ model where the width is $w$.

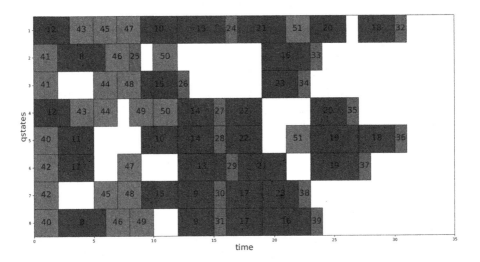

**Fig. 2.** Solution of the instance $n.8$ of the $N = 8$ benchmark set, with makespan $mk = 31$. In the plot, PS gates are depicted in green, SWAP gates in yellow, and MIX gates in blue. (Color figure online)

The difference between $TO$ and $FTO$ can be summarised as follows. Suppose that the operation $op$ is rewarded as a component of a good solution $S^*$ and that $op$ was executed at time $t_0$ in $S^*$. In the $TO$ model the pheromone value is only affected at $t_0$, while in $FTO$ also the time steps near $t_0$ can exploit of the value. The strength of influence of the pheromone value at time $t$ depends on the difference $|t - t_0|$: the smaller the distance, the higher the influence. A similar way of using pheromone has already been introduced in [1] (called Fuzzy Level Action model).

## 4    Empirical Evaluation

The QCC-ACO algorithm has been implemented in Java standard (OpenJDK, v.11.0.9.1), without using any external library. In particular, we have used the Java built-in pseudo random generator. All the experiments were run on an Intel Xeon E312 machine equipped with 16 GB of RAM.

### 4.1    Tuning

QCC-ACO has many parameters to be set: $\alpha$, $\beta$, $\rho$, the model $\mathcal{M}$ of the pheromone (the choice is among $TO$ and $FTO_w$, for some reasonable values of $w$), the number of ants $n_a$, $W$, and $L$. Some of the parameters, namely $n_a$, $W$, and $L$ were chosen after some preliminary runs, resulting that QCC-ACO is not affected in a sensitive way by the values of those parameters. The choice is $n_a = 20, W = 10, L = 10$.

Hence, we have decided to perform an extensive tuning procedure to find the best configuration for QCC-ACO in terms of $\alpha$, $\beta$, $\rho$ and $\mathcal{M}$. We have created 5 new instances of the QCC problem for each value of $N = 8, 21, 40$ only for the tuning phase, in order to avoid to use the same instances for the tuning and for the test phases. Both the parameters $\alpha$ and $\beta$ have been varied in the set $\{0, 1, 2, 3, 4\}$, while the possible values of $\rho$ were $0.1, 0.2, 0.3$. Finally, $TO$ competed against $FTO_1, \ldots, FTO_5$.

For each parameter configuration, QCC-ACO was run 10 times on the 15 tuning instances, using as termination criterion the time budget of 60 s for the instances with $N = 8$ and 300 s for the other instances.

The results of the tuning phase clearly indicates that the combinations ($\alpha = 1, \beta = 0, \rho = 0.3, \mathcal{M} = FTO3$) outperformed the others in terms of the average relative percentage difference. In line with the results obtained in [21], the best value for $\beta$ resulted to be 0, i.e. QCC-ACO works better without using the heuristic function $\eta$ in the probabilistic operation selection (formula 5). However, the heuristic function still plays an important role because it is used to prune the operations to select (see Algorithm 3) and therefore to reduce the branching factor.

Some preliminary experiments showed that the pruning stage is important because the performances of QCC-ACO greatly deteriorate if the operation selection works with $\Omega = \Omega'$, i.e. if all the eligible operations at the given time $t$ can be selected.

A second, smaller set of tuning experiments have been conducted in order to see if some small value of $\beta$ can improve the QCC-ACO performances. The parameter $\beta$ have been varied in the set $\{0, \frac{1}{4}, \frac{1}{3}, \frac{1}{2}\}$, while the other parameters have been fixed to the value of the best configuration. The second experiment confirmed that $\beta = 0$ is the best choice. Among the pheromone values, all the fuzzified versions of $TO$ outperformed the crisp model $TO$, confirming that the fuzzification of $TO$ works better. In particular, the best value for the width $w$ is 3, which is an intermediate value among all the possible value for that parameter.

Hence, we decided to adopt in the test phase the configuration ($\alpha = 1, \beta = 0, \rho = 0.3, \mathcal{M} = FTO3$).

## 4.2 Results

QCC-ACO has been tested on the 150 instances of the QCC problem (50 instances for each size $N = 8, 21, 40$) with $u = 1.0$ and $P = 2$. The termination criterion used in our experiments is the same that has been used in [5]: all runs for the $N = 8$ instances were limited to 60 s, while for the $N = 21$ and $N = 40$ instances all runs were limited to 300 s. For each instance QCC-ACO has been executed 10 times.

The results are depicted in Table 1, where the average value of the makespan, the standard deviation and the best value obtained by QCC-ACO are listed. The columns labelled *Best RH* contain the best values obtained in [5] through their Rollout heuristic. The columns labelled $\Delta$ show the difference between our results and those of our competitor. The best results obtained from either procedure are shown in bold.

For $N = 8$, QCC-ACO outperformed the Rollout heuristic (RH) in 14/50 instances, obtained the same result in 18/50 instances, and was beaten in 18/50 instances. The Wilcoxon signed rank test on QCC-ACO results against the RH results does not show any significant difference, because its p-value is large (0.2031). Despite in the $N = 8$ benchmark there is no clear winner, it is interesting to note that the average improvement obtained by ACO over RH on the improved solutions ($\Delta$ column) is equal to 1.64, versus an average worsening of 1.11. It is also interesting to see that, in 44 instances, QCC-ACO produced the same or equivalent solutions in all the 10 executions, as the standard deviation is 0.

For $N = 21$, QCC-ACO outperformed the Rollout heuristic in 35 instances, in only 7 instances QCC-ACO was outperformed by its competitor, while in the remaining 8 instances there was a tie. The average improvement obtained by ACO over RH on the improved solutions is equal to 2.60, versus an average worsening of 1.85. In this case, the Wilcoxon signed rank test has a very small p-value ($3.887 \cdot 10^{-9}$), hence the results of QCC-ACO are significantly better than those of RH.

The most impressive results were obtained for $N = 40$: in all 50 instances QCC-ACO found better solutions than the Rollout heuristic. In particular, the largest value of $\Delta$ was $-19$, obtained in three instances; remarkably, the average improvement obtained by ACO over RH is equal to 12.12. It is worth to notice that in all the instances except the instance #9, also the average makespans obtained by QCC-ACO are better than the best values obtained by the Rollout heuristic. As expected, also the p-value of Wilcoxon signed rank test is even smaller than the case of $N = 21$, i.e. $8.277 \cdot 10^{-10}$.

Figure 2 shows the plot of the solution of problem instance $n.8$ belonging to the $N = 8$ benchmark set, while Fig. 3 shows the plot of the solution of problem instance $n.24$ belonging to the $N = 40$ benchmark set (left), together with a graph of the average convergence times obtained on the 10 runs (right). Despite

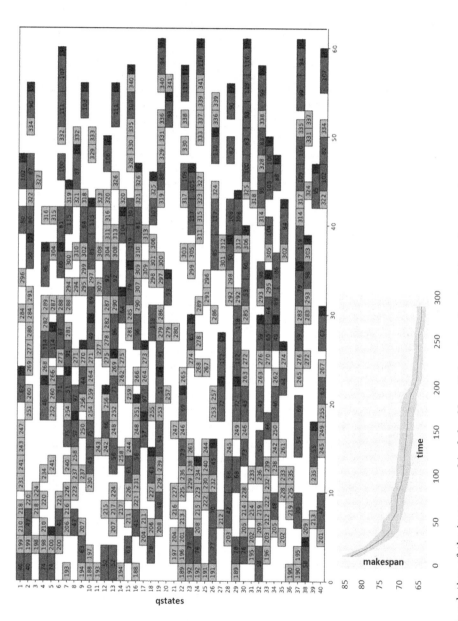

**Fig. 3.** Left: solution of the instance n.24 of the $N = 40$ benchmark set, with makespan $mk = 61$. Right: average plot of convergence times on 10 runs, the shaded area shows 95% confidence interval.

**Table 1.** Experimental results for all benchmarks

| Inst | N = 8 | | | | | N = 21 | | | | | N = 40 | | | | |
|---|---|---|---|---|---|---|---|---|---|---|---|---|---|---|---|
| | Avg. ACO | SD ACO | Best ACO | Best RH | Δ | Avg. ACO | SD ACO | Best ACO | Best RH | Δ | Avg. ACO | SD ACO | Best ACO | Best RH | Δ |
| 1 | 35.0 | 0.0 | 35 | 35 | 0 | 48.2 | 0.4 | **48** | 49 | −1 | 61.6 | 1.9 | **58** | 65 | −7 |
| 2 | 34.0 | 0.0 | **34** | 36 | −2 | 50.3 | 0.5 | 50 | 50 | 0 | 65.8 | 2.3 | **62** | 74 | −12 |
| 3 | 32.0 | 0.0 | 32 | 31 | 1 | 44.1 | 1.7 | **41** | 42 | −1 | 63.6 | 1.6 | **61** | 71 | −10 |
| 4 | 33.0 | 0.0 | 33 | 32 | 1 | 43.2 | 0.6 | **42** | 44 | −2 | 69.1 | 3.2 | **65** | 74 | −9 |
| 5 | 27.0 | 0.0 | 27 | 27 | 0 | 47.4 | 0.7 | **46** | 52 | −6 | 71.4 | 2.3 | **68** | 78 | −10 |
| 6 | 34.0 | 0.0 | **34** | 35 | −1 | 48.9 | 0.3 | **48** | 50 | −2 | 72.4 | 1.9 | **69** | 81 | −12 |
| 7 | 32.0 | 0.0 | 32 | 31 | 1 | 53.9 | 0.9 | **52** | 55 | −3 | 73.2 | 3.4 | **66** | 79 | −13 |
| 8 | 32.6 | 0.7 | **31** | 34 | −3 | 48.5 | 0.8 | **47** | 49 | −2 | 65.6 | 1.0 | **64** | 68 | −4 |
| 9 | 35.0 | 0.0 | 35 | 35 | 0 | 50.4 | 0.8 | **49** | 54 | −5 | 67.0 | 1.2 | **65** | 66 | −1 |
| 10 | 38.0 | 0.0 | 38 | 38 | 0 | 53.5 | 1.4 | **50** | 54 | −4 | 70.9 | 1.4 | **69** | 80 | −11 |
| 11 | 38.0 | 0.0 | 38 | 38 | 0 | 44.6 | 0.8 | **44** | 47 | −3 | 63.0 | 2.1 | **61** | 68 | −7 |
| 12 | 35.0 | 0.0 | 35 | 33 | 2 | 53.3 | 0.5 | **53** | 56 | −3 | 69.5 | 2.9 | **66** | 74 | −8 |
| 13 | 32.0 | 0.0 | 32 | 32 | 0 | 44.0 | 0.0 | 44 | 43 | 1 | 60.2 | 3.2 | **56** | 62 | −6 |
| 14 | 32.0 | 0.0 | 32 | 32 | 0 | 46.2 | 0.6 | 46 | 46 | 0 | 69.8 | 3.0 | **66** | 74 | −8 |
| 15 | 34.0 | 0.0 | **34** | 35 | −1 | 43.5 | 1.3 | **43** | 46 | −3 | 68.7 | 2.8 | **64** | 78 | −14 |
| 16 | 32.0 | 0.0 | 32 | 32 | 0 | 57.6 | 0.5 | 57 | 57 | 0 | 68.2 | 1.8 | **66** | 77 | −11 |
| 17 | 37.0 | 0.0 | 37 | 36 | 1 | 51.6 | 0.5 | 51 | 50 | 1 | 71.7 | 1.9 | **68** | 78 | −10 |
| 18 | 30.0 | 0.0 | 30 | 29 | 1 | 54.5 | 1.4 | **52** | 54 | −2 | 75.9 | 3.1 | **71** | 79 | −8 |
| 19 | 32.0 | 0.0 | 32 | 32 | 0 | 52.6 | 1.1 | **51** | 56 | −5 | 62.8 | 2.1 | **60** | 70 | −10 |
| 20 | 31.2 | 0.4 | 31 | 31 | 0 | 50.3 | 0.7 | **49** | 50 | −1 | 73.1 | 1.4 | **71** | 78 | −7 |
| 21 | 28.1 | 0.3 | 28 | 27 | 1 | 50.3 | 0.5 | **50** | 51 | −1 | 66.3 | 1.1 | **64** | 77 | −13 |
| 22 | 40.0 | 0.0 | 40 | 39 | 1 | 51.0 | 0.0 | **51** | 54 | −3 | 65.1 | 1.8 | **62** | 76 | −14 |
| 23 | 36.0 | 0.0 | 36 | 35 | 1 | 49.0 | 0.0 | 49 | 48 | 1 | 58.9 | 1.5 | **57** | 63 | −6 |
| 24 | 33.0 | 0.0 | 33 | 32 | 1 | 49.3 | 1.1 | **48** | 50 | −2 | 65.4 | 2.5 | **61** | 80 | −19 |
| 25 | 37.2 | 0.4 | **37** | 38 | −1 | 50.9 | 0.3 | 50 | 50 | 0 | 66.7 | 1.7 | **65** | 71 | −6 |
| 26 | 29.0 | 0.0 | 29 | 29 | 0 | 44.0 | 0.0 | **44** | 46 | −2 | 67.2 | 1.6 | **63** | 81 | −18 |
| 27 | 34.0 | 0.0 | 34 | 34 | 0 | 59.7 | 1.3 | **58** | 61 | −3 | 68.6 | 3.1 | **63** | 81 | −18 |
| 28 | 32.0 | 0.0 | 32 | 32 | 0 | 45.8 | 0.8 | **45** | 47 | −2 | 72.2 | 1.5 | **69** | 88 | −19 |
| 29 | 36.0 | 0.0 | 36 | 35 | 1 | 46.5 | 0.5 | **46** | 47 | −1 | 63.7 | 1.4 | **62** | 77 | −15 |
| 30 | 31.0 | 0.0 | 31 | 31 | 0 | 52.9 | 1.0 | **51** | 53 | −2 | 64.1 | 1.9 | **62** | 72 | −10 |
| 31 | 33.0 | 1.1 | 32 | 32 | 0 | 51.0 | 0.7 | **50** | 52 | −2 | 66.7 | 2.3 | **64** | 69 | −5 |
| 32 | 36.0 | 0.0 | 36 | 35 | 1 | 46.3 | 0.5 | **46** | 52 | −6 | 56.4 | 1.5 | **53** | 62 | −9 |
| 33 | 40.0 | 0.0 | **40** | 42 | −2 | 50.0 | 0.0 | **50** | 52 | −2 | 64.6 | 1.3 | **63** | 73 | −10 |
| 34 | 33.0 | 0.0 | **33** | 35 | −2 | 53.1 | 1.3 | 51 | 51 | 0 | 62.2 | 1.6 | **59** | 68 | −9 |
| 35 | 35.0 | 0.0 | **35** | 38 | −3 | 46.6 | 0.8 | 45 | 45 | 0 | 64.1 | 2.4 | **59** | 70 | −11 |
| 36 | 29.0 | 0.0 | 29 | 28 | 1 | 51.1 | 0.9 | 50 | 49 | 1 | 72.0 | 2.8 | **68** | 80 | −12 |
| 37 | 36.0 | 0.0 | 36 | 35 | 1 | 51.2 | 0.4 | 51 | 51 | 0 | 65.2 | 2.4 | **62** | 73 | −11 |
| 38 | 30.0 | 0.0 | 30 | 29 | 1 | 52.4 | 0.5 | **52** | 53 | −1 | 60.7 | 1.5 | **58** | 72 | −14 |
| 39 | 29.0 | 0.0 | **29** | 30 | −1 | 48.8 | 1.0 | **47** | 50 | −3 | 72.3 | 2.5 | **69** | 82 | −13 |
| 40 | 38.0 | 0.0 | 38 | 37 | 1 | 51.3 | 0.5 | 51 | 48 | 3 | 65.7 | 2.0 | **64** | 69 | −5 |
| 41 | 34.0 | 0.0 | **34** | 35 | −1 | 49.0 | 1.6 | **47** | 49 | −2 | 70.8 | 1.8 | **68** | 76 | −8 |
| 42 | 33.0 | 0.0 | 33 | 33 | 0 | 49.9 | 0.7 | **49** | 50 | −1 | 62.1 | 1.4 | **60** | 65 | −5 |
| 43 | 32.0 | 0.0 | 32 | 32 | 0 | 45.2 | 1.0 | **44** | 47 | −3 | 63.4 | 1.5 | **61** | 72 | −11 |
| 44 | 39.0 | 0.0 | 39 | 39 | 0 | 48.2 | 0.9 | 47 | 47 | 0 | 65.6 | 2.8 | **62** | 68 | −6 |
| 45 | 36.9 | 0.3 | **36** | 38 | −2 | 45.6 | 0.7 | 44 | 40 | 4 | 68.4 | 2.5 | **64** | 69 | −5 |
| 46 | 33.0 | 0.0 | **33** | 34 | −1 | 41.4 | 0.7 | **40** | 42 | −2 | 63.5 | 2.3 | **59** | 78 | −19 |
| 47 | 36.0 | 0.0 | **36** | 38 | −2 | 47.7 | 0.5 | **47** | 52 | −5 | 72.1 | 3.3 | **66** | 78 | −12 |
| 48 | 32.0 | 0.0 | **32** | 33 | −1 | 45.0 | 0.0 | **45** | 43 | 2 | 66.4 | 2.4 | **63** | 75 | −12 |
| 49 | 38.0 | 0.0 | 38 | 36 | 2 | 55.0 | 0.9 | **53** | 54 | −1 | 69.3 | 2.3 | **67** | 73 | −6 |
| 50 | 31.0 | 0.0 | 31 | 30 | 1 | 50.7 | 0.7 | **49** | 53 | −4 | 69.6 | 1.4 | **67** | 74 | −7 |

the descending trend shows the first signs of a plateau, there is still some room for a further solution improvement should the time limit of 300 s be increased.

## 5    Concluding Remarks and Future Work

In this work, we propose a novel Ant Colony Optimization (ACO) algorithm for the realization (compilation) of nearest-neighbor compliant quantum circuits. In particular, our ACO algorithm (QCC-ACO) introduces a novel pheromone model and leverages a heuristic-based priority rule inspired by the priority rules proposed by [5,17] in the recent literature to control the iterative selection of quantum gates to be inserted in the solution.

Table 1 reports the overall and direct comparison of our ACO approach with the state-of-the-art represented, to the best of our knowledge, by the experimental results proposed in [5]. According to our experimental results, QCC-ACO scales quite well on the size of the QCC problem ($N = 8, 21, 40$): overall, QCC-ACO was able to produce a total of 99 better solutions out of the considered 150 QCC instances; in particular, for $N = 40$, QCC-ACO improved over all the given 50 instances.

The priority rule used in [5] extends the one proposed in [17], using a gate selection strategy targeted at minimizing the insertion of SWAP gates, based on (i) the $D_{sum}^S$ and $D_{min}^S$ values, and on (ii) an increasing value of start time $t$ used as a scheduling criterion for iteratively inserting the operations in the solution. It is therefore reasonable to conjecture that the reasons of the better performance of [5] with respect to those obtained in [17] may reside on both previous components. In the future, it would be very interesting to perform an analysis to determine which component plays the leading role in such improvement.

Three further possible directions of future work are also worth being pursued: the first one is to apply our ACO algorithm to the case of QCC problems with cross-talk constraints [3]; the second one is to explore the feasibility of our evolutionary approach to the case of compilation of a quantum algorithms for graph coloring [8]; finally, the current ACO version may be further improved with local search procedures, though an efficient local search implementation for the QCCP seems to require a careful and not straightforward design.

## References

1. Baioletti, M., Milani, A., Poggioni, V., Rossi, F.: Experimental evaluation of pheromone models in ACOPlan. Ann. Math. Artif. Intell. **62**(3), 187–217 (2011). https://doi.org/10.1007/s10472-011-9265-7
2. Baioletti, M., Milani, A., Santucci, V.: A new precedence-based ant colony optimization for permutation problems. In: Shi, Y., et al. (eds.) SEAL 2017. LNCS, vol. 10593, pp. 960–971. Springer, Cham (2017). https://doi.org/10.1007/978-3-319-68759-9_79
3. Booth, K.E.C., Do, M., Beck, C., Rieffel, E., Venturelli, D., Frank, J.: Comparing and integrating constraint programming and temporal planning for quantum circuit compilation. In: Proceedings of the 28$^{th}$ International Conference on Automated Planning & Scheduling, ICAPS 2018, pp. 366–374 (2018)

4. Brierley, S.: Efficient implementation of quantum circuits with limited qubit interactions. Quantum Inf. Comput. **17**(13–14), 1096–1104 (2017). http://dl.acm.org/citation.cfm?id=3179575.3179577

5. Chand, S., Singh, H.K., Ray, T., Ryan, M.: Rollout based heuristics for the quantum circuit compilation problem. In: 2019 IEEE Congress on Evolutionary Computation (CEC), pp. 974–981 (2019)

6. Cirac, J.I., Zoller, P.: Quantum computations with cold trapped ions. Phys. Rev. Lett. **74**, 4091–4094 (1995). https://doi.org/10.1103/PhysRevLett.74.4091

7. Cormen, T.H., Leiserson, C.E., Rivest, R.L., Stein, C.: Introduction to Algorithms, 2nd edn. MIT Press, Cambridge (2001)

8. Do, M., Wang, Z., O'Gorman, B., Venturelli, D., Rieffel, E., Frank, J.: Planning for compilation of a quantum algorithm for graph coloring. ArXiv abs/2002.10917 (2020)

9. Dorigo, M., Stützle, T.: Ant Colony Optimization. Bradford Company, USA (2004)

10. Farhi, E., Goldstone, J., Gutmann, S.: A quantum approximate optimization algorithm. arXiv preprint arXiv:1411.4028. November 2014

11. Guerreschi, G.G., Park, J.: Gate scheduling for quantum algorithms. arXiv preprint arXiv:1708.00023, July 2017

12. Hart, J., Shogan, A.: Semi-greedy heuristics: an empirical study. Oper. Res. Lett. **6**, 107–114 (1987)

13. Herrera-Martí, D.A., Fowler, A.G., Jennings, D., Rudolph, T.: Photonic implementation for the topological cluster-state quantum computer. Phys. Rev. A **82**, 032332 (2010). https://doi.org/10.1103/PhysRevA.82.032332

14. Merkle, D., Merkle, M., Schmeck, H.: Ant colony optimization for resource-constrained project scheduling. IEEE Trans. Evol. Comput. **6**(4), 333–346 (2002)

15. Nau, D., Ghallab, M., Traverso, P.: Automated Planning: Theory & Practice. Morgan Kaufmann Publishers Inc., San Francisco (2004)

16. Nielsen, M.A., Chuang, I.L.: Quantum Computation and Quantum Information: 10th Anniversary Edition, 10th edn. Cambridge University Press, New York (2011)

17. Oddi, A., Rasconi, R.: Greedy randomized search for scalable compilation of quantum circuits. In: van Hoeve, W.-J. (ed.) CPAIOR 2018. LNCS, vol. 10848, pp. 446–461. Springer, Cham (2018). https://doi.org/10.1007/978-3-319-93031-2_32

18. Rasconi, R., Oddi, A.: An innovative genetic algorithm for the quantum circuit compilation problem. In: Proceeding of the Thirty-Third Conference on Artificial Intelligence, AAAI 2019, pp. 7707–7714. AAAI Press (2019)

19. Resende, M.G., Werneck, R.F.: A hybrid heuristic for the p-median problem. J. Heuristics **10**(1), 59–88 (2004). https://doi.org/10.1023/B:HEUR.0000019986.96257.50

20. Sete, E.A., Zeng, W.J., Rigetti, C.T.: A functional architecture for scalable quantum computing. In: 2016 IEEE International Conference on Rebooting Computing (ICRC), pp. 1–6, October 2016. https://doi.org/10.1109/ICRC.2016.7738703

21. Stützle, T.: An ant approach to the flow shop problem. In: Proceedings of the 6th European Congress on Intelligent Techniques & Soft Computing, EUFIT 1998, Aachen, Germany, pp. 1560–1564 (1998)

22. Venturelli, D., Do, M., Rieffel, E., Frank, J.: Temporal planning for compilation of quantum approximate optimization circuits. In: Proceedings of the Twenty-Sixth International Joint Conference on Artificial Intelligence, IJCAI 2017, pp. 4440–4446 (2017). https://doi.org/10.24963/ijcai.2017/620

23. Yao, N.Y., et al.: Quantum logic between remote quantum registers. Phys. Rev. A **87**, 022306 (2013). https://doi.org/10.1103/PhysRevA.87.022306

# Hybridization of Racing Methods with Evolutionary Operators for Simulation Optimization of Traffic Lights Programs

Christian Cintrano$^{(\boxtimes)}$ ⓘ, Javier Ferrer ⓘ, Manuel López-Ibáñez ⓘ, and Enrique Alba ⓘ

University of Malaga, Bulevar Louis Pasteur 35, 29010 Malaga, Spain
{cintrano,ferrer,manuel.lopez-ibanez,eat}@lcc.uma.es

**Abstract.** In many real-world optimization problems, like the traffic light scheduling problem tackled here, the evaluation of candidate solutions requires the simulation of a process under various scenarios. Thus, good solutions should not only achieve good objective function values, but they must be robust (low variance) across all different scenarios. Previous work has revealed the effectiveness of IRACE for this task. However, the operators used by IRACE to generate new solutions were designed for configuring algorithmic parameters, that have various data types (categorical, numerical, etc.). Meanwhile, evolutionary algorithms have powerful operators for numerical optimization, which could help to sample new solutions from the best ones found in the search. Therefore, in this work, we propose a hybridization of the elitist iterated racing mechanism of IRACE with evolutionary operators from differential evolution and genetic algorithms. We consider a realistic case study derived from the traffic network of Malaga (Spain) with 275 traffic lights that should be scheduled optimally. After a meticulous study, we discovered that the hybrid algorithm comprising IRACE plus differential evolution offers statistically better results than conventional algorithms and also improves travel times and reduces pollution.

**Keywords:** Hybrid algorithms · Evolutionary algorithms · Simulation optimization · Uncertainty · Traffic light planning

## 1 Introduction

In many real-world optimization problems, the evaluation of candidate solutions requires the simulation of a process under various scenarios that represent uncertainty about the real-world. Good solutions should not only achieve good objective function values but also show robustness, i.e., low variance across scenarios. To assess the robustness of solutions, it is often required to simulate each solution a number of times using different data, starting conditions or random numbers. For example, when planning the traffic light schedules within a city[1],

---

[1] Legal and technical limitations may make real-time traffic light control infeasible.

© Springer Nature Switzerland AG 2021
C. Zarges and S. Verel (Eds.): EvoCOP 2021, LNCS 12692, pp. 17–33, 2021.
https://doi.org/10.1007/978-3-030-72904-2_2

it is desirable to find a schedule that works well under many different traffic conditions [3,5,7,8,13,15–18,20,21]. A common approach is to simulate each candidate solution under a number of scenarios generated from real traffic data. However, there is a trade-off between the number of scenarios used for evaluating each solution and the number of candidate solutions evaluated.

Previous work [6] has shown that IRACE [11] is able to find high-quality and low-variance traffic light schedules by dynamically adjusting the number of simulations performed per solution. The elitist iterated racing algorithm implemented by IRACE has been traditionally used for the configuration of parameters in machine learning and optimization algorithms, where each configuration must be evaluated on a number of training instances of a problem and the algorithm themselves are often stochastic. The algorithm implemented in IRACE uses a learning mechanism inspired by reinforcement learning to sample new solutions from the best ones previously found. Although this approach tends to work well for configuring a mix of categorical and numerical parameters with dependencies and constraints among them, other operators may perform better when the problem consists only of numerical decision variables.

In this paper, we propose a hybridization of evolutionary operators from evolutionary algorithms (EAs) and the elitist iterated racing of IRACE. We evaluate its performance on the traffic light optimization problem and compare it with previous results from the literature. The idea of previous approaches to hybridizing EAs and racing was performing independent races to carry out the evaluation step within an EA [9], whereas our proposal replaces the sampling mechanism in IRACE, which is not simply a sequence of independent races, with evolutionary operators.

Besides, in order to add value to our experimentation, we use an instance based on real data from the city of Malaga, Spain. We also use a traffic simulator, SUMO [1,10], to evaluate each of the traffic light schedules generated by the algorithms. With this, we not only seek to analyze which algorithm is better but also to solve a real problem of the city.

In summary, the main contributions of this work are:

- We propose new hybrid algorithms that combine racing strategies with evolutionary operators. Thus obtaining powerful and robust algorithms.
- We optimize the traffic light plan of a real city like Malaga (Spain) using detailed micro-simulations.
- We offer an in-depth analysis of our hybrid algorithms and compare them with well-known EAs such as a genetic algorithm (GA) and a differential evolution (DE).
- We study which algorithm presents the greatest improvement to the city according to different measures about traffic quality and emission reduction.

The rest of this article is organized as follows: Sect. 2 presents a description of the Traffic Light Scheduling Problem. Section 3 describes the main contribution of this work, the hybridization between IRACE and EAs. Section 4 outlines the main aspects of our experimentation. We discuss the results obtained in Sect. 5. Finally, Sect. 6 presents conclusions and future work.

## 2   Problem Description

Traffic flow in large smart cities has become one of the most severe problems that large cities face. In some cases, this problem is further aggravated due to the high amount of traffic jams, traffic accidents, or even injured people or deaths. Therefore traffic must be regulated with some elements such as traffic lights. The larger the metropolitan area, the higher the number of traffic lights needed to regulate the traffic flow. Optimal management of traffic might be beneficial to minimize journey times, reduce fuel consumption and harmful emissions.

Traffic lights are coordinated in phases: green, yellow and red. In this way, when some traffic lights of the same intersection are in green, some others must be in red. Besides, the different pre-defined phases for an intersection are sequences repeated over time, we call traffic light program (TLP) to each of those sequences.

The large number of program combinations that appear in traffic light schedules of large cities require automatic tools to generate optimal TLP, which motivates the Traffic Light Scheduling Problem (TLSP) [8, 15, 16]. The main objective in this problem is to find optimized TLP for all the traffic lights located in the intersections of an urban area with the aim of reducing journey time, emissions, and fuel consumption.

Let us define the TLSP as follows. Let $P = \{I_1, \dots, I_n\}$ be a candidate TLP, where each $I_i$ corresponds to a different intersection defined as a set of predefined valid phases $I_i = \{\varphi_{i1}, \dots, \varphi_{im_i}\}$, where $m_i = |I_i|$ and each $\varphi_{ij} \in \mathbb{N}^+$ represents the duration (in seconds) of phase $j$ in intersection $I_i$, that is, the duration of each valid phase of light colors (e.g., "$rr\ yyg\ rr\ gyyy$"). The objective is to find a TLP $P'$ that minimizes a fitness function $f \colon \Gamma \to \mathbb{R}$ such that:

$$P' = \arg\min_{P \in \Gamma}\{f(P)\} \tag{1}$$

where $\Gamma$ is the space of all possible TLPs.

In order to define the fitness function, we need to explain some previous concepts used in the definition. The evaluation of a solution is performed using a traffic simulator that provides information regarding the flow of vehicles. Vehicles travel from a starting position to a destination position, then the travel time ($t_v$) of a vehicle $v$ is the number of simulation steps (1 second per simulation step) in which its speed was above 0.1 m/s, while its waiting time ($w_v$) is the number of simulation steps in which its speed was below 0.1 m/s.

Long phase duration may lead to a collapse of the intersection. TLPs should prioritize those phases with more green lights on the directions with a high number of vehicles circulating. So, we should maximize the following ratio measure:

$$GR(P) = \sum_{i=1}^{n} \sum_{j=1}^{|I_i|} \varphi_{ij} \cdot \frac{G_{ij}}{R_{ij}} \tag{2}$$

where $G_{ij}$ is the number of traffic lights in green, and $R_{ij}$ is the number of traffic lights in red in phase $j$ of intersection $i$ and $\varphi_{ij}$ is the duration of the phase. The minimum value of $R_{ij}$ is 1 in order to avoid a division by 0.

Finally, we define the following fitness function that should be minimized:

$$f(P) = \frac{V^{\mathrm{rem}}(P) \cdot t^{\mathrm{sim}} + \sum\limits_{v=1}^{V(P)} t_v(P) + w_v(P)}{V(P)^2 + GR(P)} \tag{3}$$

where the presence of vehicles with incomplete journeys $V^{\mathrm{rem}}(P)$ penalizes the fitness of a solution $P$ proportionally to the simulation time $t^{\mathrm{sim}}$. The number of vehicles that arrive at their destinations is squared $(V(P)^2)$ to prioritize this criterion over the rest. This fitness function has been successfully used in [7,8].

## 3  Hybridization of IRACE and Evolutionary Algorithms

There are many definitions of hybrid algorithms, yet the general idea is to combine components or concepts from different techniques to exploit desirable characteristics of those components to tackle problems with particular features [2]. In this work, we combine the elitist iterated racing strategy from IRACE with evolutionary operators to obtain an algorithm that performs well on numerical optimization problems where the fitness of each solution is uncertain and must be evaluated using multiple simulations. The elitist iterated racing strategy of IRACE decides how many simulations should be performed per solution, how solutions are compared, and which solutions should be discarded at each iteration. The evolutionary operators are responsible for generating new solutions from the surviving population of solutions. Next, we will briefly explain the base algorithm, IRACE, and the different characteristics of the hybrid algorithm.

### 3.1  IRACE

IRACE [11] is a well-known tool for automatic (hyper-)parameter configuration of optimization and machine learning algorithms. In the context of automatic parameter configuration, decision variables correspond to algorithmic parameters, candidate solutions correspond to potential configurations of an algorithm, and evaluating the fitness of a solution requires running the algorithm with a particular parameter configuration on multiple training data or problem instances. However, IRACE can be seen as an optimization method for mixed-integer black-box problem under uncertainty, and, hence, it may be used to tackle simulation-optimization problems, such as the TLSP [6].

Algorithm 1 briefly presents IRACE applied to the TLSP. Initially, a set of solutions are sampled uniformly at random. Then a race is performed to identify the best solutions among the initial set. Within a race, each solution is simulated multiple times on different traffic scenarios until there is enough evidence to eliminate it because it is performing worse than the best solution found so far. In the TLSP, we use the pairwise paired Student's $t$-test as the elimination test. The race stops once a minimum number of solutions remains alive in the race, the budget assigned to the race is exhausted, or multiple elimination tests

**Algorithm 1.** Pseudocode of IRACE

**Input:** Network data and training traffic scenarios.
**Output:** Best solution (TLP) found.

1: $t \leftarrow 1$
2: $\Theta_t \leftarrow$ SampleUniformRandomPopulation
3: $\Theta^{\text{elite}} \leftarrow$ Race$(\Theta_t)$
4: **while** $evals < totalEvals$ **do**
5:     $t \leftarrow t + 1$
6:     $\mathcal{M} \leftarrow$ Update$(\Theta^{\text{elite}})$
7:     $\Theta^{\text{new}} \leftarrow$ Sample$(\mathcal{M})$
8:     $\Theta_t \leftarrow \Theta^{\text{new}} \cup \Theta^{\text{elite}}$
9:     $\Theta^{\text{elite}} \leftarrow$ Race$(\Theta_t)$
10: **end while**
11: **Output:** best solution from $\Theta^{\text{elite}}$

fail to eliminate any solution. The solutions that remain alive after the race are called elite. These elite solutions are used to update a sampling model in a similar fashion as reinforcement learning, from which new solutions are generated. New and elite solutions together form a new population that is raced again. In elitist racing, results from previous races are re-used in subsequent races and elite solutions cannot be eliminated from the race until the contender has been evaluated in as many scenarios as the elite solution. This process is iterated until a maximum budget of simulations is exhausted. The main benefit of the racing strategy is that poor solutions are discarded quickly to avoid wasting simulations, while good solutions are simulated on many scenarios to provide a good estimate of their fitness. Moreover, the elimination test takes into account not only the mean value over multiple simulations but also the variance and the number of simulations performed so far.

## 3.2   Hybrid Algorithms

Once we have described how IRACE works, let us analyze our hybrid algorithms. In line 7 of Algorithm 1, the function Sample$(\mathcal{M})$ generates a new set of candidate solutions to the problem. In our hybrid algorithms, we replace that function with operators taken from two EAs: Genetic Algorithm (GA) and Differential Evolution (DE). We call IRACE+GA and IRACE+DE, respectively, to these new hybrid algorithms. These EAs have already demonstrated their effectiveness in solving the TLSP [6], so we consider them to hybridize with IRACE. In this way, the racing step remains intact but the sampling of new solutions is carried out by these EAs. In IRACE, the Sample$(\mathcal{M})$ procedure is equivalent to the mating selection and variation steps of an EA, i.e., selecting parents, generating new individuals from them (crossover), making some modification to the new solutions (mutation), and returning the new set of solutions. IRACE works with both numerical and categorical parameters. TLSP only has numerical parameters, so, in this paper, we do not have to deal with categorical parameters.

The set of elite solutions $\Theta^{\text{elite}}$ contains the best solutions found by IRACE after the race performed at each iteration (line 9). In our hybrid algorithm, the parents used by the evolutionary operators are selected from $\Theta^{\text{elite}}$. However, the size of $\Theta^{\text{elite}}$ may vary each iteration and may be insufficient for the number of parents required by the evolutionary operators. We handle this situation by generating additional parents by random uniform sampling (as in line 2). This mechanism also introduces more diversity to the set of parent solutions. Because we use several evolutionary operators, the number of selected parents differs from one algorithm to another. IRACE+GA needs two parents for the operator execution, while IRACE+DE needs four. The restriction in the number of parents is given by the operators used by each algorithm, because each operator requires a different number of solutions to generate a new one.

In this work, we have implemented two variants of the proposed hybrid algorithm with the following operators:

- IRACE+GA: uniform crossover [19] and integer polynomial mutation [4]
- IRACE+DE: the "DE/best/1/bin" strategy [14].

The evaluation of the new solutions returned after the Sample($\mathcal{M}$) phase is computed by performing several simulations, as carried out by IRACE. After the evaluation phase, we merge this set of new solutions $\Theta^{\text{new}}$ with set of elite solutions $\Theta^{\text{elite}}$ to execute the racing. This returns a new set of elite solutions, which will be used in the next iteration of the hybrid algorithms.

## 4   Experimental Setup

We describe here the experimental protocol followed in this work. First, we describe the real-world case study of TLSP that is the main motivation of our research. After that, we provide details about the experiments carried out. We will analyze these experiments in the next section.

### 4.1   Real World Case Study

We consider a realistic scenario derived from the traffic network of Malaga [18], which encompasses an area of about $3\,\text{km}^2$ with 58 intersections controlled by 275 traffic lights (Fig. 1). Our network model was created from real data about traffic rules, traffic element locations, road directions, streets, intersections, etc.

Once we have a realistic traffic network of a city, we need the routes and vehicles circulating and their speeds. This information was collected from sensorized points in certain streets measuring traffic density at various time intervals. From the sensed data extracted, we have applied the Flow Generator Algorithm (FGA) [18] to generate 60 different traffic scenarios with an average of 4,827 vehicles (or different vehicle routes) per scenario. In order to evaluate the reliability of a candidate solution, we split the generated traffic scenarios into two equal sets of 30 scenarios each. One (training) set is exclusively used for optimization, that is, for identifying optimal TLSP solutions. The other (testing) set of scenarios is used for comparing the solutions found during optimization.

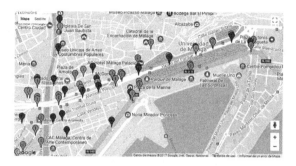

**Fig. 1.** Locations of traffic lights considered in the case of study. The colors show large (red), medium (yellow) and small (green) differences between two different solutions. (Color figure online)

### 4.2 Case Study Constraints

Real-world instances of the TLSP often present additional constraints. In our case, we consider the constraints recommended by the City Council of Malaga (Spain). Phases containing any yellow signals are called *fixed phases* because they have a predetermined duration and the set of such phases will be denoted by $Y$. These fixed phases correspond to pedestrian crosses, which last for a fixed time of $4 \times number\ of\ lanes$ seconds. Non-fixed phases have a minimum duration of $\varphi_{\min} = 15$ s. Moreover, the total program time $(Tp_i)$ within each intersection $I_i$, which is computed as the sum of its phase durations:

$$Tp_i = \sum_{\varphi_{ij} \in I_i, j=1}^{|I_i|} \varphi_{ij} \tag{4}$$

is constrained within $[Tp_{\min}, Tp_{\max}]$. For the City Council of Malaga (Spain), $Tp_{\min} = 60$ and $Tp_{\max} = 120$ s.

By default, the first programs of all intersections start at the same time. However, we also optimize an offset time at each intersection $(To_i)$ that represents a shift in seconds of the starting time of the program at the start of the simulation. If the offset value of an intersection is negative, then program start time is shifted back that number of seconds and the program actually starts on a phase before the first one; whereas if the offset is positive, the program begins as if that number of seconds has already passed, i.e., skipping those seconds from the duration of the first phase and, maybe, of later phases. Offset times enable the emergence of series of coordinated traffic lights that produce a continuous traffic flow over several intersections in one main direction. Offset values are constrained within the time interval $To_i \in [To_{\min}, To_{\max}] = [-30, 30]$.

### 4.3 Repair Procedure

The TLSP is subject to some constraints we have just explained in Sect. 4.2. To ensure that candidate solutions are valid, we propose a repair procedure that is used by all the algorithms before the simulation. The value of each phase duration $\varphi_{ij}$ is already constrained within a range that is larger than the minimum phase duration $\varphi_{\min}$. However, we need to ensure that the total program time $Tp_i$ is within $[Tp_{\min}, Tp_{\max}]$. Here we can distinguish two different cases.

In the first case, if the total program time for intersection $I_i$ is smaller than $Tp_{\min}$, then we replace each non-fixed phase (those that do not contain a yellow signal, i.e., $\varphi_{ij} \notin Y$) with

$$\varphi_{ij} = \left\lceil \varphi_{ij} \cdot \frac{Tp_{\min} - Tp_i^Y}{Tp_i - Tp_i^Y} \right\rceil \tag{5}$$

where $Tp_i^Y = \sum_{\varphi_{ij} \in I_i \cap Y} \varphi_{ij}$ is the sum of the fixed phase durations within intersection $I_i$.

In the second case, if the total program time is larger than $Tp_{\max}$, then we replace each non-fixed phase ($\varphi_{ij} \notin Y$) with

$$\varphi_{ij} = \varphi_{\min} + \left\lfloor (\varphi_{ij} - \varphi_{\min}) \cdot \frac{Tp_{\max} - Tp_i^Y - \varphi_{\min} \cdot |I_i \setminus Y|}{Tp_i - Tp_i^Y - \varphi_{\min} \cdot |I_i \setminus Y|} \right\rfloor \tag{6}$$

where $|I_i \setminus Y|$ is the number of non-fixed phases within intersection $I_i$ and $Tp_i^Y$ is the total duration of the fixed phases within intersection $I_i$.

## 4.4   Simulator: SUMO

The quality of a solution (traffic light program) is evaluated through the Simulator of Urban Mobility (SUMO) [1,10], which is a microscopic road traffic simulator that provides detailed information about vehicles like velocity, fuel consumption, emissions, journey time, waiting time, etc. The study of realistic scenarios according to real patterns of mobility of the target city is possible due to the fine-grained realistic micro-simulations provided by SUMO.

All simulations were performed with SUMO version 0.22. Since we already introduce variability by means of the different traffic scenarios, we fix the random seed used by SUMO to zero in all simulations. This means that, given a traffic scenario and a candidate solution, the simulation is deterministic. In all experiments, we stop each run of an algorithm, either a variant of IRACE or otherwise, after executing 30 000 calls to the SUMO simulator. Given that each solution is simulated on a number of different scenarios, the number of solutions evaluated per run is often much lower.

## 4.5   Algorithms

In our experiments we compare IRACE with the two hybrid variants described above, namely, IRACE+GA and IRACE+DE. In addition, to assess the contribution of the elitist racing mechanism, we also evaluate the classical GA and DE algorithms. Here, we describe the implementation details of these algorithms.

Following the conclusions from a previous work on the TLSP [6], we use default settings for IRACE and the hybrids, except the following. The population size is fixed to 10 individuals (also for GA and DE), the minimum number of traffic scenarios simulated per candidate solution is set to two ($T^{first} = 2$) and we enable the *deterministic* option that tells IRACE that the only source of uncertainty are the different scenarios and not the simulations themselves. The evolutionary algorithms use a fixed number of simulations per candidate solution. Each solution is simulated on five different training scenarios and its fitness is computed as the mean fitness over the five simulations. In [6], the authors already compared IRACE with differential evolution, genetic algorithm, particle swarm optimization, and a random search; and showed that IRACE obtained the best results with GA and DE being a close second, therefore, we focus here in the comparison of IRACE, GA, DE, and the hybrids.

Our GA implementation uses a ranking method for parent selection and elitist replacement for the next population, that is, the two best individuals of the current population are included in the next one. The operators used are uniform crossover and integer polynomial mutation with 1.0 of probability of crossover and 0.1 of probability of mutation. These parameter settings were found by additional experiments carried out in previous studies [3] to produce a search behavior that is more exploitative rather than explorative, which is more appropriate for the TLSP. Our DE implementation uses a "best/1/bin" strategy with difference factor $F = 0.5$ and probability of crossover 0.5. These are the default parameter values in jMetal [12]. Finally, IRACE+GA and IRACE+DE use the same parameter settings as the GA and DE, respectively.

The GA and DE are implemented in Java using jMetal 5.0 [12]. IRACE and the hybrids are implemented in R.[2] We used IRACE version 2.3 as the baseline.[3]

### 4.6   Experimental Details

As mentioned above, we generated 60 traffic scenarios from real sensor data and we split these scenarios into two sets of size 30. One set (training set) is used when running the algorithms to find TLSP solutions, while the other set (testing set) is used for evaluating the fitness and reliability of these solutions and comparing the various strategies analyzed in this paper. During optimization, the traffic is simulated up to a predefined time horizon (1 h plus 10 min of warm-up, in our case) in order to simulate the peak period in our real-world case study. For the constraints of the TLSP, we apply the same repair method as in [6].

The algorithms presented in this paper are non-deterministic algorithms, so we performed 30 independent runs for a fair comparison between them. After the executions, we applied the non-parametric Kruskal-Wallis test with a confidence level of 95% ($p$-value $< 0.05$) with Holms's $p$-value correction to check if the observed differences are statistically significant. In the cases where Kruskal-Wallis test rejects the null hypothesis, we run a single factor ANOVA post hoc

---

[2] The source code is available at https://github.com/NEO-Research-Group/irace-ea.
[3] Available at https://cran.r-project.org/package=irace.

test for pairwise comparisons. To properly interpret the results of statistical tests, it is always advisable to report effect size measures. For that purpose, we have also used the non-parametric effect size measure $\hat{A}_{12}$ statistic proposed by Vargha and Delaney [22]. In the case of minimization problems, such as the TLSP, higher $\hat{A}_{12}$ values suggest that algorithm 2 has a higher probability of obtaining a better result than algorithm 1, e.g., $\hat{A}_{12} = 0.3$ indicates that algorithm 2 gets better values than algorithm 1 in 30% of the runs.

The experiments were run on a cluster of 16 machines with Intel Core2 Quad processors Q9400 at 2.66 GHz and 4 GB memory and 3 machines equipped with three Intel Xeon CPU (E5-2670 v3) at 2.30 GHz and 64 GB memory. The cluster was managed by HTCondor 8.2.7, which allowed us to perform parallel independent executions to reduce the overall experimentation time.

## 5    Results

To give an in-depth view of the performance of our hybrid algorithms against the standard ones, we will analyze their performance in several sets of scenarios (training and testing). With this, we want to present the competitiveness of our proposal and give a solution to the TLSP.

### 5.1    Training Set

During the training phase, each algorithm performs a maximum of 30,000 simulations. Figure 2 shows the best fitness obtained, over the number of simulations, averaged over 30 runs of each algorithm. We can see that, in general, up to 10,000 simulations, all the algorithms improve significantly the quality of their

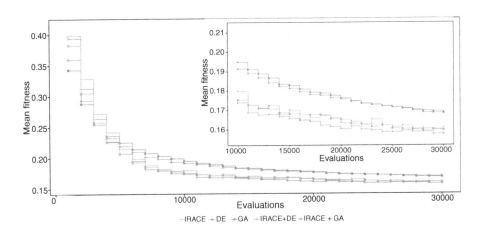

**Fig. 2.** Mean fitness of the best solutions found so far within each run, as estimated by each algorithm at each moment of its execution on traffic scenarios from the training set. Results in the range [10,000, 30,000] are magnified.

**Table 1.** Results of the $\widehat{A}_{12}$ test for the evaluation of the last solutions found over training scenarios. Probability that the algorithm (column) is better than another algorithm (row). We highlight in bold the values when the algorithm in the column is better than the algorithm in the row.

|          | IRACE  | IRACE+DE | IRACE+GA | GA       | DE     |
|----------|--------|----------|----------|----------|--------|
| IRACE    | –      | **0.6711** | **0.6100** | **0.6067** | 0.4933 |
| IRACE+DE | 0.3289 | –        | 0.3956   | **0.5122** | 0.3556 |
| IRACE+GA | 0.3900 | **0.6044** | –        | **0.5478** | 0.4267 |
| GA       | 0.3933 | 0.4878   | 0.4522   | –        | 0.3811 |
| DE       | **0.5067** | **0.6444** | **0.5733** | **0.6189** | —      |

**Table 2.** Statistics of each algorithm from the best solutions obtained in the 30,000 simulation of the training. We mark in bold the lower value of each metric.

| Algorithm | Mean   | Median   | STD Dev. |
|-----------|--------|----------|----------|
| IRACE+DE  | **0.1585** | 0.1563 | 0.0101   |
| IRACE+GA  | 0.1597 | 0.1590   | 0.0076   |
| IRACE     | 0.1621 | 0.1615   | **0.0064** |
| GA        | 0.1684 | **0.1562** | 0.0364   |
| DE        | 0.1689 | 0.1596   | 0.0215   |

solutions, but after this number of simulations, the improvement slows down. Although the GA and DE obtain the best results up to 5,000 simulations, they are quickly overtaken by IRACE and its hybrid variants. Figure 2 also shows in more detail the differences, starting from 10,000 simulations, between IRACE, IRACE+DE and IRACE+GA. We can notice that IRACE+DE consistently obtains the lowest mean fitness, while IRACE and IRACE+GA show a similar result. The plot also shows that the fitness reported by the IRACE hybrids sometimes increases due to the racing procedure performing additional simulations to refine the estimation of the fitness.

We have performed an $\widehat{A}_{12}$ test at the end of the training execution (30,000 simulations) to check if IRACE+DE is indeed better than the other algorithms. Table 1 shows the results of the $\widehat{A}_{12}$ test among the different algorithms, where each value gives the probability of the algorithm in the column returning a better solution than the one in the row. The test indicates that GA is better than the rest of the algorithms. However, we also look at the other statistics shown in Table 2. Although GA has better median than IRACE+DE's only by $10^{-4}$, the standard deviation of IRACE+DE is 3.6 times less than GA. Thus, we conclude that IRACE+DE is more robust than GA. EAs obtain lower mean and median, while IRACE reports solutions with smaller variability. These results support our approach to hybridizing IRACE with EAs to obtain good quality robust solutions. Particularly, IRACE+DE looks like a good option if we want to apply these features.

## 5.2   Testing Set

The above reported statistics were obtained after evaluating the final solutions on the same scenarios used during optimization, but the training scenarios will never arise exactly in the real-world. We evaluate again the solutions on the 30 testing scenarios to properly assess their quality in unseen scenarios. Figure 3 shows the boxplots of each independent execution in each algorithm. EAs have the highest variability, while the other three algorithms have more robust boxplots.

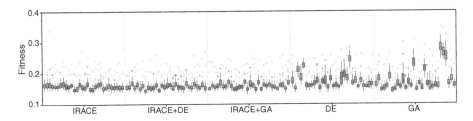

**Fig. 3.** Fitness of the solutions obtained by the 5 algorithms. Each boxplot shows the distribution of fitness values of one solution on the 30 traffic scenarios in the test set.

**Table 3.** Statistics of each algorithm from the best solutions obtained in the testing phase. We mark in bold the lower value of each metric.

| Algorithm | Mean | Median | STD Dev |
|---|---|---|---|
| IRACE+DE | **0.1607** | **0.1571** | **0.0175** |
| IRACE+GA | 0.1620 | 0.1577 | 0.0184 |
| IRACE | 0.1623 | 0.1581 | 0.0184 |
| GA | 0.1793 | 0.1630 | 0.0407 |
| DE | 0.1769 | 0.1676 | 0.0275 |

**Table 4.** Wilcoxon Test $p$-value of the testing set with Holm correction.

|  | IRACE | DE | IRACE+DE | GA |
|---|---|---|---|---|
| DE | $<2e-16$ | — | — | — |
| IRACE+DE | **0.0237** | $<2e-16$ | — | — |
| GA | $3.5e-11$ | **0.0011** | $<2e-16$ | — |
| IRACE+GA | 0.3284 | $<2e-16$ | 0.2122 | $5.3e-13$ |

To better compare the different algorithms, we summarize the mean, median and standard deviation (see Table 3) between the different independent runs. IRACE+DE gets the best results in each of these metrics, followed by IRACE and IRACE+GA, which are very similar, and the last ones, the EAs. This is a great result for IRACE+DE because, as it was proved in the training results, remarks the competitiveness of the algorithm also in the testing phase.

Anyway, the algorithms using IRACE obtain very similar results. This makes us wonder if there are significant differences between them. To study this, we perform a Wilcoxon rank-sum test between the algorithms to check if there are significant differences. Table 4 shows the $p$-values reported by the test. As we expected, IRACE+DE has significant differences compared to IRACE and EAs. This result support our working hypothesis: including EAs (specifically a DE) into IRACE can improve the performance. IRACE+GA and IRACE do not offer significant differences between them, which is not a bad result either, since at least the hybrid algorithm reaches a similar performance to IRACE. Lastly, EAs have significant differences with the others algorithms.

Finally, we perform an $\widehat{A}_{12}$ test to see if our hybrid algorithms (especially IRACE+DE) effectively beat the other competitors. Table 5 shows the results for the $\widehat{A}_{12}$ test. We observe that IRACE+DE is better than standard IRACE 53.62% of the time, and 66.17% better than evolutionary ones. While IRACE+GA is 51.33% of the time better than IRACE and 64.34% better than the evolutionary ones. These differences are in favour of our approach. After this experimentation, we can conclude that hybridizing IRACE with evolutionary algorithms is a viable and competitive option. With this idea, we join the best of both types of algorithms obtaining a powerful and robust algorithm, which allows us to find better solutions for TLSP than the commonly used algorithms.

**Table 5.** Results of the $\widehat{A}_{12}$ test for testing. Probability that the algorithm (column) is better than another algorithm (row). We highlight in bold the values when the algorithm in the column is better than the algorithm in the row.

|          | IRACE  | IRACE+DE | IRACE+GA | GA     | DE     |
|----------|--------|----------|----------|--------|--------|
| IRACE    | —      | **0.5362** | **0.5133** | 0.4066 | 0.3198 |
| IRACE+DE | 0.4638 | —        | 0.4780   | 0.3820 | 0.2946 |
| IRACE+GA | 0.4867 | **0.5220** | —        | 0.3985 | 0.3147 |
| GA       | **0.5934** | **0.6180** | **0.6015** | —      | 0.4506 |
| DE       | **0.6802** | **0.7054** | **0.6853** | **0.5494** | —      |

## 5.3 Impact in Real World

The previous analysis has focused on the fitness function, an approximation which encompasses some knowledge of traffic flow to guide the search, however, it is quite complex to extract useful information for the domain's expert. Therefore in this section, we study the main traffic and environmental indicators which give the domain's expert more information about the solution.

In a real-world problem, it is desirable to analyze the impact that a representative solution of the different algorithms would have in a real environment. We choose one solution from each algorithm, as a typical traffic light plan as follows: (i) we calculate the mean of the fitness obtained in the 30 scenarios of the testing set by each of the 30 solutions of each algorithm, (ii) we order upwards these mean fitness for each algorithm, (iii) we select, as the representative, the solution whose fitness value is at the 16th position, that is, immediately following the median solution. We cannot select the median because there are an even number of solutions (30).

We simulate again each of the representative solutions in the test scenarios but allowing all the vehicles to reach their destination. This means that the fitness values are not penalized, hence, they are smaller than those reported in the previous boxplots. With these new simulations, we obtain 34 different traffic and environmental measures of the 30 testing scenarios. Figure 4 shows some of the most important measures for each algorithm. In all measures, hybrid algorithms get the best results. IRACE+DE obtains the best average values in *MeanTravelTime* and *MeanWaitingTime*, while IRACE+GA has the lowest *MaxTravelTime* and *MaxWaitingTime*. In practice, if we implement the IRACE+DE solution, citizens would complete the journeys in less time (329.60 s) and with less waiting time at intersections (88.84 s). If IRACE+GA solution were implemented, the *MeanTravelTime* is higher than in IRACE+DE solution, but in the worst case (*MaxTravelTime* and *MaxWaitingTime*), IRACE+GA obtains the minimum values. On the complete opposite side, we have GA and DE, with the worst results of the comparison.

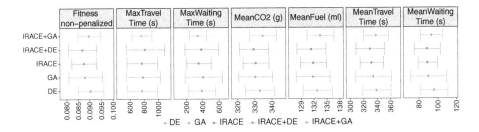

**Fig. 4.** Traffic measures per vehicle. Mean values (and standard deviation) over 30 test traffic scenarios of the median solutions for the five algorithms.

Regarding the environmental impact (fuel consumption and CO2 emissions), IRACE+DE gives the most eco-friendly solutions. Nowadays, pollution is a serious issue in many cities, so offering solutions that reduce emissions and fuel is of vital importance in today's cities.

With all these results, we can confirm that better TLPs result in less CO2 emissions, less fuel consumption, and less journey time for the citizen. Our hybrid proposals, specially IRACE+DE, not only offer competitive solutions from a scientific point of view, but it would also have a positive impact in the city at multiple levels both environmental and for the quality of life of the citizens.

# 6    Conclusions

In this article, we have proposed new hybrid algorithms combining IRACE with two evolutionary algorithms: GA and DE. These new hybrid algorithms are ideally suited for black-box numerical optimization problems under uncertainty, by using evolutionary operators designed for numerical optimization to generate better solutions, while handling uncertainty by means of the elitist racing strategy in IRACE. We have used these hybrid algorithms (IRACE+DE and IRACE+GA), IRACE, a GA, and a DE, to solve the TLSP using the real instance of Málaga, Spain, and the SUMO traffic simulator to evaluate the solutions. The results obtained in the experiments confirm the competitiveness of the hybridization strategy. Both hybrid algorithms offer better results than GA (60% of the time) and DE (70% of the time) on realistic traffic scenarios. Particularly, IRACE+DE returns the best results during the testing, being also competitive during the training. Besides, we have seen the impact that the solutions would have on the city. Our hybridization strategies obtain the best results in travel times, fuel consumption, $CO_2$ emissions, etc. These results reinforce our algorithmic proposal and show the efficiency that IRACE+DE and IRACE+GA obtain when solving a real-world problem.

As future work, we will consider other algorithms and operators that have proven to be effective in numerical optimization problems for hybridization with IRACE. Although preliminary experiments hybridizing IRACE with JADE [23], a well-known variant of DE, did not improve the results over the IRACE+DE proposed in this paper, we plan to perform a deeper analysis of IRACE+JADE to extract any insights about the behavior of the new hybrid algorithms. Also, we plan to test our hybrid algorithms on other black-box numerical optimization problems under uncertainty to further validate our results.

**Acknowledgements.** This research was partially funded by the University of Málaga, Andalucía Tech and the project TAILOR Grant #952215, H2020-ICT-2019-3. C. Cintrano is supported by a FPI grant (BES-2015-074805) from Spanish MINECO. M. López-Ibáñez is a "Beatriz Galindo" Senior Distinguished Researcher (BEAGAL 18/00053) funded by the Ministry of Science and Innovation of the Spanish Government. J. Ferrer is supported by a postdoc grant (DOC/00488) funded by the Andalusian Ministry of Economic Transformation, Industry, Knowledge and Universities.

# References

1. Behrisch, M., Bieker, L., Erdmann, J., Krajzewicz, D.: SUMO - simulation of urban mobility: an overview. In: SIMUL 2011, The Third International Conference on Advances in System Simulation, ThinkMind, Barcelona, Spain, pp. 63–68 (2011)
2. Blum, C., Raidl, G.R.: Hybrid metaheuristics-powerful tools for optimization. In: Artificial Intelligence: Foundations, Theory, and Algorithm. Springer, Heidelberg (2016). https://doi.org/10.1007/978-3-319-30883-8

3. Bravo, Y., Ferrer, J., Luque, G., Alba, E.: Smart mobility by optimizing the traffic lights: a new tool for traffic control centers. In: Alba, E., Chicano, F., Luque, G. (eds.) Smart-CT 2016. LNCS, vol. 9704, pp. 147–156. Springer, Cham (2016). https://doi.org/10.1007/978-3-319-39595-1_15

4. Deb, K., Agrawal, S.: A niched-penalty approach for constraint handling in genetic algorithms. In: Dobnikar, A., Steele, N.C., Pearson, D.W., Albrecht, R.F. (eds.) Artificial Neural Nets and Genetic Algorithms (ICANNGA-99), pp. 235–243. Springer, Vienna (1999). https://doi.org/10.1007/978-3-7091-6384-9_40

5. Ferrer, J., García-Nieto, J., Alba, E., Chicano, F.: Intelligent testing of traffic light programs: validation in smart mobility scenarios. Math. Probl. Eng. **2016**, 1–19 (2016)

6. Ferrer, J., López-Ibáñez, M., Alba, E.: Reliable simulation-optimization of traffic lights in a real-world city. Appl. Soft Comput. **78**, 697–711 (2019)

7. García-Nieto, J., Alba, E., Olivera, A.C.: Swarm intelligence for traffic light scheduling: application to real urban areas. Eng. Appl. Artif. Intell. **25**(2), 274–283 (2012)

8. García-Nieto, J., Olivera, A.C., Alba, E.: Optimal cycle program of traffic lights with particle swarm optimization. IEEE Trans. Evol. Comput. **17**(6), 823–839 (2013)

9. Heidrich-Meisner, V., Igel, C.: Hoeffding and Bernstein races for selecting policies in evolutionary direct policy search. In: Danyluk, A.P., Bottou, L., Littman, M.L. (eds.) Proceedings of the 26th International Conference on Machine Learning, ICML 2009, pp. 401–408. ACM Press, New York (2009)

10. Krajzewicz, D., Erdmann, J., Behrisch, M., Bieker, L.: Recent development and applications of SUMO - Simulation of Urban MObility. Int. J. Adv. Syst. Meas. **5**(3–4), 128–138 (2012)

11. López-Ibáñez, M., Dubois-Lacoste, J., Pérez Cáceres, L., Stützle, T., Birattari, M.: The irace package: iterated racing for automatic algorithm configuration. Oper. Res. Perspect. **3**, 43–58 (2016)

12. Nebro, A.J., Durillo, J.J., Vergne, M.: Redesigning the jMetal multi-objective optimization framework. In: Laredo, J.L.J., Silva, S., Esparcia-Alcázar, A.I. (eds.) GECCO (Companion), pp. 1093–1100. ACM Press, New York (2015)

13. Péres, M., Ruiz, G., Nesmachnow, S., Olivera, A.C.: Multiobjective evolutionary optimization of traffic flow and pollution in Montevideo Uruguay. Appl. Soft Comput. **70**, 472–485 (2018)

14. Price, K., Storn, R.M., Lampinen, J.A.: Differential Evolution: A Practical Approach to Global Optimization, p. 539. Springer, Heidelberg (2005). https://doi.org/10.1007/3-540-31306-0

15. Sánchez, J., Galán, M., Rubio, E.: Applying a traffic lights evolutionary optimization technique to a real case: "Las Ramblas" area in Santa Cruz de Tenerife. IEEE Trans. Evol. Comput. **12**(1), 25–40 (2008)

16. Sánchez-Medina, J.J., Galán-Moreno, M.J., Rubio-Royo, E.: Traffic signal optimization in "La Almozara" district in Saragossa under congestion conditions, using genetic algorithms, traffic microsimulation, and cluster computing. IEEE Trans. Intell. Transp. Syst. **11**(1), 132–141 (2010). ISSN 1524-9050

17. Stolfi, D.H., Alba, E.: Red swarm: reducing travel times in smart cities by using bio-inspired algorithms. Appl. Soft Comput. **24**, 181–195 (2014)

18. Stolfi, D.H., Alba, E.: An evolutionary algorithm to generate real urban traffic flows. In: Puerta, J., et al. (eds.) CAEPIA 2015. LNCS (LNAI), vol. 9422, pp. 332–343. Springer, Cham (2015). https://doi.org/10.1007/978-3-319-24598-0_30

19. Syswerda, G.: Uniform crossover in genetic algorithms. In: Schaffer, J.D. (ed.) Proceedings of the Third International Conference on Genetic Algorithms, pp. 2–9. Morgan Kaufmann Publishers, San Mateo (1989)
20. Teklu, F., Sumalee, A., Watling, D.: A genetic algorithm approach for optimizing traffic control signals considering routing. Comput. Aided Civ. Infrastruct. Eng. **22**(1), 31–43 (2007)
21. Teo, K.T.K., Kow, W.Y., Chin, Y.K.: Optimization of traffic flow within an urban traffic light intersection with genetic algorithm. In: Proceedings - 2nd International Conference on Computational Intelligence, Modelling and Simulation, CIMSim 2010, pp. 172–177. IEEE Press (2010)
22. Vargha, A., Delaney, H.D.: A critique and improvement of the CL common language effect size statistics of McGraw and Wong. J. Educ. Behav. Stat. **25**(2), 101–132 (2000)
23. Zhang, J., Sanderson, A.C.: JADE: adaptive differential evolution with optional external archive. IEEE Trans. Evol. Comput. **13**(5), 945–958 (2009)

# Decomposition-Based Multi-objective Landscape Features and Automated Algorithm Selection

Raphaël Cosson[1(✉)], Bilel Derbel[1], Arnaud Liefooghe[1], Hernán Aguirre[2], Kiyoshi Tanaka[2], and Qingfu Zhang[3]

[1] Univ. Lille, CNRS, Inria, Centrale Lille, UMR 9189 CRIStAL, 59000 Lille, France
{raphael.cosson.etu,bilel.derbel,arnaud.liefooghe}@univ-lille.fr
[2] Faculty of Engineering, Shinshu University, Nagano, Japan
{ahernan,ktanaka}@shinshu-u.ac.jp
[3] City University of Hong Kong, Kowloon Tong, Hong Kong
qingfu.zhang@cityu.edu.hk

**Abstract.** Landscape analysis is of fundamental interest for improving our understanding on the behavior of evolutionary search, and for developing general-purpose automated solvers based on techniques from statistics and machine learning. In this paper, we push a step towards the development of a landscape-aware approach by proposing a set of landscape features for multi-objective combinatorial optimization, by decomposing the original multi-objective problem into a set of single-objective sub-problems. Based on a comprehensive set of bi-objective $\rho$mnk-landscapes and three variants of the state-of-the-art MOEA/D algorithm, we study the association between the proposed features, the global properties of the considered landscapes, and algorithm performance. We also show that decomposition-based features can be integrated into an automated approach for predicting algorithm performance and selecting the most accurate one on blind instances. In particular, our study reveals that such a landscape-aware approach is substantially better than the single best solver computed over the three considered MOEA/D variants.

## 1 Introduction

**Context.** Evolutionary algorithms have been proven extremely effective for solving a broad range of optimization problems. In the last decades, the community has gained a deep understanding on the key components underlying the design of a successful evolutionary approach for a given problem. However, one of the main challenge remains to *automate* the process of choosing the most suitable algorithm or configuration for the instance under consideration. In fact, it is well known that the structural properties of an optimization problem highly impact the dynamics and performance of search algorithms, leading to the requirement of adopting a *landscape-aware algorithm selection and configuration* methodology for the success of evolutionary problem solving. On the one hand, landscape

C. Zarges and S. Verel (Eds.): EvoCOP 2021, LNCS 12692, pp. 34–50, 2021.
https://doi.org/10.1007/978-3-030-72904-2_3

analysis [15] provides a principled approach for studying and analyzing the relation between the underlying search space structure and algorithm behavior. On the other hand, machine learning techniques can be leveraged to perform sophisticated tasks, such as predicting algorithm performance, identifying the best algorithm configuration, or selecting the best algorithm [7]. As a byproduct, and since the pioneer work of Rice [14], landscape-aware algorithm configuration has emerged as an appealing approach for increasing the effectiveness and efficiency of evolutionary algorithms. In this paper, we contribute to the development of such an approach when specifically dealing with *multi-objective* and *combinatorial* optimization problems.

**Related Work.** Independently of whether the target problems and algorithms aim at optimizing a single or multiple objectives, and of whether they have a combinatorial or continuous nature, every landscape-aware methodology needs to address the following two research challenges: (i) the design of a set of informative and high-level *landscape features*, and (ii) the development of automated recommendation systems integrating the so-designed features on the basis of *statistical or machine learning prediction models*. Looking at the specialized literature, a large amount of work has been conducted for single-objective optimization since pioneering works in the field [4]. For instance, in single-objective continuous optimization, the exploratory landscape analysis (ELA) constitutes one major achievement made by the community to collect and combine existing features under a common tool and methodology [7]. Similarly, a number of features for single-objective combinatorial optimization have been developed over the years, and recent studies integrate them into sophisticated automated approaches for algorithm selection and configuration [1,9]. Those single-objective features are either based on problem-specific characteristics such as the maximum cost between two cities in the traveling salesperson problem [13], or on general descriptors from the underlying landscape. In the latter case, this is achieved by relying on a neighborhood relation over the search space in order to define a (combinatorial) landscape, and by studying its properties and characteristics in terms of multimodality, ruggedness, or neutrality [15].

Despite the significant progress made in the last decades, the existing literature on the development of a unified landscape-aware approach targeting multi-objective optimization problems is more scarce [6,8]. Although landscape features can in principle be applied to the multi-objective case, the statistical and machine learning models considered in the single-objective case are still to be studied and validated when turning into a multi-objective setting. For multi-objective combinatorial optimization, one can refer to the recent study in [8], providing a comprehensive analysis of problem-independent multi-objective landscape features, and showing their effectiveness in predicting algorithm performance and in selecting from an algorithm portfolio. The features described there are mostly based on *dominance* and (hypervolume) *indicator*. In fact, they were designed to grasp the landscape characteristics, but also to capture the search behavior of dominance-based multi-objective algorithms. In a subsequent study [10], those previous features were shown to provide useful information about the performance of different classes of multi-objective evolutionary algorithms (MOEAs). However, their correlation with the performance of decomposition-based MOEAs reveal to be less significant [10]. Getting inspiration from the fact that successful MOEAs may rely on different search paradigms, our work considers applying other mechanisms, such as decomposition, to design new multi-objective landscape features, hence pushing one step further the development of a landscape-aware automated evolutionary approach.

**Methodology and Contribution.** We rely on the concept of decomposition [16] to develop a new set of general-purpose multi-objective landscape features and to study their effectiveness when integrated into an automated algorithm selection task. Our interest in the concept of decomposition stems from the fact that it provides a state-of-the-art framework, represented by the MOEA/D algorithm [18], by simply decomposing the multi-objective problem into a number of single-objective sub-problems. We view the decomposition paradigm as an opportunity to leverage existing single-objective features for multi-objective landscape analysis. Our contributions can be summarized as follows.

(i) We propose a new set of high-level multi-objective landscape features based on decomposition. Intuitively, we attempt to capture the multi-objective landscape by aggregating the characteristics from the single-objective sub-problem landscapes obtained by decomposition. As such, the proposed features are obtained by first defining single-objective features for each sub-problem, and then aggregating them by means of descriptive statistics.

(ii) We consider the task of predicting performances of three variants of the state-of-the-art MOEA/D algorithm using a tree-based regression model to study the effectiveness of a decomposition-based landscape-aware methodology for automatically selecting the best performing algorithm for a given instance.

(iii) Throughout an extensive set of experiments using $\rho$mnk-landscapes with two objectives as a case study, we conduct a systematic analysis on the association between the designed features and the benchmarked landscapes, as well as the association between features and algorithm performance. Our findings reveal that the designed features are able to capture the benchmark parameters ($\rho$ and $k$), and to substantially improve the so-called single best solver when integrated into a landscape-aware algorithm selection approach.

**Outline.** In Sect. 2, we describe the proposed multi-objective landscape features. In Sect. 3, we study the association of features with benchmark parameters. In Sect. 4, we investigate the integration of the proposed features into an automated landscape-aware approach. In Sect. 5, we conclude the paper.

# 2 From Single- to Multi-objective Features Based on Decomposition

## 2.1 Multi-objective Optimization

A multi-objective combinatorial optimization problem can be defined by a set of $\mathtt{m}$ objective functions $f = (f_1, f_2, \ldots, f_{\mathtt{m}})$, and a discrete set $X$ of feasible solutions in the *decision space*. Let $Z = f(X) \subseteq \mathbb{R}^{\mathtt{m}}$ be the set of feasible outcome vectors in the *objective space*. To each solution $x \in X$ is assigned an objective vector $z \in Z$, on the basis of the vector function $f : X \to Z$. In a maximization context, an objective vector $z \in Z$ is *dominated* by a vector $z' \in Z$ iff $\forall m \in \{1, \ldots, \mathtt{m}\}$, $z_m \leqslant z'_m$ and $\exists m \in \{1, \ldots, \mathtt{m}\}$ s.t. $z_m < z'_m$. A solution $x \in X$ is dominated by a solution $x' \in X$ iff $f(x)$ is dominated by $f(x')$. A solution $x^\star \in X$ is *Pareto optimal* if there does not exist any other solution $x \in X$ such that $x^\star$ is dominated by $x$. The set of all Pareto optimal solutions is the *Pareto set*. Its mapping in the objective space is the *Pareto front*. Our goal is to identify a good Pareto set approximation, for which multi-objective evolutionary algorithms (MOEAs) constitute a popular effective option [3].

## 2.2 Rationale, Methodology and Features Overview

In this work, we get inspiration from the so-called MOEAs based on decomposition [16] in order to design new high-level multi-objective features. This algorithm class is based on searching for good-performing solutions in multiple regions of the Pareto front by *decomposing* the original multi-objective problem into a number of scalarized *single-objective sub-problems*. Each sub-problem is obtained by a different parameterization of the same underlying scalarizing function. This is typically what the state-of-the-art MOEA/D algorithm [18] performs, while introducing a cooperation among sub-problem solving. In particular, this offers much flexibility for integrating existing single-objective search operators and solvers, which is actually one of the main reasons for the success of decomposition-based MOEAs. Let us however recall that our main goal is *not* to design a new multi-objective algorithm, but to design new multi-objective features that can feed the design of a general-purpose landscape-aware approach.

Therefore, we propose to rely on the simple observation that *each* of the so-defined sub-problems also implies a single-objective landscape that we can attempt to analyze and characterize. In other words, by studying the single-objective landscape implied by the sub-problems, we should be able to extract some knowledge about the original multi-objective problem. More precisely, the methodology that we adopt in the reminder of this paper consists in: (i) defining a number of single-objective landscapes using decomposition, (ii) extracting single-objective features for each sub-problem landscape, and (iii) aggregating those single-objective features into new multi-objective features. These steps are detailed below.

**Sub-problem Landscape Definition.** Firstly, we define $\mu$ scalarized single-objective sub-problems, where both the scalarizing function and the $\mu$ value are user-defined parameters. Among the different scalarizing functions that may be used, the Chebyshev function is one of the most effective since it can be shown that any Pareto optimal solution can be achieved by solving a well-parameterized Chebyshev sub-problem. In the rest of this paper, we should hence use the Chebyshev scalarizing function: $g(x|\omega) = \max_{i \in \{1,\ldots,m\}} \omega_i \cdot |z_i^\star - f_i(x)|$, such that $x \in X$, $\omega = (\omega_1, \ldots, \omega_m)$ is a positive weight vector, and $z^\star = (z_1^\star, \ldots, z_m^\star)$ is a reference point such that $z_i^\star > f_i(x) \ \forall x \in X$, $i \in \{1, \ldots, m\}$. It should be clear that $\mu$ different sub-problems can be obtained by choosing $\mu$ different weight vectors, denoted by $\omega^j$, $j \in \{1, \ldots, \mu\}$.

It is worth noticing that we do not make any assumption about the original (black-box) multi-objective problem, so that we have no information about what region every sub-problem is actually mapping to. Hence, the value of $\mu$ as well as the procedure to generate the weight vector can be a critical issue. This is studied in more details later when reporting our empirical results.

**Single-Objective Landscape Features.** Next, we define a landscape for every single-objective sub-problem $j \in \{1, \ldots, \mu\}$, for which we compute a number of underlying high-level single-objective features. Following the standard literature on single-objective landscape analysis [15], the landscape of sub-problem $j$ can be defined as a triplet $(X, \mathcal{N}, g(\cdot|\omega^j))$, such that $\mathcal{N} : X \longrightarrow 2^X$ is a neighborhood relation defined among solutions for the considered problem; e.g., 1-bit-flips for binary strings, or swaps for permutations.

The considered sub-problem features are based on sampling the so-defined landscape to compute some statistics. Following the standard literature [5,15], we consider two simple sampling strategies, namely random walks (rws) and adaptive walks (aws). Generally speaking, a walk is an ordered sequence of solutions $(x_0, x_1, \ldots, x_\ell)$ such that $x_0 \in X$, and $x_t \in \mathcal{N}(x_{t-1}) \ \forall t \in \{1, \ldots, \ell\}$. In a random walk, $x_t$ being the current solution, the next solution $x_{t+1}$ is simply a random neighbor under $\mathcal{N}$. The length of a random walk is a user-defined parameter. In an adaptive walk, the next solution $x_{t+1}$ is selected to be an improving neighbor with respect to the single-objective scalarizing function $g(\cdot|\omega^j)$. Consequently, the length of an adaptive walk is the number of steps performed until no further improvement is possible, $x_\ell$ is then a local optimum. Notice that the reference point $z^*$ required for computing the scalar fitness values is updated on the basis of the best objective values seen so far during the walk.

Given a sub-problem $j \in \{1, \ldots, \mu\}$ and a walk $(x_0, x_1, \ldots, x_\ell)$, we consider the following four classes of single-objective features, as summarized in Table 1:

- **Fitness value (fv_*).** In the first class, we compute some statistics informing about the distribution of fitness values observed along the walk. More precisely, we consider the mean (avg) and standard deviation (sd) of the fitness values of collected solutions. We also consider three additional statistics, namely the first auto-correlation coefficient (r1), the kurtosis (kr), and the skewness (sk) of fitness values. The kurtosis and the skewness are standard measures in statistical analysis, while the first auto-correlation coefficient is mostly used in the landscape analysis literature. Denoting by $\bar{g}(\cdot|\omega^j)$ the average fitness value of solutions in the walk, the first auto-correlation coefficient is defined as follows [5]:

$$r1 = \frac{\sum_{t=0}^{\ell-1} \left(g(x_t|\omega^j) - \bar{g}(\cdot|\omega^j)\right) \cdot \left(g(x_{t+1}|\omega^j) - \bar{g}(\cdot|\omega^j)\right)}{\left(\sum_{t=0}^{\ell-1} g(x_t|\omega^j) - \bar{g}(\cdot|\omega^j)\right)^2}$$

- **Fitness difference (fd_*).** In the second class, we compute the average fitness difference with the neighboring solutions for every $x_i$, $i \in \{1,\ldots,\ell\}$: $\frac{1}{|\mathcal{N}(x_i)|} \sum_{x \in \mathcal{N}(x_i)} \left(g(x_i|\omega^j) - g(x|\omega^j)\right)$. Similarly, we consider the mean (avg), standard deviation (sd), minimum (min) and maximum (max) fitness difference over solutions from the walk.
- **Improving neighbors (in_*).** In the third class, we compute the proportional number of improving neighbors for each solution $x_i$, $i \in \{0,\ldots,\ell\}$. Then, we consider the mean, standard deviation, minimum, and maximum of this measure. It is worth noticing that the second and third classes of features require to evaluate the fitness value of neighbors from all solutions from the walk.
- **Length of the adaptive walk (law_*).** The fourth class only contains features extracted from the adaptive walk. In particular, we consider the length of the adaptive walk as a feature to characterize the sub-problem landscape. This length was shown to provide an estimation of the number of local optima in single-objective landscape analysis [5].

**Table 1.** A summary of the proposed landscape features.

| Description | Sub-problem features | | MO features |
|---|---|---|---|
| | Random walk | Adaptive walk | |
| Fitness values | fv_rws_s | fv_aws_s | fv_rws_s_r; fv_aws_s_r |
| | $s \in \{\text{avg}, \text{sd}, \text{r1}, \text{kr}, \text{sk}\}$ | | $r \in \{\text{avg}, \text{sd}, \text{c1}, \text{c2}\}$ |
| Fitness differences | fd_rws_s | fd_aws_s | fd_rws_s_r; fd_aws_s_r |
| | $s \in \{\text{avg}, \text{sd}, \text{min}, \text{max}\}$ | | $r \in \{\text{avg}, \text{sd}, \text{c1}, \text{c2}\}$ |
| Improving neighbors | in_rws_s | in_aws_s | in_rws_s_r; in_aws_s_r |
| | $s \in \{\text{avg}, \text{sd}, \text{min}, \text{max}\}$ | | $r \in \{\text{avg}, \text{sd}, \text{c1}, \text{c2}\}$ |
| Walk length | – | law | law_r |
| | | | $r \in \{\text{avg}, \text{sd}, \text{c1}, \text{c2}\}$ |

**Aggregated Multi-objective Features.** The features described above are computed for each sub-problem $j \in \{1,\ldots,\mu\}$, and then have a dimension $\mu$.

For $\mu > 1$, we need to aggregate these $\mu$-dimensional single-objective features into 1-dimensional multi-objective features. To do so, we use two standard statistics, namely the mean (avg) and the standard deviation (sd). In addition, we use a polynomial regression in order to fit each single-objective feature as a function of the weight vector $\omega^j$ of sub-problem $j$. The coefficient of the polynomial model are then used as additional aggregated features. In this study, since we experiment bi-objective optimization problems, we consider a second order polynomial regression and propose to use the first (c1) and the second (c2) coefficients as additional multi-objective features. As summarized in Table 1, we end up with 4 aggregation statistics over respectively 5, 4, and 4 fv, fd, and in features, in addition to 4 aggregated features on the length of adaptive walk. This amounts to a total of 108 decomposition-based multi-objective features.

## 3    A Preliminary Exploratory Analysis

As a first step, we analyze the relevance of the proposed features in capturing the characteristics of multi-objective optimization problems regardless of any particular evolutionary search algorithm. Therefore, we conduct a preliminary exploratory analysis in order to highlight the association between the designed features and the properties of well-established benchmark landscapes.

### 3.1    Experimental Setup

Following previous work [8], we consider $\rho$mnk-landscapes [17] as a problem-independent model used for constructing multi-objective multimodal landscapes with objective correlation. Candidate solutions are binary strings of size $n$. The objective function vector $f = (f_1, \ldots, f_i, \ldots, f_m)$ is defined as $f : \{0,1\}^n \mapsto [0,1]^m$ such that each objective $f_i$ is to be maximized. The objective value $f_i(x)$ of a solution $x = (x_1, \ldots, x_j, \ldots, x_n)$ is an average value of the individual contributions associated with each variable $x_j$. Given objective $f_i$, and each variable $x_j$, a component function $f_{ij} : \{0,1\}^{k+1} \mapsto [0,1]$ assigns a real-valued contribution for every combination of $x_j$ and its $k$ *epistatic interactions* $\{x_{j_1}, \ldots, x_{j_k}\}$. These $f_{ij}$-values are uniformly distributed in $[0,1]$. The objective functions to be maximized can written as: $f_i(x) = \frac{1}{n} \sum_{j=1}^{n} f_{ij}(x_j, x_{j_1}, \ldots, x_{j_k})$, $\forall i \in \{1, \ldots, m\}$. In this work, the $k$ epistatic interactions are set uniformly at random among the $(n-1)$ variables other than $x_j$. By increasing the value of $k$ from 0 to $(n-1)$, problem instances can be gradually tuned from smooth to rugged. The $f_{ij}$-values additionally follow a multivariate uniform distribution of dimension $m$, defined by an $m \times m$ positive-definite symmetric covariance matrix $(c_{pq})$ s.t. $c_{pp} = 1$ and $c_{pq} = \rho$ for all $p \neq q$ where $\rho > \frac{-1}{m-1}$ defines the correlation among the objectives.

    In our work, we focus on bi-objective landscapes, i.e., $m = 2$. We use a latin hypercube sampling to generate a set of 1 000 balanced instances spanning parameters ranges: $n \in \{50, 51, \ldots, 200\}$, $k \in \{0, 1, 2, \ldots, 8\}$ and $\rho \in\, ]-1, 1]$. The random walk length is set to $\ell = 1\,000$ across all problem sizes. A unique random walk is performed for *all* sub-problems, whereas one adaptive walk is performed for *each* sub-problem. The number of sub-problems is set to $\mu = 20$ and weight vectors are distributed uniformly; i.e., $\omega^j = ((j-1)/(\mu-1), 1 - (j-1)/(\mu-1))$. The neighborhood relation is the standard 1-bit-flip.

For our analysis, we first conduct an exploratory analysis to better visualize and understand the proposed features. Next, we construct a regression model to study the accuracy of features in grasping the global properties of $\rho$mnk-landscapes, and we analyze the correlation among features.

## 3.2 Visual Analysis of Single-Objective Features

In Fig. 1, we report the values of *single-objective* features as a function of sub-problems. Due to space restriction, we report a single representative feature for each of the four classes. The blue curves correspond to $\rho$mnk-landscapes with $k = 0$, the green ones to $k = 2$ and the red ones to $k = 4$. The color scales from red to orange, green to cyan, and blue to purple, respectively, and correspond to the objective correlation parameter $\rho$ varying from 1 to $-1$; i.e., from highly correlated to highly conflicting objectives. For example, the standard deviation of fitness values from a random walk (fv_rws_sd, bottom left), the average fitness difference from a random walk (fd_rws_avg, second line, second column) and the standard deviation of improving neighbors (in_rdw_sd, second line, third column) gives a clear differentiation between landscapes with different k-values. The lower the benchmark parameter k, the lower the standard deviation of fitness values and the average fitness difference. Similarly, the standard deviation of fitness values from an adaptive walk (vf_aws_sd, top left) and the length of an adaptive walk (law_aws_avg, top right) seems to be clearly associated with parameter $\rho$. The flatter the curve rendering the evolution of these two features as a function of weight vectors, the higher the objective correlation $\rho$.

From our visual inspection, we can conclude that the landscape features are representative of the different *global* benchmark parameters, which are *unknown* in a black-box optimization scenario. However, this first analysis considers the single-objective features and not the aggregated 1-dimensional multi-objective features, which are discussed next.

## 3.3 Correlation Analysis of Features and Landscape Parameters

Investigating the accuracy of the designed multi-objective features, we consider a typical machine learning task consisting in predicting the value of the (unknown) global benchmark parameters k and $\rho$. We respectively construct a random forest classification model and a random forest regression model [2], using the whole set of multi-objective features computed over all considered $\rho$mnk-landscapes. Random forest has the nice property of providing a measure of feature importance for model fitting. In Fig. 2, we report the relative importance of each feature extracted from the random forest models, using the Gini impurity as a measure of quality. Values are averaged over 10 independent repetitions of model fitting.

The first notable observation is that feature importance is different depending on whether we aim at predicting benchmark parameter $\rho$ or k. The objective

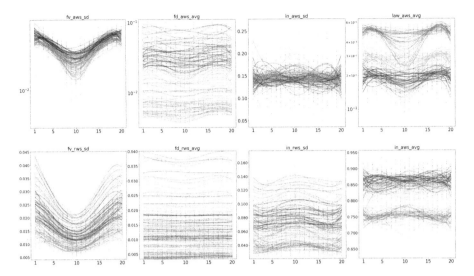

**Fig. 1.** Feature values of $\rho$mnk-landscapes decomposed into 20 sub-problems ($n = 25$, each color correspond to a particular configuration of $\rho$ and $k$). (Color figure online)

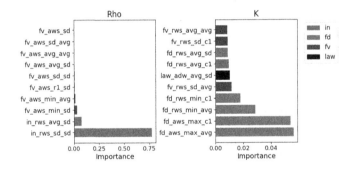

**Fig. 2.** Relative importance of features to predict $\rho$ (left) and $k$ (right). (Color figure online)

correlation $\rho$ appears mostly related to a single feature: the standard deviation of the number of improving neighbors (in_rws_sd). By contrast, deciding on parameter $k$ is related to multiple different features, that mostly correspond to the fitness difference computed from adaptive walks, in particular the maximum fitness difference (fd_aws_max). To push the analysis further, we show in Fig. 3 the Spearman correlation matrix among a subset of features as well as benchmark parameters $\rho$ and $k$. It is worth noticing that we do not show all 108 features, but only a subset of 60 features with representative correlation values due to space restriction. Interestingly, we can distinguish four clusters, denoted $C_1$, $C_2$, $C_3$ and $C_4$ in the figure. The largest cluster $C_1$ contains more than 30 features. There is a positive correlation between the first auto-correlation coefficient of fitness values vf_r1 and k, of at least 0.25. This cluster also contains a small subset of 2 features related to the number of improving neighbors in_*, with a particularly high correlation value (>0.8) (the $3 \times 3$ red points at the center of the figure). In the second cluster $C_2$, we observe a large subset of features being negatively correlated with k (the dark blue columns/lines intersecting in a completely red square in the middle-right of the figure). For both clusters related to k ($C_1$ and $C_2$), features are not extracted from a unique class. This means that features from different classes can be used to characterize k.

The remaining 2 clusters ($C_3$ and $C_4$) are associated with the second benchmark parameter $\rho$. In particular, besides the average number of improving neighbors (in_rws_avg), cluster $C_3$ also contains the standard deviation (in_rws_sd), both computed from a random walk. This latter feature, that was found to be the most important to predict $\rho$ in Fig. 2, has a nearly perfect negative correlation with $\rho$, equals to $-1$. The last cluster $C_4$ contains features based on the fitness values computed over adaptive walks (fv_aws), which appear to be the most positively correlated with $\rho$ ($3 \times 3$ red points at the bottom right of the figure), with a correlation higher than 0.7.

To summarize, we find that all feature classes are useful to characterize the (unknown) global benchmark parameter k, rendering the degree of non-linearity of the problem. However, the fitness distribution (fv) and the number of improving neighbors (in) classes have more importance than the fitness difference (fd) and the length of adaptive walks (law) for characterizing the (unknown) global benchmark parameter $\rho$, relating to objective correlation.

## 4   Landscape-Aware MOEA/D Selection

In this section, we conduct a second set of experiments in order to study the accuracy of the designed features when integrated into an automated algorithm selection approach. We consider the more sophisticated task of selecting the best performing algorithm among an algorithm portfolio. More precisely, we consider three variants of the well-established MOEA/D algorithm [12,18] as a case study. In the following, we start by briefly describing the considered portfolio before addressing our main target task.

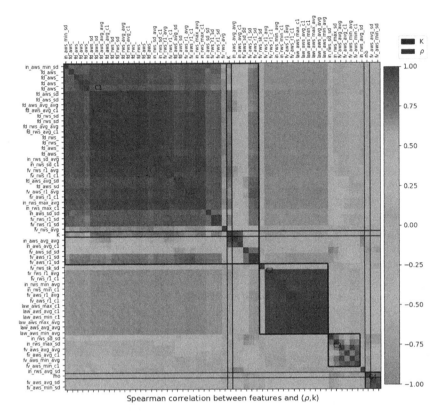

**Fig. 3.** Pairwise correlation among a subset of features. (Color figure online)

## 4.1 Algorithm Portfolio

As mentioned earlier, the MOEA/D algorithm is based on a flexible decomposition-based framework that can be configured in different manners. In its baseline variant [18], MOEA/D first decomposes the problem into a number of scalarized sub-problem, as discussed previously. Then, a solution is evolved for each sub-problem in a cooperative way. The algorithm iterates over the sub-problems and, at each iteration, an offspring is generated by means of crossover and mutation on the basis of parent solutions selected from the so-called $T$-neighborhood; i.e., the sub-problems corresponding to the $T$ closest weights in the objective space. The new offspring can then replace any of the sub-problem solutions in the $T$-neighborhood of the current sub-problem. This corresponds to a standard evolutionary process, where selection and replacement are performed iteratively over sub-problems. In [12], it is shown that the selection and replacement underlying the standard MOEA/D framework are key algorithm components that highly impact performance. Several generational variants are proposed therein, allowing to tune the selection and replacement underlying the

MOEA/D framework from fully cooperative (i.e., among *all* sub-problems) to fully selfish (i.e., independently of any other sub-problem).

Interestingly, it was found that no variant outperforms the other independently of the global benchmark parameter values $\rho$ and k for the considered $\rho$mnk-landscapes. Since such parameters are unknown in a black-box optimization scenario, the study presented in [12] leaves open the challenging question of which variant to choose in an automated manner. In addition, this constitutes a perfect and typical setting for the main automated algorithm selection task addressed in this paper. We consider, in the following, three representative variants of the MOEA/D framework, exposing different degrees of cooperation among sub-problem solving. For reproducibility, and in order to be consistent with the notations from [12], these variants are denoted as follows: (i) MOEA/D, referring to the standard variant [18], (ii) MOEA/D-SC, a generational variant where selection is performed in a selfish manner for every sub-problem whereas replacement is performed in a cooperative manner, and (iii) MOEA/D-SS, a (selfish) generational variant exposing the lower degree of cooperation among sub-problems. Besides population size, the three variants have the same set of parameters: $\delta = 1$, $nr = 2$, $p_{mut} = \frac{1}{n}$ and $p_{cr} = 1$. Due to space restriction, we refer to [12] for a full description of these MOEA/D variants.

In order to highlight the relevance of this portfolio in studying the accuracy of our features when integrated into a landscape-aware algorithm selection approach, we briefly report their relative performance using exactly the same set of $\rho$mnk-landscapes as in the previous section. Every algorithm is executed 20 times on each instance, using a population size equals to n, and a budget of $10^6$ evaluations. The performance of an algorithm is computed as its hypervolume relative deviation w.r.t. the best-found approximation set for each instance. The hypervolume measures the area covered by an approximation set and enclosed by a reference point [19]. For a given instance, the reference point is set to the best value seen across all runs for each objective. We then count the number of times an algorithm is statistically outperformed using a two-sided Mann-Whitney test with a p-value of 0.05 and a Bonferroni correction. Results are reported in Table 2.

**Table 2.** Performance matrix of the three MOEA/D variants. The diagonal reports the number of times where the corresponding algorithm is statistically outperformed by another one (the lower the better). The other cells report how many instances (out of 1000) the algorithm in the corresponding line is statistically better than the algorithm in the corresponding column (the higher the better).

|  | MOEA/D-SC | MOEA/D | MOEA/D-SS |
|---|---|---|---|
| MOEA/D-SC | 205 | 18 | 310 |
| MOEA/D | 85 | 137 | 312 |
| MOEA/D-SS | 120 | 119 | 622 |

We clearly see that each algorithm is outperformed by another one on a subset of instances. The basic MOEA/D variant seems to have a reasonably good behavior, since it is less-often outperformed than the two other variants overall (see diagonal). A more detailed analysis, omitted due to space restriction, shows that there is a complex interaction between algorithm performance and the benchmark parameters $\rho$ and k which can be summarized as follows: (i) the smaller k and $\rho$, the better MOEA/D and MOEA/D-SC against MOEA/D-SS, (ii) the larger $\rho$ (highly correlated), the better MOEA/D-SS, and (iii) the larger k and the smaller $\rho$, the better MOEA/D-SC. Of course, this general trend has some exceptions, but it shows the impact of the unknown benchmark parameters on the relative performance of algorithms.

## 4.2  Automated Algorithm Selection

**Task and Experimental Methodology.** We study the accuracy of the proposed features by investigating the selection of the best performing MOEA/D variant. For this purpose, we adopt the following standard supervised-learning approach. We first train three models in order to predict the performance of every considered MOEA/D variant. We use the average hypervolume deviation as defined in the previous section as a measure of algorithm performance on a given instance, which then corresponds to our output prediction variable. Considering an unseen test instance, the landscape features are first computed, the performance of each algorithm is then predicted on the basis of the trained models, and the algorithm having the best prediction is selected as the recommended one. We consider the same set of $\rho$mnk-landscapes described in the previous section. We adopt a standard validation methodology where an instance is selected for training with probability 0.9 and for testing with probability 0.1. We use a set of 100 random regression trees to learn and predict the expected relative hypervolume deviation. Reported values are computed over 50 independent runs.

We recall that the proposed multi-objective features rely on some weight vectors $\mu$. We consider a variable number of weight vectors in the range $\mu \in \{1, 2, 3, 4, 5, 6, 10, 20\}$. Additionally, we consider two alternatives for generating weight vectors, namely uniform or random. In a random setting, a weight vector is generated uniformly at random. In a uniform setting, the weight vector are evenly distributed in the objective space. In particular, for $\mu = 1$, the weight vector selected in the uniform setting is $(0.5, 0.5)$; i.e., the "middle" of the objective space. In this case, the multi-objective features are simply the same than the corresponding single-objective features from the single scalarized sub-problem. For $\mu = 2$, our uniform setting corresponds to weight vector $(1, 0)$ and $(0, 1)$. This means that our features are obtained by aggregating the single-objective features computed independently for each objective of the original multi-objective problem. The impact of this setting is carefully analyzed in our experiments.

**Prediction Accuracy.** In order to assess the prediction accuracy, we compute three complementary measures. The first two directly relate to the prediction error: the percentage of times the selected algorithm does not have the best hypervolume deviation in average, and the percentage of times the selected algorithm is statistically outperformed by at least one other algorithm. The third indicator, which is a straightforward adaptation from [11], measures the gap between: (i) the performance of the single best solver (SBS) having the best performance in average over the training set (without model training), and (ii) the performance of the virtual best solver (VBS), obtained by a model that would make perfect predictions. More precisely, let $I_{train}$ and $I_{test}$ be the set of training and testing instances, respectively, and let $\mathrm{rhf}(A, i)$ be the average relative hypervolume deviation of a given algorithm $A \in \mathcal{A} = \{\text{MOEA/D}, \text{MOEA/D-SS}, \text{MOEA/D-SC}\}$ on instance $i \in I_{train} \cup I_{test}$. For every algorithm $A \in \mathcal{A}$ and instance subset $J$, let $\overline{\mathrm{rhf}}(A, J) = \frac{1}{|J|} \sum_{i \in J} \mathrm{rhf}(A, i)$. We define SBS as the algorithm having the best $\overline{\mathrm{rhf}}$ value on the training set $I_{train}$, i.e., $SBS = \arg\min_{A \in \mathcal{A}} \{\overline{\mathrm{rhf}}(A, I_{train})\}$. We define VBS as the virtual 'algorithm' obtained by a perfect prediction model (an oracle); i.e., the algorithm with the best $\mathrm{rhf}(\cdot, i)$ value for each $i \in I_{test}$. Finally, let Recommended Solver (RS) be the algorithm predicted by the actual trained model. The *merit* indicator is:

$$M = \frac{\overline{\mathrm{rhf}}(RS, I_{test}) - \overline{\mathrm{rhf}}(VBS, I_{test})}{\overline{\mathrm{rhf}}(SBS, I_{test}) - \overline{\mathrm{rhf}}(VBS, I_{test})}$$

It should be clear that: (i) A merit of 0 indicates that the model does not make any error, (ii) A merit in the range $[0, 1[$ indicates that the model is more efficient than the SBS but worse than the VBS, (iii) a merit greater than 1 indicates that the model is worse than the SBS. Achieving a merit value of 0 is clearly a very challenging task and one seeks for a merit value below 1 (better than the SBS) and as close as possible to 0 (the VBS).

**Experimental Results and Discussion.** Our main results are summarized in Fig. 4 showing the accuracy indicators as a function of the number of weight vectors $\mu$ and their type (random or uniform). For completeness, we also show the $R^2$ coefficient obtained by the training models.

We first clearly see that the choice of the weight vector distribution is of critical importance. In fact, a random choice does not obtain a good accuracy, except when the number of weights $\mu$ is substantially large. By contrast, a uniform strategy appears to perform reasonably well, even when the number of weights is low. Interestingly, for uniform weights, the worst accuracy is obtained with $\mu = 2$. Notice that such a setting is even substantially outperformed by a random choice of weight vectors. This indicates that computing single-objective features independently for each objective is not a recommended strategy. By contrast, computing features for decomposed sub-problems is effective even when using a very low number of weights. This indicates that a decomposition-based approach for multi-objective landscape analysis contains a valuable information about algorithm performance. Surprisingly, we found that a uniform choice of

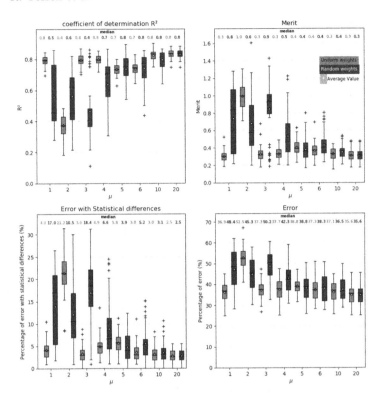

**Fig. 4.** $R^2$ (top left), merit (top right) and error rates (bottom) according to the number of weights $\mu$ and their distribution. (Color figure online)

few weight vectors with $\mu \in \{1, 3\}$ performs reasonably well, although increasing $\mu > 3$ allows to obtain a better accuracy. The relatively good results achieved with $\mu = 1$ are however to be interpreted very carefully, taking into account that the shape of the Pareto front for $\rho$mnk-landscapes, although having different magnitude in the objective space, is convex, symmetric and centered in the middle of the objective space, regardless of the values of $\rho$ and $\mathbf{k}$ [17]. Although this is a recurrent observation for many multi-objective combinatorial optimization problems, one might need to carefully choose $\mu$ when tackling problems with different Pareto front shapes. Such considerations are left for future investigations.

At last, in order to further show the accuracy of the proposed multi-objective landscape features, we experiment a baseline random forest model using the (unknown) global benchmark parameters $\rho$ and $\mathbf{k}$ as input variables to predict algorithm performance. Contrary to a black-box scenario where the knowledge about $\rho$ and $\mathbf{k}$ is not available, the accuracy of such a 'white-box' model should highlight the relevance and reliability of the proposed black-box features. We found that such a model trained with $\rho$ and $\mathbf{k}$ obtains an average merit of 0.41. Comparatively, black-box features obtain an average merit of 0.31, 0.37 and 0.29 respectively, for $\mu \in \{1, 3, 20\}$ uniform weight vectors. This once again indicates that the proposed approach is very effective, and that the designed high-level black-box features seem to provide more accurate prediction models than the global benchmark parameters, hence allowing to substantially improve over the single best solver, and also to get closer to the ideal virtual best solver.

## 5 Conclusion and Open Issues

In this paper, we push a step towards the development of automated landscape-aware selection and configuration approaches by proposing a set of multi-objective landscape features and analyzing their effectiveness in grasping the global properties of black-box multi-objective combinatorial optimization problems, together with their efficiency in selecting the best performing algorithm among three variants of the MOEA/D state-of-the art algorithm. The proposed features are based on the simple idea of aggregating the single-objective features extracted from a number of sub-problems obtained by decomposition. Our empirical analysis on a wide range of $\rho$mnk-landscapes provides insights into the accuracy of the proposed approach. However, it also raises some interesting research questions.

For instance, problems with more than two objectives, where the distribution of weight vector is expected to play an even more important role, as well as other real-like problems, are to be studied. Moreover, we excluded the cost of feature computation from our analysis, which can be an issue when the total affordable budget for the optimization task is restricted. An interesting observation is that we only need to extract features from a relatively low number of sub-problems, as considering a single sub-problem is already shown to provide a reasonably good performance. Additionally, solely one random walk is required to compute the considered single-objective features, so that one can eventually end up with negligible cost features. In this respect, a more fine-grained analysis of the cost-vs-importance of features is to be conducted in our future investigations. Finally, an interesting question is to conduct a systematic analysis of the proposed decomposition-based features w.r.t. existing dominance- and indicator-based features, and to analyze their relative cost and accuracy. Given that decomposition-based features were shown to be effective and does not need any (costly) dominance- and indicator-based computations, it is our hope that unifying the two classes of features would allow us to end up with efficient high-level state-of-the-art features for landscape-aware multi-objective optimization.

## References

1. Beham, A., Wagner, S., Affenzeller, M.: Algorithm selection on generalized quadratic assignment problem landscapes. In: GECCO 2018, pp. 253–260 (2018)
2. Breiman, L.: Random forests. Mach. Learn. **45**(1), 5–32 (2001). https://doi.org/10.1023/A:1010933404324
3. Deb, K.: Multi-Objective Optimization Using Evolutionary Algorithms. Wiley, Hoboken (2001)

4. Grefenstette, J.J.: Predictive models using fitness distributions of genetic operators. Found. Genet. Algorithms **3**, 139–161 (1995)

5. Kauffman, S.A.: The Origins of Order. Oxford University Press, Oxford (1993)

6. Kerschke, P., Trautmann, H.: The R-package FLACCO for exploratory landscape analysis with applications to multi-objective optimization problems. In: CEC, pp. 5262–5269 (2016)

7. Kerschke, P., Hoos, H.H., Neumann, F., Trautmann, H.: Automated algorithm selection: survey and perspectives. Evol. Comput. **27**(1), 3–45 (2019)

8. Liefooghe, A., Daolio, F., Verel, S., Derbel, B., Aguirre, H., Tanaka, K.: Landscape-aware performance prediction for evolutionary multi-objective optimization. IEEE Trans. Evol. Comput. **24**(6), 1063–1077 (2020)

9. Liefooghe, A., Derbel, B., Verel, S., Aguirre, H., Tanaka, K.: Towards landscape-aware automatic algorithm configuration: preliminary experiments on neutral and rugged landscapes. In: Hu, B., López-Ibáñez, M. (eds.) EvoCOP 2017. LNCS, vol. 10197, pp. 215–232. Springer, Cham (2017). https://doi.org/10.1007/978-3-319-55453-2_15

10. Liefooghe, A., Verel, S., Derbel, B., Aguirre, H., Tanaka, K.: Dominance, indicator and decomposition based search for multi-objective QAP: landscape analysis and automated algorithm selection. In: Bäck, T., et al. (eds.) PPSN 2020. LNCS, vol. 12269, pp. 33–47. Springer, Cham (2020). https://doi.org/10.1007/978-3-030-58112-1_3

11. Lindauer, M., van Rijn, J.N., Kotthoff, L.: The algorithm selection competition series 2015–17. CoRR abs/1805.01214 (2018)

12. Marquet, G., Derbel, B., Liefooghe, A., Talbi, E.-G.: Shake them all!. In: Bartz-Beielstein, T., Branke, J., Filipič, B., Smith, J. (eds.) PPSN 2014. LNCS, vol. 8672, pp. 641–651. Springer, Cham (2014). https://doi.org/10.1007/978-3-319-10762-2_63

13. Mersmann, O., Bischl, B., Trautmann, H., Wagner, M., Bossek, J., Neumann, F.: A novel feature-based approach to characterize algorithm performance for the traveling salesperson problem. Ann. Math. Artif. Intell. **69**, 151–182 (2013). https://doi.org/10.1007/s10472-013-9341-2

14. Rice, J.R.: The algorithm selection problem. Adv. Comput. **15**, 65–118 (1976)

15. Richter, H., Engelbrecht, A. (eds.): Recent Advances in the Theory and Application of Fitness Landscapes. Emergence Complexity and Computation. Springer, Heidelberg (2014). https://doi.org/10.1007/978-3-642-41888-4

16. Trivedi, A., Srinivasan, D., Sanyal, K., Ghosh, A.: A survey of multiobjective evolutionary algorithms based on decomposition. IEEE TEVC **21**(3), 440–462 (2017)

17. Verel, S., Liefooghe, A., Jourdan, L., Dhaenens, C.: On the structure of multiobjective combinatorial search space: MNK-landscapes with correlated objectives. Eur. J. Oper. Res. **227**(2), 331–342 (2013)

18. Zhang, Q., Li, H.: MOEA/D: a multiobjective evolutionary algorithm based on decomposition. IEEE TEVC **11**, 712–731 (2008)

19. Zitzler, E., Thiele, L., Laumanns, M., Fonseca, C.M., da Fonseca, V.G.: Performance assessment of multiobjective optimizers: an analysis and review. IEEE TEVC **7**(2), 117–132 (2003)

# MATE: A Model-Based Algorithm Tuning Engine
## A Proof of Concept Towards Transparent Feature-Dependent Parameter Tuning Using Symbolic Regression

Mohamed El Yafrani[1(✉)], Marcella Scoczynski[2], Inkyung Sung[1], Markus Wagner[3], Carola Doerr[4], and Peter Nielsen[1]

[1] Operations Research Group, Aalborg University, Aalborg, Denmark
mey@mp.aau.dk
[2] Federal University of Technology Paraná (UTFPR), Curitiba, Brazil
[3] Optimisation and Logistics Group, The University of Adelaide, Adelaide, Australia
[4] Sorbonne Université, CNRS, LIP6, Paris, France

**Abstract.** In this paper, we introduce a Model-based Algorithm Tuning Engine, namely MATE, where the parameters of an algorithm are represented as expressions of the features of a target optimisation problem. In contrast to most *static* (feature-independent) algorithm tuning engines such as *irace* and *SPOT*, our approach aims to derive the best parameter configuration of a given algorithm for a specific problem, exploiting the relationships between the algorithm parameters and the features of the problem. We formulate the problem of finding the relationships between the parameters and the problem features as a symbolic regression problem and we use genetic programming to extract these expressions in a human-readable form. For the evaluation, we apply our approach to the configuration of the $(1 + 1)$ EA and RLS algorithms for the OneMax, LeadingOnes, BinValue and Jump optimisation problems, where the theoretically optimal algorithm parameters to the problems are available as functions of the features of the problems. Our study shows that the found relationships typically comply with known theoretical results – this demonstrates (1) the potential of model-based parameter tuning as an alternative to existing static algorithm tuning engines, and (2) its potential to discover relationships between algorithm performance and instance features in human-readable form.

**Keywords:** Parameter tuning · Model-based tuning · Genetic programming

## 1 Motivation

The performance of many algorithms is highly dependent on tuned parameter configurations made with regards to the user's preferences or performance criteria [4], such as the quality of the solution obtained in a given CPU cost, the

C. Zarges and S. Verel (Eds.): EvoCOP 2021, LNCS 12692, pp. 51–67, 2021.
https://doi.org/10.1007/978-3-030-72904-2_4

smallest CPU cost to reach a given solution quality, the probability to reach a given quality, with given thresholds, and so on. This configuration task can be considered as a second layer optimisation problem [19] relevant in the fields of optimisation, machine learning and AI in general. It is a field of study that is increasingly critical as the prevalence of the application of such methods is expanded. Over the years, a range of automatic parameter tuners have been proposed, thus leaving the configuration to a computer rather than manually searching for performance-optimised settings across a set of problem instances. These tuning environments can save time and achieve better results [2].

Among such automated algorithm configuration (AAC) tools, we cite GGA [2], ParamILS [23], SPOT [3] and irace [30]. These methods have been successfully applied to (pre-tuned) state-of-the-art solvers of various problem domains, such as mixed integer programming [21], AI planning [16], machine learning [33], or propositional satisfiability solving [24]. Figure 1 illustrates the abstract standard architecture adopted by these tools.

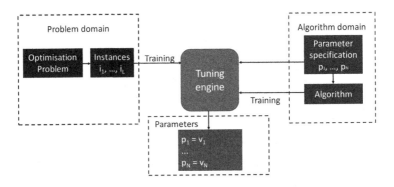

**Fig. 1.** Standard architecture of tuning frameworks.

However, the outcomes of these tools are static (or feature-independent), which means an algorithm configuration derived by any of these tools is not changed depending on an instance of a target optimisation problem. This leads to a significant issue as theoretical and empirical studies on various algorithms and problems have shown that parameters of an algorithm are highly dependent on features of a specific instance of a target problem [12] such as the problem size [6,35].

A possible solution to this issue is to cluster problem instances into multiple sub-groups by their size (and other potential features), then use curve fitting to map features to parameters [15,31]. A similar approach is also found in [29] that first partitions problem instances based the values of their landscape features and selects an appropriate configuration of a new problem instance based on its closeness to the partitions. However, the former approach does not scale well to multiple features and parameters, and the latter faces over-fitting issues due to the nature of the partitioning approach, making it difficult to assign an unseen instance to a specific group.

Some works have incorporated problem features in the parameter tuning process. SMAC [22] and PIAC [28] are examples of model-based tools that consider instance features to define parameter values by applying machine learning techniques to build the model. However, an issue of these approaches is the low explainability of the outcome. For instance, while machine learning techniques such as random forest and neural networks can be used to map the parameters to problem features with a high accuracy, they are considered as black-boxes, i.e., the outcome is virtually impossible to understand or interpret. Explainability is an important concept, as not only it allows us to understand the relationships between input and output [32], but in the context of parameter tuning, it can provide an outcome that can be used to inspire fundamental research [17,18].

To tackle these issues, we propose an offline algorithm tuning approach that extracts relationships between problem features and algorithm parameters using a genetic programming algorithm framework. We will refer to this approach as MATE, which stands for Model-based Algorithm Tuning Engine. The main contributions in this work are as follows:

1. We formulate the model-based parameter tuning problem as a symbolic regression problem, where knowledge about the problem is taken into account in the form of problem features;
2. We implement an efficient Genetic Programming (GP) algorithm that configures parameters in terms of problem features; and
3. In our empirical investigation, we rediscover asymptotically-correct theoretical results for two algorithms (1+1-EA and RLS) and four problems (One-Max, LeadingOnes, BinValue, and Jump). In these experiments, MATE shows its potential in algorithm parameter configuration to produce models based on instance features.

## 2  Background

Several methods have tried to tackle the dependence between the problem features and the algorithm parameters. The Per Instance Algorithm Configuration (PIAC) [28], for example, can learn a mapping between features and best parameter configuration, building an Empirical Performance Model (EPM) that predicts the performance of the algorithm for sample (instance, algorithm/configuration) pairs. PIAC methodology has been applied to several combinatorial problems [20,25,36] and continuous domains [5].

Sequential Model-based Algorithm Configuration (SMAC) [22] is also an automated algorithm configuration tool which considers a model, usually a random forest, to design the relationship between a performance metric (e.g. the algorithm runtime) and algorithm parameter values. SMAC can also include problem features within the tuning process as a subset of input variables.

Table 1 presents a summary for some state-of-the-art methods including the approach proposed in this paper. The term 'feature-independent' means that

**Table 1.** Summary of the state-of-the-art related works

| Approach name | Algorithm | Characteristics | Ref |
|---|---|---|---|
| GGA | Genetic algorithm | Feature-independent, model-free | [2] |
| ParamILS | Iterated Local Search | Feature-independent, model-free | [23] |
| irace | Racing procedure | Feature-independent, model-free | [30] |
| SPOT | Classical regression, tree-based, random forest and Gaussian process | Feature-independent, model-based | [3] |
| PIAC | Regression methods | Feature-dependent, model-based | [28] |
| SMAC | Random forest | Feature-dependent, model-based | [22] |
| MATE | Genetic programming | Feature-dependent, model-based, explainable | |

the corresponding approach does not consider instance features. 'Model-based' approaches use a trained model (e.g. machine learning, regression, etc.) to design parameter configurations. Model-free approaches generally rely on an experimental design methodology or optimisation method to find parameter settings of an algorithm that optimise a cost metric on a given instance set.

The main differences between MATE and the other related approaches are:

1. A transparent machine learning method (GP) is utilised to enable human-readable configurations (in contrast to, e.g., random forests, neural networks, etc.).
2. The training phase is done on one specific algorithm and one specific problem in our approach – the model is less instance-focused but more problem-domain focused by abstracting via the use of features. For example, the AAC experiments behind [17,18] have guided the creation of new heavy-tailed mutation operators that were beating the state-of-the-art. Similarly, the AAC and PIAC experiments in [34] showed model dependencies on easily-deducible instance features.

Lastly, our present paper is much aligned with the recently founded research field "Data Mining Algorithms Using/Used-by Optimisers (DUO)" [1]. There, data miners can generate models explored by optimisers; and optimisers can adjust the control parameters of a data miner.

# 3   The MATE Framework

## 3.1   Problem Formulation and Notation

Let us denote an optimisation problem by $\mathcal{B}$ whose instances are characterised by the problem-specific features $\mathcal{F} = \{f_1, \ldots, f_M\}$. A target algorithm $\mathcal{A}$ with its parameters $\mathcal{P} = \{p_1, \ldots, p_N\}$ is given to address the problem $\mathcal{B}$. A set of instances $\mathcal{I} = \{i_1, \ldots, i_L\}$ of the problem $\mathcal{B}$ and a $L \times M$ matrix $\mathcal{V}$, whose element value $v_{i,j}$ represents the $j$th feature value of the $i$th problem instance, are given.

Under this setting, we define the model-based parameter tuning problem as the problem of deriving a list of mappings $\mathcal{M} = \{m_1, \ldots, m_N\}$ where each mapping $m_j : \mathbb{R}^M \to \mathbb{R}$, which we will refer to as a *parameter expression*, returns a value for the parameter $p_j$ given feature values of an instance of the problem $\mathcal{B}$. Specifically, the objective of the problem is to find a parameter expression set $\mathcal{M}^*$, such that the performance of the algorithm $\mathcal{A}$ across all the problem instances in $\mathcal{I}$ is optimised.

## 3.2   Architecture Overview

In this section, we introduce our approach for parameter tuning based on the problem features. Figure 2 illustrates the architecture of the MATE tuning engine. In contrast to static methods, we consider the features of the problem. These feature are to be used in the training phase in addition to the instances, the target algorithm and the parameter specifications. Once the training is finished, the model can be used on unseen instances to find the parameters of the algorithm in terms of the problem feature values of the instance.

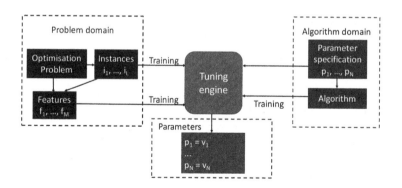

**Fig. 2.** Architecture of the proposed MATE framework

For example, a desired outcome of applying the MATE framework can be:

– Mutation probability of an evolutionary algorithm in terms of the problem size;

– Perturbation strength in an iterated local search algorithm in terms of the ruggedness of the instance and the problem size; and
– Population size of an evolutionary algorithm in terms of the problem size.

Note that all the examples include the problem size as a problem feature. In both theory and practice, the problem size is among the most important problem features, and it is usually known prior to the optimisation, without any need for a pre-processing step. More importantly, an extensive number of theoretical studies showed that the optimal choice of parameters is usually expressed in terms of the problem size (see, e.g. [6,12,35]).

### 3.3  The Tuning Algorithm

We use a tree-based Genetic Programming system as the tuning algorithm. It starts with a random population of trees, where each tree represents a potential parameter expression. Without loss of generality, we assume that the target problem is always a maximisation problem[1].

**The Score Function and Bias Reduction.** The score function is expressed as the weighted sum of the obtained objective values on each instance in the training set $\mathcal{I}$. Using the notations previously introduced, the score is defined in Eq. (1):

$$S(t) = \frac{1}{L} \Sigma_{i \in I} \frac{z_{\mathcal{A}}(m_1(v_{i,1}, \ldots, v_{i,M}), \ldots, m_N(v_{i,1}, \ldots, v_{i,M}), i)}{R_i} \tag{1}$$

where:

– $S(.)$ is the GP score function,
– $z_{\mathcal{A}}(\varphi_1, \ldots, \varphi_N, i)$ is a function measuring the goodness of applying the algorithm $A$ with the parameter values $\varphi_1, \ldots, \varphi_N$ to instance $i$,
– $R_i$ is the best known objective value for instance $i$.

The weights are used as a form of normalisation to reduce the bias some instances might induce. A solution to address this issue would be to use the optimal value or a tight upper bound. However, since we assume the such values are unknown (the problem itself can be unknown), we use the best known objective value ($R_i$) as a reference instead. In order to always ensure that score is well contained, the reference values are constantly updated whenever possible during the tuning process.

---

[1] The current MATE implementation is publicly available at https://gitlab.com/yafrani/mate.

**Table 2.** Summary of problems

| Problem | Features | Training set |
|---------|----------|--------------|
| OneMax($n$) | $n$: number of bits | $n = 10, 20, 50, 100, 200, 500$ |
| BinValue($n$) | $n$: number of bits | $n = 10, 20, 50, 100, 200, 500$ |
| LeadingOnes($n$) | $n$: number of bits | $n = 10, 20, 50, 100, 200, 500$ |
| Jump($m, n$) | $m$: width of region with bad fitness values | (m,n)=(2,10), (3,10), (4,10), (5,10), |
| | $n$: number of bits | (2,20), (3,20), (4,20), |
| | | (2,50), (3,50), |
| | | (2,100), (3,100), |
| | | (2,200) |

**Table 3.** MATE setup

| Attribute/Parameter | Value/Content |
|---------------------|---------------|
| Terminals | $\{1, 2, -1, -2\} \bigcup \mathcal{F}$ |
| Functions | Arithmetic operators |
| Number of GP generations | 100 |
| Population size | 20 |
| Tournament size | 5 |
| Replacement rate | $<75\%$ |
| Initialisation | Grow (50%) and full (50%) methods |
| Mutation operator | Random mutations |
| Mutation probability | 0.2 |
| Crossover operator | Sub-tree gluing |
| Crossover rate | 80% |
| Number of independent runs of target algorithm | 10 |
| $p$-value for the Wilcoxon ranksum test | 0.02 |

**Replacement Strategy – Statistical Significance and Bloat Control.** As the target algorithm can be stochastic, it is mandatory to perform multiple runs to ensure statistical significance (refer to Table 3). Thus, the replacement of trees is done based on the Wilcoxon rank-sum test.

Another aspect to take into account during the replacement process is bloat control. In our implementation, we use a simple bloat minimisation method based on the size of tree (number of nodes).

Given a newly generated tree $(Y)$, we compare it against each tree $(X)$ in the current population starting from the ones with the lowest scores using the following rules:

- If $Y$ is deemed to be significantly better than $X$ (using the Wilcoxon test). then we replace $X$ with $Y$ irrespective of the sizes.
- If there is no statistical significance between $X$ and $Y$, but $Y$ has a smaller size than $X$, then we replace $X$ with $Y$.
- Otherwise, we do not perform the replacement.

## 4    Computational Study

### 4.1    Experimental Setting

To evaluate our framework, we consider two target algorithms, the $(1 + 1)$ EA$(\mu)$ and RLS$(k)$. The $(1 + 1)$ EA$(\mu)$ is a simple hill-climber which uses standard bit mutation with mutation rate $\mu$. RLS$(k)$ differs from the $(1 + 1)$ EA$(\mu)$ only in that it uses the mutation operator that always flips $k$ uniformly chosen, pairwise different bits. That is, the mutation strength $k$ is deterministic in RLS, whereas it is binomially distributed in case of the $(1 + 1)$ EA$(\mu)$, Bin$(n, \mu)$, where $n$ is the number of bits.

We use MATE to configure the two algorithms for the four different problems with different time budgets as summarised in Table 2. In the table, the features of the problems used to tune the algorithm parameters and the different feature values chosen to generate problem instances of the problems are also presented. These problems have been chosen because they are among the best studied benchmark problems in the theory of evolutionary algorithms [13]. The details of our GP implementation for the experiments are presented in Table 3. Based on Table 3 and the set of features, our GP method uses a minimalistic set of 6 terminals at most: $m$, $n$ and $\{1, 2, -1, -2\}$.

It is worth noting that we are focusing in this paper on tuning algorithms with a single parameter. This is done to deliver a first prototype that is validated on algorithms and problems extensively studied by the EA theory community. An extension to tuning several algorithm parameters forms an important direction for future work.

For example, given a budget of $(1 + o(1))en \ln(n)$, it is known that the $(1+1)$EA$(1/n)$ optimises the OneMax function as well as any other linear functions with a decent probability. It is also known that the $1/n$ is asymptotically optimal [27]. Note, though, that such *fixed-budget results* are still very sparse [26], since the theory of EA community largely focuses on expected optimisation times. Since these can nevertheless give some insight into the optimal parameter settings, we note the following:

- OneMax and BinValue: the $(1+1)$EA$(1/n)$ optimises every linear function in expected time $en \ln(n)$, and no parameter configuration has smaller expected running time, apart from possible lower order terms [35]. For RLS, it is not difficult to see that $k = 1$ yields an expected optimisation time of $(1 + o(1))n \ln(n)$, and that this is the optimal (static) mutation strength;

– LeadingOnes: on average, RLS(1) needs $n^2/2$ steps to optimise LeadingOnes. This choise also minimises the expected optimisation time. For the (1+1) EA, $\mu \approx 1.59/n$ minimises the expected optimisation time, which is around $0.77n^2$ for this setting [6]. The standard mutation rate $\mu = 1/n$ requires $0.86n^2$ evaluations, on average, to locate the optimum, of the LeadingOnes function. For LeadingOnes, it is known that the optimal parameter setting drastically depends on the available budget. This can be inferred from the proofs in [6,9]; and

– Jump: mutation rate $m/n$ minimises the expected optimisation time of the (1+1) EA on Jump($m, n$), which is nevertheless $\Theta((e/m)^m n^m)$ [12].

## 4.2   Performance Analysis

**Training Phase.** The experimental study is conducted by running MATE ten times on each algorithm, problem and budget combination (refer to Table 4 for the list of budgets). This results in an elite population of 20 individuals for each setting, from which we select the top 5 expressions in terms of the score. These results are then merged and the 3 most frequent expressions are selected. For instance, the expression $2/n$ for OneMax with $0.5enln(n)$ appears 92 times over the 200 individuals (population size (20) × runs (10)).

In the current implementation, expression types (integers and non-integers) are not taken into account during the evolution. Therefore, the resulting expressions are converted into integers in the case of RLS by merging all real numbers $r$ using $\lfloor r \rfloor$ (e.g. $k = 3/2$ will be replaced by $k = 1$). On the other hand, expressions are simplified for EA by eliminating additive constants (e.g. $\mu = 1/(n+1)$ is replaced by $\mu = 1/n$).

**Evaluation Phase I.** To assess the performance of MATE, we evaluate for each problem-budget combination each of the top 3 most frequent expressions, by running them 100 independent times on each training dimension. We then normalise the outputs as in Eq. (1). The results are shown in the box plots in Table 4.

*Comparison Amongst the Top 3 Configurations.* When comparing the top 3 ranked configurations, we observe the following from Table 4 while we compare medians.

– OneMax: For (1+1) EA, $\mu = 1/n$, which ranked second for budgets $0.5en \ln n$ and $en \ln n$ and first for budget $2en \ln n$ performs better than $\mu = 1/2 * n$; while for RLS, the expression $k = 1$ appears at least on 94%, providing the best results;

– BinValue: $\mu = 1/n$ represents 18% on $en \ln n$ for (1+1) EA experiments, and a similar performance with $\mu = 2/n$ and $\mu = 3/n$; while on $0.5en \ln n$ case the $\mu = 1/n$ expression provides better results than $\mu = 1/2$ and $\mu = 1/3$; on the same way the expression $k = 1$ corresponds to 60% of the cases on RLS with the budget of $2n \ln n$ with a better performance than $k = 2$ and $k = n$;

**Table 4.** Results for 20 settings.

| | 1+1-EA | | RLS | |
|---|---|---|---|---|
| | **Budget** | **Result** | **Budget** | **Result** |
| **OneMax** | $0.5en\ln(n)$ | 2/n [46%], 1/n [32%], 1/(2*n) [10%], 3/n [4%], 5/(2*n) [2%], 3/(2*n) [0%]  (x-axis: 0.85 0.90 0.95) | $n\ln(n)^*$ | 1 [98%], 3 [2%], 2 [0%]  (x-axis: 0.85 0.90 0.95 1.00) |
| | $en\ln(n)^*$ | 2/n [46%], 1/n [26%], 1/(2*n) [14%], 3/n [12%], 5/(2*n) [2%], 3/(2*n) [0%]  (x-axis: 0.95 0.96 0.97 0.98 0.99 1.00) | $2n\ln(n)^{**}$ | 1 [94%], 3 [4%], 2 [2%]  (x-axis: 0.90 0.92 0.94 0.96 0.98 1.00) |
| | $2en\ln(n)^{**}$ | 1/n [44%], 2/n [26%], 1/(2*n) [12%], 3/n [8%], 3/(2*n) [8%], 5/(2*n) [0%]  (x-axis: 0.975 0.980 0.985 0.990 0.995 1.000) | | |
| **Bin Value** | $0.5en\ln(n)$ | 1/2 [36%], 1/n [26%], 1/3 [6%]  (x-axis: 0.92 0.94 0.96 0.98 1.00) | $0.5n\ln(n)$ | n [42%], 2 [32%], 1 [22%]  (x-axis: 0.80 0.85 0.90 0.95 1.00) |
| | $en\ln(n)^*$ | 2/n [44%], 1/n [18%], 3/n [14%]  (x-axis: 0.990 0.992 0.994 0.996 0.998 1.000) | $n\ln(n)^*$ | 2 [40%], n [36%], 1 [14%]  (x-axis: 0.90 0.92 0.94 0.96 0.98 1.00) |
| | $2en\ln(n)^{**}$ | 2/n [48%], 3/n [18%], 1/n [12%]  (x-axis: 0.9990 0.9992 0.9994 0.9996 0.9998 1.0000) | $2n\ln(n)^{**}$ | 1 [60%], 2 [32%], n [6%]  (x-axis: 0.980 0.985 0.990 0.995 1.000) |
| **LeadingOnes** | $0.5n^2$ | 1/n [52%], 2/n [28%], 4/n [20%], 3/n [0%], 5/(2*n) [0%], 3/(2*n) [0%]  (x-axis: 0.6 0.7 0.8 0.9) | $0.5n^{2*}$ | 2 [70%], 1 [26%], 3 [4%]  (x-axis: 0.7 0.8 0.9 1.0) |
| | $0.8n^{2**}$ | 1/n [62%], 2/n [18%], 3/n [16%], 4/n [2%], 5/(2*n) [0%], 3/(2*n) [0%]  (x-axis: 0.70 0.75 0.80 0.85 0.90 0.95 1.00) | $0.75n^{2**}$ | 1 [88%], 2 [12%], 3 [0%]  (x-axis: 0.75 0.80 0.85 0.90 0.95 1.00) |
| | $0.9n^{2**}$ | 1/n [48%], 2/n [28%], 3/n [18%], 5/(2*n) [4%], 4/n [0%], 3/(2*n) [0%]  (x-axis: 0.75 0.80 0.85 0.90 0.95 1.00) | | |
| **Jump** | $n^m$ | 1/n [36%], 2/n [32%], m/n [12%]  (x-axis: 0.875 0.900 0.925 0.950 0.975 1.000) | $n^m$ | m [34%], 2*m [24%], 3 [20%]  (x-axis: 0.850 0.875 0.900 0.925 0.950 0.975 1.00) |
| | $en^{m**}$ | 1/n [68%], 2/n [22%], m/n [6%]  (x-axis: 0.92 0.94 0.96 0.98 1.00) | $2n^m$ | m [42%], 2*m [20%], 3 [18%]  (x-axis: 0.850 0.875 0.900 0.925 0.950 0.975 1.000) |

† The $y$-axis show the best found expressions with its frequency between square brackets, and the $x$-axis represents the normalised fitness.

- LeadingOnes: $\mu = 1/n$ is the most frequent expression among all considered budgets on (1+1) EA and $\mu = 2/n$ presents the best performance amongst the top 3 expressions for all budget cases; $k = 1$ represents 88% on RLS cases with $0.75n^2$ iterations and performs better than $k = 2$ and $k = 3$ for both considered budgets.
- Jump: $\mu = 2/n$ and $\mu = m/n$ present similar results for both budget cases; $\mu = 1/n$ appears on 36% and 68% of the cases on (1+1) EA on the considered budgets respectively, and performs worse than the other two $\mu$ configurations; for RLS experiments $k = m$ is the most frequent expression and performs better than $k = 2 * m$ and $k = 3$.

*Comparison of Top 3 Configurations Against Other Parameter Settings.* For a fair assessment of our results, we add to this comparison some expressions that were not ranked in the top 3. These are $\mu = i/n$ with $i \in \{1, 3/2, 2, 5/2, 3, 4\}$ for (1+1) EA($\mu$) for OneMax and LeadingOnes. For readability purposes, the top 3 expressions are complemented with 3 of these additional expressions in the same order they are shown. We can observe in Table 4 that these additional expressions present low frequencies, $\mu = 3/n$ being the highest case with 12% with the budget $en \ln n$, while expressions $\mu = 3/(2n)$ and $\mu = 5/(2n)$ are the lowest cases among the considered budgets. Note that the frequencies do not necessarily sum up to 100% as other expressions not reported here might have occurred.

*Comparison with Theoretical Results.* As we have mentioned in the beginning of this section, one should be careful when comparing theoretical results that have been derived either in terms of running time or in terms of asymptotic convergence analysis, as typically done in runtime analysis. It is well known that optimal parameter settings for concrete (typically, comparatively small) dimensions can be different from the asymptotically optimal ones [7,8]. We nevertheless see that the configurations that minimise the expected running times (again, in the classical, asymptotic sense) also show up in the top 3 ranked configurations. In Table 4, we highlight the asymptotically optimal best possible running time by an asterisk*. Budgets exceeding this bound are marked by two asterisks**. As for the individual problems, we note the following:

- OneMax: It is interesting to note here that the performance is not monotonic in $k$, i.e., $k = 2$ performs worse than $k = 1$ and $k = 3$. This is caused by a phenomenon described in [11, Section 4.3.1], which states that, regardless of the starting point, the expected progress is always maximised by an uneven mutation strength. MATE correctly identifies this and suggests uneven mutation strengths in almost all cases.
- BinValue: We observe that it is very difficult here to distinguish the performance of the different configurations. This is in the nature of BinValues, as setting the first bit correctly already ensures 50% of the optimal fitness values. We very drastically see this effect in the recommendation to use $k = n$ for the RLS cases. With this configuration, the algorithm evaluates only two points: the random initial point $x$ and its pairwise complement $\bar{x}$, regardless of the

**Table 5.** Results for larger OneMax and LeadingOnes instances

budget. As can be seen in Table 4, the performance of this simple strategy is quite efficient, and hard to beat

- LeadingOnes: As mentioned earlier, for the (1+1) EA, the optimal mutation rate in terms of minimising the expected running time is around $\mu = 1.59/n$. We see that $\mu = 3/(2n)$, which did not show in the top 3 ranked configurations performs better than any of the suggestions by MATE.
- Jump: as discussed, mutation rate $\mu = m/n$ minimises the expected optimisation time. MATE recognises it as a good configuration in some of the runs. However, we see that $\mu = 2/n$, which equals $\mu = m/n$ for 5 out of our 12 training sets, shows comparable performance, and in the $en^m$ budget case even slightly better performance.

**Evaluation Phase II.** To properly assess the performance of MATE, we conducted experiments for OneMax and LeadingOnes instances of larger sizes that were not considered in the training phase. The goal of this experiment is to empirically demonstrate that our approach generalises well for large and unseen instances. These results are presented in Table 5 where 100 runs were performed for OneMax with $n \in \{1000, 2000, 5000\}$ and LeadingOnes with $n \in \{750, 1000\}$. We can observe the following:

- There is less overlap amongst the confidence intervals especially for smaller budgets, which means there is a higher level of separability amongst the performances of the different expressions.
- By comparing these results with the ones from Table 4, we can observe that the results of the top 3 expressions on large instances are statistically better in the majority of cases.
- OneMax: For (1+1) EA, in contrast to the results in Table 4 where $\mu = 1/n$ and $\mu = 3/(2n)$ show a similar performance, here $\mu = 1/n$ performs better than the other expressions. For RLS, the best performing expression is $k = 1$, which was ranked first.
- LeadingOnes: For (1+1) EA the best expressions are $\mu = 2/n$, which was ranked second, and $\mu = 3/(2n)$, which was not ranked among the top 3 expressions. For RLS, $k = 1$, ranked first and second, is the best performing expression.

### 4.3   Comparative Study

Herein, we compare the performance of MATE with irace and SMAC. The goal is to investigate the sensitivity of the obtained parameters on unseen instances. For a fair comparison, we run irace and SMAC with 2000 maximum experiments (which we believe is equivalent to the 100 GP generations with a population size of 20 individuals in MATE) considering the training instances presented in Table 2. We report the best elite parameter values returned by irace (2 candidates), SMAC (1 candidate) and MATE (most frequent expressions) in the columns $\mu$ and $k$ in Table 6, while the score (Eq. 1) is shown in column *Score* with the standard deviation as a subscript. These parameter values are then applied over 100 runs performed for OneMax with $n \in \{1000, 2000, 5000\}$ and LeadingOnes with $n \in \{750, 1000\}$.

Table 6 shows that MATE significantly outperforms irace and SMAC for (1+1) EA. On the other hand, the three methods show a similar performance on RLS. This is due to the fact that the parameter $\mu$ in (1+1) EA is highly sensitive to the problem feature $n$. In contrast, the parameter $k$ in RLS is independent from $n$ and its best value ($k = 1$) was identified by the three methods for both OneMax and LeadingOnes.

**Table 6.** Results for MATE, irace and SMAC for OneMax and LeadingOnes instances.

| | 1+1-EA | | | | | | RLS | | | | | | |
|---|---|---|---|---|---|---|---|---|---|---|---|---|---|---|
| | Budget | MATE μ | MATE Score | irace μ | irace Score | SMAC μ | SMAC Score | Budget | MATE k | MATE Score | irace k | irace Score | SMAC k | SMAC Score |
| OneMax | $\frac{e\ln(n)}{2}$ | $\frac{2}{n}$ | $0.99_{0.001}$ | 0.258 | $0.57_{0.002}$ | 0.009 | $0.8_{0.003}$ | $n\ln(n)$ | 1 | $1_0$ | 1 | $1_0$ | 1 | $1_0$ |
| | | $\frac{1}{n}$ | $0.99_{0.001}$ | 0.216 | $0.58_{0.002}$ | | | | 3 | $0.96_{0.002}$ | | | | |
| | | $\frac{1}{2n}$ | $0.98_{0.002}$ | | | | | | 2 | $0.94_{0.003}$ | | | | |
| | $e\ln(n)$ | $\frac{1}{n}$ | $1_0$ | 0.009 | $0.82_{0.002}$ | 0.016 | $0.76_{0.003}$ | $2n\ln(n)$ | 1 | $1_0$ | 1 | $1_0$ | 1 | $1_0$ |
| | | $\frac{2}{n}$ | $1_0$ | 0.013 | $0.79_{0.003}$ | | | | 3 | $0.98_{0.001}$ | | | | |
| | | $\frac{1}{2n}$ | $1_0$ | | | | | | 2 | $0.97_{0.002}$ | | | | |
| | $2e\ln(n)$ | $\frac{2}{n}$ | $1_0$ | 0.594 | $0.54_{0.002}$ | 0.008 | $0.86_{0.002}$ | | | | | | | |
| | | $\frac{1}{n}$ | $1_0$ | 0.589 | $0.54_{0.002}$ | | | | | | | | | |
| | | $\frac{1}{2n}$ | $1_0$ | | | | | | | | | | | |
| LeadingOnes | $0.5n^2$ | $\frac{1}{n}$ | $0.7_{0.025}$ | 0.430 | $0.03_{0.002}$ | 0.024 | $0.29_{0.007}$ | $0.5n^2$ | 2 | $0.86_{0.014}$ | 1 | $0.98_{0.026}$ | 1 | $0.98_{0.02}$ |
| | | $\frac{2}{n}$ | $0.8_{0.021}$ | 0.409 | $0.03_{0.002}$ | | | | 1 | $0.97_{0.027}$ | 5 | $0.61_{0.01}$ | | |
| | | $\frac{4}{n}$ | $0.71_{0.015}$ | | | | | | 3 | $0.75_{0.013}$ | | | | |
| | $0.8n^2$ | $\frac{1}{n}$ | $0.95_{0.023}$ | 0.255 | $0.05_{0.002}$ | 0.005 | $0.83_{0.017}$ | $0.75n^2$ | 1 | $1_0$ | 1 | $1_0$ | 1 | $1_0$ |
| | | $\frac{2}{n}$ | $0.99_{0.012}$ | 0.258 | $0.05_{0.002}$ | | | | 2 | $0.95_{0.009}$ | | | | |
| | | $\frac{3}{n}$ | $0.91_{0.018}$ | | | | | | 3 | $0.82_{0.01}$ | | | | |
| | $0.9n^2$ | $\frac{1}{n}$ | $0.99_{0.011}$ | 0.158 | $0.07_{0.003}$ | 0.006 | $0.75_{0.013}$ | | | | | | | |
| | | $\frac{2}{n}$ | $1_0$ | 0.153 | $0.07_{0.006}$ | | | | | | | | | |
| | | $\frac{3}{n}$ | $0.95_{0.014}$ | | | | | | | | | | | |

# 5    Conclusions and Future Directions

With this article, we have presented MATE as a model-based algorithm tuning engine: its human-readable models map instance features to algorithm parameters. Our experiments showed that MATE can find known asymptotic relationships between the feature values and algorithm parameters. We also compared the performance of MATE with iRace and SMAC investigating the sensitivity of the obtained parameters on unseen instances of larger size. With its scalable models, MATE performed best. It is worth noting that MATE can be a useful guideline tool for theory researchers due to its white-box nature, similarly to how results in [14] inspired the analysis of a generalised one-fifth success rule in [10]. But MATE can also be extended to be used as a practical toolbox for feature-based algorithm configuration.

In the future, we intend to explore, among other, the following three avenues. First, the design of MATE itself will be subject to extensions, e.g. to better handle performance differences between instances via ranks or racing. Second, while our proof-of-concept study here was motivated by theoretical insights, we will investigate more realistic problems for which instance features are readily available, such as the travelling salesperson problem and the assignment problem. Third, we will investigate approaches to extend MATE to handle multiple parameters to demonstrate its ability to tune more sophisticated algorithms.

**Acknowledgements.** M. Martins acknowledges CNPq (Brazil Government). M. Wagner acknowledges the ARC Discovery Early Career Researcher Award DE160100850. C. Doerr acknowledges support from the Paris Ile-de-France Region. Experiments were performed on the AAU's CLAUDIA compute cloud platform.

# References

1. Agrawal, A., Menzies, T., Minku, L.L., Wagner, M., Yu, Z.: Better software analytics via "duo": data mining algorithms using/used-by optimizers. Empirical Softw. Eng. **25**(3), 2099–2136 (2020)
2. Ansótegui, C., Sellmann, M., Tierney, K.: A gender-based genetic algorithm for the automatic configuration of algorithms. In: Gent, I.P. (ed.) CP 2009. LNCS, vol. 5732, pp. 142–157. Springer, Heidelberg (2009). https://doi.org/10.1007/978-3-642-04244-7_14
3. Bartz-Beielstein, T., Flasch, O., Koch, P., Konen, W., et al.: SPOT: a toolbox for interactive and automatic tuning in the R environment. In: Proceedings, vol. 20, pp. 264–273 (2010)
4. Belkhir, N., Dréo, J., Savéant, P., Schoenauer, M.: Feature based algorithm configuration: a case study with differential evolution. In: Handl, J., Hart, E., Lewis, P.R., López-Ibáñez, M., Ochoa, G., Paechter, B. (eds.) PPSN 2016. LNCS, vol. 9921, pp. 156–166. Springer, Cham (2016). https://doi.org/10.1007/978-3-319-45823-6_15
5. Belkhir, N., Dréo, J., Savéant, P., Schoenauer, M.: Per instance algorithm configuration of CMA-ES with limited budget. In: Genetic and Evolutionary Computation Conference. GECCO 2017, pp. 681–688. ACM (2017)
6. Böttcher, S., Doerr, B., Neumann, F.: Optimal fixed and adaptive mutation rates for the leadingones problem. In: Schaefer, R., Cotta, C., Kołodziej, J., Rudolph, G. (eds.) PPSN 2010. LNCS, vol. 6238, pp. 1–10. Springer, Heidelberg (2010). https://doi.org/10.1007/978-3-642-15844-5_1
7. Buskulic, N., Doerr, C.: Maximizing drift is not optimal for solving onemax. In: Genetic and Evolutionary Computation Conference, GECCO 2019, pp. 425–426. ACM (2019). http://arxiv.org/abs/1904.07818
8. Chicano, F., Sutton, A.M., Whitley, L.D., Alba, E.: Fitness probability distribution of bit-flip mutation. Evol. Comput. **23**(2), 217–248 (2015)
9. Doerr, B.: Analyzing randomized search heuristics via stochastic domination. Theor. Comput. Sci. **773**, 115–137 (2019)
10. Doerr, B., Doerr, C., Lengler, J.: Self-adjusting mutation rates with provably optimal success rules. In: Proceeding of Genetic and Evolutionary Computation Conference (GECCO 2019), pp. 1479–1487. ACM (2019). https://doi.org/10.1145/3321707.3321733, https://arxiv.org/abs/1902.02588
11. Doerr, B., Doerr, C., Yang, J.: Optimal parameter choices via precise black-box analysis. Theor. Comput. Sci. **801**, 1–34 (2020)
12. Doerr, B., Le, H.P., Makhmara, R., Nguyen, T.D.: Fast genetic algorithms. In: Genetic and Evolutionary Computation Conference, GECCO 2017, pp. 777–784. ACM (2017)
13. Doerr, B., Neumann, F.: Theory of evolutionary computation. In: Recent Developments in Discrete Optimization. Springer, Cham (2020)
14. Doerr, C., Wagner, M.: Simple on-the-fly parameter selection mechanisms for two classical discrete black-box optimization benchmark problems. In: Proceeding of Genetic and Evolutionary Computation Conference (GECCO 2018), pp. 943–950. ACM (2018). https://doi.org/10.1145/3205455.3205560
15. El Yafrani, M., Ahiod, B.: Efficiently solving the traveling thief problem using hill climbing and simulated annealing. Inf. Sci. **432**, 231–244 (2018)
16. Fawcett, C., Helmert, M., Hoos, H., Karpas, E., Röger, G., Seipp, J.: Fd-autotune: domain-specific configuration using fast downward. In: ICAPS 2011 Workshop on Planning and Learning, pp. 13–17 (2011)

17. Friedrich, T., Göbel, A., Quinzan, F., Wagner, M.: Heavy-tailed mutation operators in single-objective combinatorial optimization. In: Auger, A., Fonseca, C.M., Lourenço, N., Machado, P., Paquete, L., Whitley, D. (eds.) PPSN 2018. LNCS, vol. 11101, pp. 134–145. Springer, Cham (2018). https://doi.org/10.1007/978-3-319-99253-2_11
18. Friedrich, T., Quinzan, F., Wagner, M.: Escaping large deceptive basins of attraction with heavy-tailed mutation operators. In: Genetic and Evolutionary Computation Conference. GECCO 2018, pp. 293–300. ACM (2018)
19. Hoos, H.H.: Programming by optimization. Commun. ACM **55**(2), 70–80 (2012)
20. Hutter, F., Hamadi, Y., Hoos, H.H., Leyton-Brown, K.: Performance prediction and automated tuning of randomized and parametric algorithms. In: Benhamou, F. (ed.) CP 2006. LNCS, vol. 4204, pp. 213–228. Springer, Heidelberg (2006). https://doi.org/10.1007/11889205_17
21. Hutter, F., Hoos, H.H., Leyton-Brown, K.: Automated configuration of mixed integer programming solvers. In: Lodi, A., Milano, M., Toth, P. (eds.) CPAIOR 2010. LNCS, vol. 6140, pp. 186–202. Springer, Heidelberg (2010). https://doi.org/10.1007/978-3-642-13520-0_23
22. Hutter, F., Hoos, H.H., Leyton-Brown, K.: Sequential model-based optimization for general algorithm configuration. In: Coello, C.A.C. (ed.) LION 2011. LNCS, vol. 6683, pp. 507–523. Springer, Heidelberg (2011). https://doi.org/10.1007/978-3-642-25566-3_40
23. Hutter, F., Hoos, H.H., Leyton-Brown, K., Stützle, T.: ParamILS: an automatic algorithm configuration framework. J. Artif. Intell. Res. **36**, 267–306 (2009)
24. Hutter, F., Lindauer, M., Balint, A., Bayless, S., Hoos, H., Leyton-Brown, K.: The configurable SAT solver challenge (CSSC). Artif. Intell. **243**, 1–25 (2017)
25. Hutter, F., Xu, L., Hoos, H.H., Leyton-Brown, K.: Algorithm runtime prediction: Methods & evaluation. Artif. Intell. **206**, 79–111 (2014)
26. Jansen, T.: Analysing stochastic search heuristics operating on a fixed budget. Theory of Evolutionary Computation. NCS, pp. 249–270. Springer, Cham (2020). https://doi.org/10.1007/978-3-030-29414-4_5
27. Lengler, J., Spooner, N.: Fixed budget performance of the (1+1) EA on linear functions. In: ACM Conference on Foundations of Genetic Algorithms, FOGA 2015, pp. 52–61. ACM (2015)
28. Leyton-Brown, K., Nudelman, E., Shoham, Y.: Learning the empirical hardness of optimization problems: the case of combinatorial auctions. In: Van Hentenryck, P. (ed.) CP 2002. LNCS, vol. 2470, pp. 556–572. Springer, Heidelberg (2002). https://doi.org/10.1007/3-540-46135-3_37
29. Liefooghe, A., Derbel, B., Verel, S., Aguirre, H., Tanaka, K.: Towards landscape-aware automatic algorithm configuration: preliminary experiments on neutral and rugged landscapes. In: Hu, B., López-Ibáñez, M. (eds.) EvoCOP 2017. LNCS, vol. 10197, pp. 215–232. Springer, Cham (2017). https://doi.org/10.1007/978-3-319-55453-2_15
30. López-Ibáñez, M., Dubois-Lacoste, J., Cáceres, L.P., Birattari, M., Stützle, T.: The irace package: iterated racing for automatic algorithm configuration. Oper. Res. Perspect. **3**, 43–58 (2016)
31. Mascia, F., Birattari, M., Stützle, T.: Tuning algorithms for tackling large instances: an experimental protocol. In: Nicosia, G., Pardalos, P. (eds.) LION 2013. LNCS, vol. 7997, pp. 410–422. Springer, Heidelberg (2013). https://doi.org/10.1007/978-3-642-44973-4_44
32. Rai, A.: Explainable AI: from black box to glass box. J. Acad. Market. Sci. **48**(1), 137–141 (2020)

33. Snoek, J., Larochelle, H., Adams, R.P.: Practical bayesian optimization of machine learning algorithms. In: Advances in Neural Information Processing Systems, pp. 2951–2959 (2012)
34. Treude, C., Wagner, M.: Predicting good configurations for github and stack overflow topic models. In: 16th International Conference on Mining Software Repositories. MSR 2019, pp. 84–95. IEEE (2019)
35. Witt, C.: Tight bounds on the optimization time of a randomized search heuristic on linear functions. Comb. Probab. Comput. **22**, 294–318 (2013)
36. Xu, L., Hutter, F., Hoos, H.H., Leyton-Brown, K.: SATzilla: portfolio-based algorithm selection for SAT. J. Artif. Intell. Res. **32**, 565–606 (2008)

# An Improvement Heuristic Based on Variable Neighborhood Search for a Dynamic Orienteering Problem

Hoang Thanh Le[1(✉)], Martin Middendorf[1], and Yuhui Shi[2]

[1] Swarm Intelligence and Complex Systems Group, Institute of Computer Science,
Leipzig University, Leipzig, Germany
{lht,middendorf}@informatik.uni-leipzig.de
[2] Computer Science and Engineering, Southern University of Science
and Technology, Shenzhen, China
shiyh@sustech.edu.cn

**Abstract.** The Dynamic Orienteering Problem (DOP) is studied where nodes change their value over time. An improvement heuristic that is based on Variable Neighborhood Search is proposed for the DOP. The new heuristic is experimentally compared with two heuristics that are based on state-of-the-art algorithms for the static Orienteering Problem. For the experiments several benchmark instances are used as well as instances that are generated from existing road networks. The results show that the new heuristic outperforms the other heuristics with respect to several evaluation criteria and different measures for run time. An additional experiment shows that the new heuristic can be easily adapted to become a standalone algorithm that does not need given initial solutions. The standalone version obtains better results than two state-of-the-art algorithms.

**Keywords:** Dynamic optimization · Neighborhood search · Anytime algorithms · Routing problems

## 1 Introduction

The Orienteering Problem (OP) is to find a simple cycle in a given undirected graph with edge weights and node values such the total edge weight of the cycle does not exceed a given threshold and the total value of all nodes in the cycle is maximal. The name of the problem comes from a corresponding sport [31]. The OP has applications in many areas, e.g., in the planning of city trips [32] and fuel delivery routes [10], in agriculture [23] or for surveillance activities in military scenarios [33]. The OP commonly occurs in applications where the nodes are locations in a city or a region and the cycle corresponds to a route on a road network.

This paper considers a dynamic version of the Orienteering Problem (DOP) where the nodes to be visited can change their value (measuring, e.g., their

C. Zarges and S. Verel (Eds.): EvoCOP 2021, LNCS 12692, pp. 68–83, 2021.
https://doi.org/10.1007/978-3-030-72904-2_5

attractiveness or urgency) during the optimization process. This is interesting because in each of the above mentioned application areas it might be possible that unexpected changes of node values occur after which a planned route has to be revised without exceeding the time limit (or violating other problem-specific constraints). For example, customers might change their order and thus their payment value, tourist spots might change their attractiveness over the course of a day or emergencies might occur that change the urgency with which locations have to be visited.

The DOP necessitates the development of algorithms that are able to quickly adapt existing solutions. For this reason, we focus on *improvement heuristics* that try to improve or adapt a given cycle (path) as opposed to heuristics that independently construct a new solution (in the following the latter are referred to as *standalone algorithms*). Improvement heuristics can be easily integrated into existing tour planning systems in order to improve precalculated solutions and adapt them if changes occur.

In particular, an improvement heuristic ($VNS_{DOP}$) is proposed that is based on Variable Neighborhood Search. This heuristic is experimentally compared with two other heuristics that use state-of-the-art algorithms for the Static Orienteering Problem. The performance of the heuristics is compared using benchmark instances [19] as well as instances that are generated from existing road networks. In addition, it is evaluated whether the proposed heuristic $VNS_{DOP}$ is also suited as a standalone algorithm that does not need given initial solutions.

The remaining sections are structured as follows. A short overview of existing works on Orienteering Problems and their dynamic variants is given in Sect. 2. A formal description of the Dynamic Orienteering Problems is presented in Sect. 3. The heuristic $VNS_{DOP}$ is introduced in Sect. 4. Experimental results are presented in Sect. 5. Conclusions are given in Sect. 6.

## 2 Related Work

In one of the earliest works on the "Orienteering Problem" (OP) [31] the name of the problem was chosen because the problem occurs in a sport with the same name. However, the OP is also known as the Selective TSP [21], the Maximum Collection Problem [17] or, due to the structural similarities to Vehicle Routing Problems, as a Vehicle Routing Problem with Profits [4]. There exist several possible formulations of the OP as a linear program (see, e.g., [15,21]).

The OP is known to be NP-hard [10] and numerous heuristics have been developed for this problem. One of them is the Greedy Randomized Adaptive Search Procedure (GRASP) which has multiple variants, such as Memetic GRASP [24] or GRASP with path relinking [5]. The latter is outperformed by a newer GRASP variant that removes path segments [18] as well as by an Evolutionary Algorithm [19]. Other heuristics include an Ant Colony Optimization for a multi-objective version of the OP [29], an Evolution-inspired algorithm with Hill Climbing [28] or an approximation algorithm for the case of an OP with a directed graph and unit values for the nodes [25].

Regarding dynamic versions of the OP, there are several OP variants which contain probabilistic elements. These problems are known as Stochastic OPs and can be interpreted as a specific type of dynamic OP. In one variant [2] it is possible that nodes randomly become unavailable after a path has been chosen so that some of the nodes have to be skipped. In this problem, the difference between the expected total value of the path and the expected length of the path is to be maximized. To solve the problem a Mixed-Integer Linear model and a Matheuristic have been proposed. In another variant the node values are normally distributed so that the probability is to be maximized that the total value of a path exceeds a given target value. For this problem an exact algorithm as well as a Genetic Algorithm have been presented in [13]. In [36] the sale of books on a university campus is presented as a Stochastic OP. It is modelled as a Markov Decision Process and solved using Approximate Dynamic Programming allowing for the route to dynamically change while the path is traversed.

There also exist (true) dynamic OP variants where the problem instance changes over time. One example is the OP with Time Windows which, e.g., represent business hours, at which certain nodes have to be visited (see, e.g., the surveys in [9, 11] or the application to the city of Tehran in [1]). Another example is studied in [8] where the weights of the edges, i.e., the travel times along the edges, change depending on the departure time from the starting node. The case where the value of a node decreases linearly depending on when it is visited is studied in [7]. However, in these works the edge weights or node values depend on the time at which a given node is visited which mostly depends on the distance previously traveled on a given path. To the best of our knowledge, there exist no works that have considered a (time-)dynamic OP in which the changes of an instance occur during the optimization process.

## 3   Problem Description

The Orienteering Problem (OP) studied in this work is specified by a tuple $(G, d, B, T, v_0, s)$ which consists of an undirected, connected, simple graph $G = (V, E)$ with node set $V$, edge set $E \subset \{\{u, v\} \mid u, v \in V\}$, a function $d : E \to \mathbb{R}_{\geq 0}$ that assigns to each edge $e \in E$ a length $d(e) > 0$ which can be interpreted as the distance to traverse that edge, a positive value $B$ ("budget"), a time interval $T$ ("time horizon"), a node $v_0 \in V$ ("depot"), and a value function $s$ that assigns to each node $v \in V$ a non-negative value $s(v)$ ("score", "profit"). It is assumed that the function $d$ satisfies the triangle inequality.

In this work, a closed path $P$ is a sequence $(v_0, v_1, \ldots, v_n, v_0)$ of nodes of $G$ where each node occurs at most once except the depot $v_0$ which occurs exactly twice and forms the start and the end of the path. The length of a closed path $P$ is defined as the sum of the distances between the nodes $v_i$ and $v_{i+1}$ ($i \in \{0, 1, 2, \ldots, n-1\}$) plus the distance between the nodes $v_n$ and $v_0$ where the distance between two nodes $u$ and $v$ is defined as the length of a shortest path between $u$ and $v$. The length of a path is the sum of the lengths of its edges. Since the triangle inequality holds for $d$ it follows that if $\{u, v\} \in E$, the length of a shortest path between $u$ and $v$ equals $d(\{u, v\})$. The length of a closed path $P$ is denoted by $Length(P)$. The value of a closed path $P$ is the sum of the values $s(v_i)$ for all nodes $v_i$ ($i \in \{0, 1, 2, \ldots, n\}$). The value of $P$ is denoted by $Value(P)$. The objective of the DOP is to find a closed path $P$ in $G$ with $Length(P) \leq B$ that maximizes $Value(P)$.

For the dynamic version of the OP which is called Dynamic Orienteering Problem (DOP), we consider the case where the values of the nodes can change over time, i.e., the value function $s$ is a (time-)dynamic function $s : V \times T \to \mathbb{R}_{\geq 0}$ where $T$ is a time interval. The term "time" in this work refers to the time during the run of an optimization algorithm. This is to be distinguished from the "time" at which a node is visited (which, for the most part, depends on the distance travelled so far on a given path $P$ as, e.g., in [7,8]). Note, that OPs where only the node values can change have the property that a valid solution $P$ whose length does not exceed the budget $B$ cannot become invalid over time. How $s$ is defined for the experiments is described in Sect. 5.3. We assume that the changes in $s$ are not known in advance to an optimization algorithm.

Since the static OP is known to be *NP*-complete [10], the *NP*-completeness of the DOP immediately follows. Algorithms for the OP need to consider two sub-problems: 1) A suitable subset $V' \subseteq V$ with a preferably high sum of node values has to be selected; 2) For the nodes in $V'$, a closed path $P$ has to be calculated with preferably small values for $Length(P)$. The latter sub-problem is similar to the well-studied Traveling Salesperson Problem (TSP). If the solution for the second sub-problem exceeds the available budget $B$ with respect to path length, the first sub-problem might need to be solved again. Due to the budget constraint, the first sub-problem is similar to the well-known Knapsack Problem (KP) such that the OP can be considered to contain an interplay of these two combinatorial optimization problems.

## 4   Variable Neighborhood Search

The principle of Neighborhood Search algorithms is to modify a given solution $P_{old}$ by selecting a new solution $P_{new}$ from a set of similar solutions (the "neighborhood" of $P_{old}$) that minimizes or maximizes an objective function $f$. Variable Neighborhood Search algorithms are based on the observation that a local extremum, i.e., a local maximum or local minimum of $f$ within one neighborhood is not necessarily a local extremum within a different neighborhood [12]. Therefore, Variable Neighborhood Search algorithms change the neighbourhood function, i.e., the set of solutions considered "similar to $P_{old}$" over the course of the algorithm. In the following, this is written as $P_{new} = \arg\min_{P \in N_k(P_{old})} f(P)$ (for the case of a minimization problem) where $N_k(P_{old})$ denotes the neighborhood of the solution $P_{old}$ that changes depending on an index $k$. As an example, the neighborhood $N_{\mathsf{remove}}(P)$ with the index $k = \mathsf{remove}$ indicates the set of paths that can be obtained by removing a non-depot node from $P$.

Since solving OP problems is considered here to entail the solution of two sub-problems with different objectives, the heuristic that is proposed in this work is based on a generalization of this principle: Given a path $P_{old}$, the algorithm selects a new solution $P_{new}$ from a neighborhood $N_k$ that maximizes or minimizes an objective function $f_\ell$ which also changes during the run of the algorithm depending on an index $\ell$:

$$P_{new} = \begin{cases} \arg \max_{P \in N_k(P_{old})} f_\ell(P, P_{old}) & \text{if } f_\ell \text{ is to be maximized} \\ \arg \min_{P \in N_k(P_{old})} f_\ell(P, P_{old}) & \text{if } f_\ell \text{ is to be minimized} \end{cases} \tag{1}$$

The combination of different neighborhoods with varying objective functions—which can also be interpreted in the sense that the structure of the neighborhood is now determined by the pair $(N_k, f_\ell)$—allows for a finer adjustment of the algorithm's behavior. In particular, using different functions $f_\ell$ allows the algorithm to focus on specific aspects of the given optimization problem. In this work, we consider the following functions:

$$f_{length}(P, P') = 1/|Length(P) - Length(P')| \tag{2}$$

$$f_{value}(P, P') = |Value(P) - Value(P')| \tag{3}$$

$$f_{ratio}(P, P') = \left| \frac{Value(P) - Value(P')}{Length(P) - Length(P')} \right| = f_{length}(P, P') \cdot f_{value}(P, P') \tag{4}$$

$$f_{random}(P, P') = r \tag{5}$$

In $f_{random}$ the number $r$ is drawn randomly with uniform probability from the interval $[0, 1]$. For this case it can be shown that Eq. (1) is equivalent to choosing $P_{new}$ as a random element from $N_k(P_{old})$. In the following, we define $\mathcal{F} = \{f_{length}, f_{value}, f_{ratio}, f_{random}\}$ as the set of considered functions and if $Length(P) = Length(P')$, we define $f_{length}(P, P') = \infty$ and $f_{ratio}(P, P') = \infty$.

As for neighborhoods, we consider two sets $N_{add}(P)$ and $N_{remove}(P)$ for a path $P$ defined as follows. $N_{add}(P)$ is the set of paths $P'$ that can be obtained by inserting an unvisited node with positive value into $P$ (regardless of whether $Length(P')$ exceeds the budget $B$) and $N_{remove}(P)$ is the set of paths that can be obtained by removing a node that is unequal to the depot $v_0$ from $P$. Search steps with respect to these two neighborhoods thus allow the algorithm to add and remove nodes from a path.

The choice of the objective functions is motivated by the structure of the OP. The objective function $f_{length}$ focuses on the path length and prefers solutions of short length. This is relevant for the aforementioned sub-problem that corresponds to the TSP. The objective function $f_{value}$ primarily considers the total value of a path without taking its length into account. This can be interpreted as solely focusing on the KP sub-problem. A third objective function $f_{ratio}$ combines $f_{length}$ and $f_{value}$ such that the selection of new paths takes both sub-problems into account. Function $f_{random}$ corresponds to a random selection of paths which is used to perturb solutions adding an exploration mechanism to the algorithm. Note that the absolute values in $\mathcal{F}$ allow the algorithm to select solutions with high total values and/or short length by appropriately choosing whether to maximize or minimize all functions in $\mathcal{F}$ when adding or removing

---

**Algorithm 1.** VNS$_{\text{DOP}}$

---

**Parameters:** initial iterations $k_{init}$, initial insertion probability $p$, initial solution $P_0$

1: $P \leftarrow P_0$
2: **for** $it = 1, \ldots, k_{init}$ **do**          ▷ initial exploration phase
3:     **for each** node $v$ not in $P$ **do**
4:        with probability $p$, insert $v$ into $P$ at a random position
5:     **end for**
6:     optimize the order of nodes in $P$ using the Chained Lin-Kernighan heuristic
7:     **while** $Length(P) > B$ **do**
8:        $P \leftarrow \arg\min_{P' \in N_{\mathsf{remove}}(P)} f_{\mathsf{ratio}}(P', P)$
9:     **end while**
10: **end for**
11: **while** termination criterion not satisfied **do**      ▷ main iterations
12:     $f_1, f_2 \leftarrow$ random elements from the set $\mathcal{F}$
13:     **repeat**
14:        $P \leftarrow \arg\max_{P' \in N_{\mathsf{add}}(P)} f_1(P', P)$
15:     **until** $Length(P) > B \lor N_{\mathsf{add}}(P) = \varnothing$
16:     optimize the order of nodes in $P$ using the Chained Lin-Kernighan heuristic
17:     **while** $Length(P) > B$ **do**
18:        $P \leftarrow \arg\min_{P' \in N_{\mathsf{remove}}(P)} f_2(P', P)$
19:     **end while**
20: **end while**
21: **return** best solution found so far

---

nodes. For the neighborhoods that are considered in this work, two neighboring paths differ by a single node. Hence, the evaluation of these functions can be performed efficiently by only considering the differing nodes.

The proposed algorithm (VNS$_{\text{DOP}}$) is shown in Algorithm 1. It starts with a short, exploration-focused *initial phase* containing a randomized procedure that for a path $P$ and a parameter $p \in [0,1]$ iterates over all unvisited nodes with a positive value and with probability $p$ inserts them into $P$ at a random position regardless of whether the length of the resulting path exceeds the budget $B$. This exploration-focused procedure aims to do a quick, cursory scan of different areas of the solution space (that also includes paths which do not contain all nodes in $V$) in order to find a promising subset of nodes from which solutions are then improved using the Chained-Lin-Kernighan heuristic [3] based on the implementation provided by the Concorde TSP solver [6]. The computation time for the Chained-Lin-Kernighan heuristic is included in the run time measurement that is described in Sect. 5.1. Afterwards and if necessary, nodes are removed based on the minimization of the function $f_{\mathsf{ratio}}$ which provides a balanced view on both sub-problems without using a large amount of time. The initial phase lasts for $k_{init}$ iterations where $k_{init}$ is given as a parameter.

Afterwards, the *main iterations* begin where the algorithm repeats three steps until a termination criterion is satisfied. First, it performs multiple search steps with respect to the neighborhood $N_{\mathsf{add}}$ that maximize a randomly chosen function from $\mathcal{F}$ starting with a given initial solution $P_0$ until the resulting path exceeds the budget $B$ after which the path is now potentially infeasible. Next, the calculated path $P$ is improved with respect to $Length(P)$ by applying the Chained-Lin-Kernighan heuristic In the third step, if the improved path $P$ still exceeds the budget $B$, the algorithm removes nodes from $P$ by performing search steps with respect to $N_{\mathsf{remove}}(P)$. In the VNS framework, this can also be interpreted as a change in neighborhood since a local optimum with respect to $N_{\mathsf{add}}(P)$ was reached where it is not possible to feasibly improve the path by adding nodes. The search steps minimize a randomly chosen function from $\mathcal{F}$ until the resulting path $P$ satisfies $Length(P) \leq B$. The random selection of functions aims to provide a proper degree of variation in each iteration of the algorithm, which in combination with $f_{\mathsf{random}}$ strengthens the exploration and allows the algorithm to flexibly adapt to changes in the problem as it constantly tests a variety of paths based on the current solution.

## 5    Computational Evaluation

In this section we describe the experiments that have been done to evaluate the proposed heuristic $\mathrm{VNS_{DOP}}$ and their results.

### 5.1    Measurement of Algorithm Performance

In the DOP it is possible that the quality of a path $P$ as well as the optimal solution, i.e., the optimal path $P^*$, changes during the time interval $T$. Thus it is not sufficient to only consider an algorithm's performance at a certain point in time, e.g., when it terminates. Instead, it is necessary to take the algorithm's optimization behavior over the entirety of its run time into account. For this reason a measurement framework with extensive logging functionalities has been implemented (the source code is available at [22]). In order to measure an algorithm's performance over time, this work uses *progress curves* as described in [34], which plot over time the quality $Value(P_{best})$ of the best solution $P_{best}$ found so far. Similar to [34], the progress curves are recorded with respect to different time measures to evaluate different aspects of an algorithm. The general formulation of progress curves allows them to be easily applied to other optimization problems if appropriate problem-specific time measures are chosen. In this work, the following two time measures are used:

1. Function evaluations ($FE$) count how often the total length $Length(P)$ or the total value $Value(P)$ of $P$ has been calculated. This measure can show how an algorithm deals with the TSP sub-problem since for a given subset of nodes $V' \subseteq V$ typically multiple paths with different lengths exist.
2. Subsets ($SS$) count how many different subsets $V' \subseteq V$ of nodes have been used for the calculation of paths. This time measure shows the effort of an algorithm to select a suitable subset of $V$ and thus the effort for solving the sub-problem that is similar to the KP.

Observe that these time measures do not depend on the hardware that is used to run an algorithm.

Since the quality criterion $Value(P)$ is to be maximized, it is preferable for the resulting progress curve to quickly reach high values as opposed to progress curves for minimization problems such as the TSP where low values are desirable. This allows us to use the sum $UB_t = \sum_{v \in V} s(v,t)$ of all node values at a time $t \in T$ as a trivial upper bound for normalizing the solution quality such that for every time point $t$ we have that $Value(P)/UB_t \in [0,1]$, similar to the "optimization accuracy" measure used for dynamic optimization problems [26]. It should be noted that besides $UB_t$ the value $Value(P_{best})$ of the best solution $P_{best}$ found so far might also change over time due to the dynamics of the problem (which are described in Sect. 5.3). Taking these factors into account, an example for such a resulting progress curve is shown in Fig. 1 left.

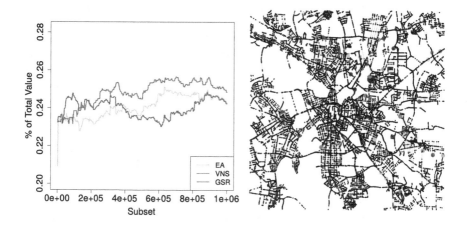

**Fig. 1.** Left: Example of a progress curve with respect to time measure $SS$ (VNS is used as an abbreviation for VNS$_{DOP}$). Right: Visualization of the OP instance Leipzig-100-80000-3 generated on basis of OpenStreetMap data for Leipzig. The blue node is the depot $v_0$ and the other colored nodes correspond to nodes with assigned values where a bright color indicates a high value. (Color figure online)

## 5.2   Choice of Algorithms for Comparison

Based on the literature review presented in Sect. 2, we selected the Evolutionary Algorithm from [19] and the Greedy Randomized Adaptive Search Procedure with Segment Remove from [18] as algorithms for comparison with our algorithm VNS$_{DOP}$. Both comparison algorithms are fairly recent algorithms that obtained favorable results when experimentally compared to other modern heuristics for the OP in the respective studies. In the following, they are abbreviated as EA and GSR, respectively. The authors of EA provide their source code on GitHub, but for the experiments in this work the algorithm has been reimplemented based on the source code and the description in [19] in order to fit and utilize our measurement framework for the DOP mentioned in Sect. 5.1. The algorithm GSR [18] has also been reimplemented as its source code is not available.

Since EA and GSR are both originally used as standalone algorithms, we also implemented an improvement heuristic variant of each algorithm, denoted as EA$'$

and GSR′ respectively. For a given initial solution $P_0$, algorithm EA′ initializes half of its population with mutated variants of $P_0$ that have been obtained with its mutation operator, after which EA′ is identical to the original EA. As for GSR′, the algorithm skips the construction phase for an initial solution if $P_0$ does not just contain the depot $v_0$ and immediately proceeds to the local search phase using $P_0$ (otherwise the algorithm is run as GSR). If the local search phase ends before the termination criterion is satisfied, the path obtained so far is modified using the Segment Remove operator proposed in [18] and algorithm GSR′ is repeated. The source code (in C++) for the proposed algorithm VNS$_{DOP}$, both reimplemented algorithms EA and GSR, and the variants EA′ and GSR′ is available on [22].

### 5.3   Problem Instances

The DOP instances that are used in this work are based on two sets of (static) OP instances. The first set is based on OPLib [20], a benchmark for Orienteering Problems. There we chose the instances brazil58, brazil48, gr48, gr120 from the gen4-subset since it contains the most difficult instances [20]. The 4 instances from the subset were chosen because they specify the distance function $d$ by a distance matrix whereas most of the other instances specify distances by using point coordinates and euclidean distances between them. However, in road networks it is not uncommon that the travel distance between two nodes differs from their euclidean distance which is why we did not consider these instances.

The second set of instances (in the following referred to as city) is intended to contain properties of road networks in cities and was generated as follows: We used map data from OpenStreetMap [27] from which extracts for two German cities, Leipzig and Berlin (as examples for a smaller and a larger city) were downloaded by using the download server from [30]. We then applied a parser [16] to extract the roads and processed the resulting files to obtain the road network as a graph. On these two graphs, we randomly selected 50, 100 and 150 nodes three times each and assigned to them a random initial value $s(v, 0) \in \{1, 2, \ldots, 10\}$ excluding one random node which was set as a depot node $v_0$ with value 0 (default value). In addition, distance matrices for these nodes were calculated so that the instance data is given in the same form as in the OPLib instances.

Regarding the budget $B$ for the city instances, we chose the value $B = 80\,000$ based on the following reasoning: A survey [14] measured that the average speed in the two aforementioned cities is 11 mph ($\approx 17.70\,\text{km/h}$). Since road transport drivers in the European Union are not allowed to drive for more than 4.5 h without a break [35], it is potentially possible to drive 49.5 miles during that time frame which, after rounding, corresponds to approximately 80 km or 80 000 m since the generated graphs measure distances in meters. Using the Concorde TSP Solver [6], it has been verified that for none of the generated instances all nodes can be reached within this budget. This set of instances contains 18 Orienteering Problems of which an example is shown in Fig. 1 right.

Based on the (static) OP instances described above, dynamic OP instances where the value function $s$ changes over time were generated as follows. For each node $v \in V$ in the graph of the problem instance, let $s(v, 0)$ be the *initial value* that was given beforehand (for the OPLib instances) or assigned in the procedure above (for the city instances). Let $V_1 = \{v \in V \mid s(v, 0) > 0 \lor v = v_0\}$ be the subset of nodes that are the depot or have a positive initial value (also called "points of interest" [9]). In the following, these nodes are set to be nodes for which the value function $s$ can potentially change its value. Additionally, since it is also desirable to generate instances where changes occur with different frequencies, we consider two dynamic levels: Formally, let the dynamic level be $L \in \{|V_1|, \lfloor |V_1|/2 \rfloor\}$ where $L = |V_1|$ (or $L = \lfloor |V_1|/2 \rfloor$) indicates that a dynamic instance with a high (or low) dynamic level is generated. The value $L = |V_1|$ was chosen since it allows for each node in $V_1$ to potentially change its value once.

Then for each time measure $tm \in \{FE, SS\}$, we consider the discrete time interval $T = \{1\,tm, 2\,tm, \ldots, 1\,000\,000\,tm\}$ and $L$ evenly distributed time points $t_i = \lfloor i \cdot |T|/(L+1) \rfloor$ tm with $i \in \{1, 2, \ldots, L\}$. These time points form a set $T' = \{t_0, t_1, t_2, \ldots, t_L\}$ where $t_0 = 0$ denotes the start time for all time measures. The time-dynamic function $s$ is now defined as follows. For each $t_i$ with $i \in \{1, 2, \ldots, L\}$, a node $v_i^* \in V_1 \backslash \{v_0\}$ is randomly chosen from all non-depot nodes in $V_1$ and we set

$$s(v, t_i) = \begin{cases} \max\{0,\, s(v, t_{i-1}) + q\} & \text{if } v = v_i^* \\ s(v, t_{i-1}) & \text{if } v \neq v_i^* \end{cases} \tag{6}$$

for all $v \in V$ where $q$ is a nonzero integer randomly drawn from the interval $[-5, 5]$. In other words, at each $t_i$ a random node in $V_1$ undergoes an absolute change in value by a nonzero integer between $-5$ and $5$, (i.e., by 50% of the highest number that can be randomly drawn for $s(v, 0)$ as described above). For all other $t \in T \backslash T'$, the function $s$ is set to be constant: $s(v, t) = s(v, t - 1\,tm)$ for all $v \in V$. Applying this procedure to the 22 static instances described above for each value of $L$ and tm led to 88 DOP instances, which are uploaded to [22].

## 5.4    Initial Solutions for the Improvement Heuristics

Improvement heuristics require an initial solution $P_0$ that is to be improved. In order to investigate how the compared algorithms deal with initial solutions of varying quality, the following five initial solutions were generated for each of the 22 (static) OP instances used in this work: two random paths $P_{rand1}, P_{rand2}$ generated by randomly selecting and inserting nodes without exceeding the budget $B$ (two paths in order to reduce the variance), a path $P_{greedy}$ constructed by a simple greedy heuristic which, based on the path $P = (v_0)$, repeatedly appends nodes that increase its length the least until the path length exceeds $B$, the best solution $P_{EA}$ of the EA [19] from Sect. 5.2 after a run time of 10 min and the solution $P_{GSR}$ calculated by the algorithm GSR [18]. The solutions $P_{rand1}, P_{rand2}$ and $P_{greedy}$ represent solutions of low and medium quality, respectively, whereas the solutions $P_{EA}$ and $P_{GSR}$ correspond to high-quality solutions generated by state-of-the-art algorithms for the OP. In total, 110 initial solutions were created which are available at [22]. For the dynamic OP instances, the same initial solution is used that was generated for the static instance it is based on.

## 5.5 Parameter Values

The proposed algorithm VNS$_{DOP}$ from Sect. 4 contains two parameters: The number of iterations in the initial phase $k_{init}$ and the probability $p$ for random insertions in the initial phase. Regarding the former, since the number of possible paths grows rapidly with increasing number of nodes in the graph, we consider it reasonable to scale the length of the initial exploration phase with the size of the instance. However, if the algorithm focuses too strongly on exploration, then there might not be enough time to refine the discovered solutions. We thus chose $k_{init} = \sqrt{|V_1|}$, i.e., the square root of the number of nodes with positive value including the depot as a compromise between these two conflicting aspects.

As for the parameter $p$, it is desirable that $p$ approximates the fraction of nodes that are contained in an optimal solution so that the insertion and subsequent optimization of the path length leads to a solution of high quality. The authors of the EA presented in [19] dealt with a similar problem by utilizing the formula $\sqrt{B/Length(P_{LK})}$ which in the following is also used for the parameter $p$. This formula, which incorporates the budget $B$ and the length of the path $P_{LK}$ obtained by applying the Chained-Lin-Kernighan heuristic on all nodes in $V_1$, provides an efficiently calculable approximation for the number of nodes in an optimal solution (in relation to all nodes). The parameters from the other algorithms EA and GSR are set as described in their respective studies [18,19].

## 5.6 Comparison of VNS$_{DOP}$ as an Improvement Heuristic with Other Metaheuristics

Each of the compared improvement heuristics was executed on each of the 110 instances described in Sect. 5.3 with each of the initial solutions and 10 repetitions over which the progress curves were averaged. The runs were performed on a cluster with 4 computers that each have eight 3.4-GHz-cores (each run being executed on one core) and 32 GB RAM. For the dynamic instances, the compared algorithms were set to terminate when the time horizon $T$ expires, i.e., after 1 000 000 function evaluations or 1 000 000 subsets are calculated, depending on the time measure tm by which the value function $s$ changes. Runs on the static instances were set to terminate after 1 000 000 time units have passed for both of the time measures.

Plotting the progress curves for each instance and initial solution leads to 660 diagrams so that an individual evaluation of each diagram is not feasible. For this reason, we calculate the percentage that the area under the progress curve ("area under curve", $AUC$) occupies from the area of a theoretical curve with constant value 1. This value, denoted $AUC^{rel}$ ("relative AUC"), satisfies $AUC^{rel} \in [0, 1]$ and provides an aggregate quality measure for the progress curves similar to [34]. These values allow us to quantify the performance on a given instance where a high value indicates that an algorithm quickly obtains solutions of high quality with respect to the time horizon $T$.

In order to compare the AUC values over different instances (similar to [26]), we normalize them with respect to the best attained value. Formally, if $AUC_{I,A}^{rel}$ denotes the relative AUC for algorithm $A$ on an instance $I$, the normalized value $AUC_{I,A}^{norm} \in [0,1]$ is calculated as $AUC_{I,A}^{norm} = AUC_{I,A}^{rel} / \max_{A'} AUC_{I,A'}^{rel}$ with $A' \in \{\text{VNS}_{\text{DOP}}, \text{EA}, \text{GSR}\}$ where a value close to 1 can be interpreted as the algorithm $A$ reaching a performance similar to the best performing algorithm for instance $I$. This type of evaluation measure can be seen as an extension of an evaluation measure also known as "collective mean fitness" that is commonly used for dynamic optimization problems [26] where instead of the best values per iteration/generation the optimization behavior over the entire run time is taken into account.

**Table 1.** Average values for $AUC_{I,A}^{norm}$ for the improvement heuristics, aggregated by different criteria, time measures and truncated to 3 decimal places. Values in bold indicate the best average value for each aggregation criterion and time measure. The left half of the table shows the AUC values calculated over the entire time horizon, whereas the right half of the table shows the values when the calculation of AUC values is restricted to the final 500 000 time units. Here "VNS" is used as an abbreviation for VNS$_{\text{DOP}}$ and the notation $P_0 = P_{\text{rand}}$ indicates the aggregation over all instances with random initial solutions ($P_0 \in \{P_{\text{rand1}}, P_{\text{rand2}}\}$).

| Aggregation over instances with | AUC over all 1 000 000 time units | | | | | | AUC over the last 500 000 time units | | | | | |
|---|---|---|---|---|---|---|---|---|---|---|---|---|
| | tm $= FE$ | | | tm $= SS$ | | | tm $= FE$ | | | tm $= SS$ | | |
| | EA$'$ | GSR$'$ | VNS | EA$'$ | GSR$'$ | VNS | EA$'$ | GSR$'$ | VNS | EA$'$ | GSR$'$ | VNS |
| Initial solution | | | | | | | | | | | | |
| $P_0 = P_{\text{rand}}$ | 0.938 | 0.821 | **1.000** | 0.960 | 0.890 | **1.000** | 0.943 | 0.872 | **1.000** | 0.959 | 0.918 | **0.999** |
| $P_0 = P_{\text{greedy}}$ | 0.944 | 0.899 | **1.000** | 0.963 | 0.940 | **1.000** | 0.950 | 0.923 | **1.000** | 0.964 | 0.949 | **0.999** |
| $P_0 = P_{\text{EA}}$ | 0.957 | 0.940 | **1.000** | 0.972 | 0.963 | **0.999** | 0.963 | 0.946 | **0.999** | 0.973 | 0.960 | **0.999** |
| $P_0 = P_{\text{GSR}}$ | 0.958 | 0.918 | **1.000** | 0.972 | 0.937 | **0.999** | 0.964 | 0.920 | **0.999** | 0.973 | 0.934 | **0.999** |
| Instance set | | | | | | | | | | | | |
| OPLib | 0.986 | 0.934 | **1.000** | 0.990 | 0.967 | **1.000** | 0.988 | 0.956 | **1.000** | 0.991 | 0.973 | **0.999** |
| city | 0.938 | 0.868 | **1.000** | 0.960 | 0.919 | **0.999** | 0.945 | 0.896 | **0.999** | 0.960 | 0.927 | **0.999** |
| Dynamic level | | | | | | | | | | | | |
| $L = |V_1|$ | 0.947 | 0.881 | **1.000** | 0.966 | 0.926 | **0.999** | 0.953 | 0.907 | **0.999** | 0.966 | 0.934 | **0.999** |
| $L = \lfloor|V_1|/2\rfloor$ | 0.947 | 0.880 | **1.000** | 0.966 | 0.930 | **0.999** | 0.954 | 0.907 | **0.999** | 0.966 | 0.938 | **0.999** |
| Static OPs | 0.946 | 0.879 | **1.000** | 0.964 | 0.927 | **0.999** | 0.952 | 0.907 | **1.000** | 0.965 | 0.935 | **0.999** |
| All instances | 0.947 | 0.880 | **1.000** | 0.965 | 0.928 | **0.999** | 0.953 | 0.907 | **0.999** | 0.966 | 0.936 | **0.999** |

Table 1 shows the average $AUC_{I,A}^{norm}$ aggregated over different subsets of the instances. The $AUC_{I,A}^{norm}$ and $AUC_{I,A}^{rel}$ values for all instances and algorithms as well as the data used for plotting the progress curves are available at [22] so that future works can be compared with these algorithms. The aggregation criteria were chosen to investigate how the different improvement heuristics perform with initial solutions of varying quality on both sets of instances with different dynamic levels. Here a high value indicates that on average the curves for the respective algorithm are highly similar to the best performing algorithm on the considered instances. In order to specifically investigate the steady-state performance of the algorithms without effects that are caused by differences during the initialization, an additional evaluation that only considers the AUC values after 500 000 time units has been performed. This evaluation shows the optimization behavior for the case that a heuristic is used long-term in a dynamic environment. The results are shown in the right half of Table 1.

It can be seen that in both sets of evaluations $VNS_{DOP}$ obtains the best results for the considered time measures and aggregation criteria. Especially for time measure $FE$ and when all time units are taken into account, its average value over all instances is 1.000 which means that it obtained the best $AUC_{I,A}^{norm}$ value on all instances showing a high performance for different initial solutions and levels of dynamic changes in the score function $s$, including static OP instances. In addition, the high values for $VNS_{DOP}$ in the right half of Table 1 also indicate that the algorithm maintains a high performance over time. Algorithm $EA'$ obtains slightly higher values in the right half of Table 1 than in the left half, whereas for $GSR'$ this is true in all but 2 cases. However, in general it can be seen that the values for $EA'$ are higher than the AUC values for $GSR'$ indicating that $EA'$ outperforms $GSR'$, but not $VNS_{DOP}$.

### 5.7 Comparison of $VNS_{DOP}$ as a Standalone Algorithm with Other Metaheuristics

In this section it is investigated whether $VNS_{DOP}$ is also suited as a standalone heuristic algorithm for the Dynamic OP that does not utilize a starting solution $P_0$. For this, the same setup and instances as in the previous section are used and the $VNS_{DOP}$ algorithm from Sect. 4 is called with the same parameters as in the previous experiment, with the exception of the initial path $P_0$ which is set as the path $P_0 = (v_0)$ that only contains the depot. The other algorithms are used in their original version, i.e., as EA and GSR which do not use user-defined initial solutions. However, since GSR in its original formulation [18] terminates after the local search phase (whereas EA allows for user-defined termination criteria), comparisons with EA and GSR are performed with two separate sets of runs. More precisely, for the comparison with EA the same termination criteria as in the previous experiment in Sect. 5.6 are used, whereas for the comparison with GSR, we first run GSR on each instance and measure the number of required time units for $FE$ and $SS$. Afterwards, the evaluation of $VNS_{DOP}$ and GSR is restricted to the same number of time units that GSR used.

The results, shown in Table 2, show that $VNS_{DOP}$ as a standalone algorithm reaches the best performance for all aggregation criteria and time measures. Similar to Sect. 5.6, it can be seen that the gap between EA and $VNS_{DOP}$ is smaller than between GSR and $VNS_{DOP}$ which indicates that EA reaches a higher performance than GSR. As for GSR, even though it obtains good results for the time measure $SS$ when compared to $FE$ (indicating that it carefully selects a subset $V' \subseteq V$ and thoroughly tests it before changing the subset) it is still outperformed by $VNS_{DOP}$ in all criteria and both time measures.

**Table 2.** Average values for $AUC_{T,A}^{norm}$ for the runs with standalone algorithms, aggregated by different criteria, time measures and truncated to 3 decimal places. Values in bold indicate the best average value for each aggregation criterion and time measure. Note that the results between EA and VNS$_{DOP}$ are separate from the runs with GSR and VNS$_{DOP}$ due to the different termination criteria between EA and GSR.

| Aggregation over instances with | Comparison with EA | | | | Comparison with GSR | | | |
|---|---|---|---|---|---|---|---|---|
| | tm = $FE$ | | tm = $SS$ | | tm = $FE$ | | tm = $SS$ | |
| | EA | VNS$_{DOP}$ | EA | VNS$_{DOP}$ | GSR | VNS$_{DOP}$ | GSR | VNS$_{DOP}$ |
| Instance set | | | | | | | | |
| OPLib | 0.936 | **0.999** | 0.956 | **0.999** | 0.772 | **1.000** | 0.868 | **0.995** |
| city | 0.990 | **0.999** | 0.992 | **0.999** | 0.886 | **1.000** | 0.943 | **1.000** |
| Dynamic level | | | | | | | | |
| $L = |V_1|$ | 0.951 | **0.999** | 0.963 | **0.998** | 0.808 | **1.000** | 0.898 | **0.990** |
| $L = \lfloor|V_1|/2\rfloor$ | 0.943 | **1.000** | 0.966 | **0.999** | 0.782 | **1.000** | 0.880 | **1.000** |
| Static OPs | 0.944 | **0.999** | 0.958 | **0.999** | 0.787 | **1.000** | 0.866 | **0.997** |
| All instances | 0.946 | **0.999** | 0.962 | **0.999** | 0.793 | **1.000** | 0.882 | **0.996** |

In addition, since both [19] and [18] measure algorithm performance using the final solution quality at the time of termination, statistical tests for the static OP instances were performed comparing the average quality of the best solution found at the end. In particular, we used the sign test for paired samples which is a non-parametric test and compared VNS$_{DOP}$ with the other two algorithms pairing by the static OP instances. Performing the test showed that VNS$_{DOP}$ obtained better results than EA and GSR on both time measures $FE$ and $SS$ with a highly significant difference ($n = 22$, $p < 0.001$ for each of these tests).

## 6    Conclusion

This work considered a Dynamic Orienteering Problem (DOP) where the value function $s$ for the nodes changes over time during the optimization process. Since dynamic problems necessitate the development of algorithms that can quickly adapt existing solutions, an improvement heuristic based on Variable Neighborhood Search (VNS$_{DOP}$) has been proposed that optimizes a given initial solution. The main idea of VNS$_{DOP}$ is to take the two interacting sub-problems of the OP into account by using different functions and neighborhoods by which new solutions are selected, but this concept can also be applied to other combinatorial optimization problems by choosing the functions and neighborhoods in accordance with their characteristics and structure.

The heuristic VNS$_{DOP}$ was compared with two improvement heuristics based on existing state-of-the-art methods for the static OP. The experimental evaluation of the algorithms, which considered the performance over the entire time horizon, showed that VNS$_{DOP}$ is able to deal with several types of instances with differing dynamic levels (including static OPs) and with initial solutions of varying quality outperforming the other algorithms with respect to several time measures. An additional experiment showed that $VNS_{DOP}$ is also suited as a standalone heuristic not requiring initial solutions as it obtained better results over time than the two aforementioned state-of-the-art methods for the considered criteria. For future research, dynamic Orienteering Problems with other dynamic levels $L$ and DOPs where other factors, such as the budget $B$ or the distances $d$ between nodes dynamically change, potentially affecting the validity of existing solutions are to be investigated.

**Acknowledgements.** This work was funded by the Deutsche Forschungsgemeinschaft (DFG, German Research Foundation) - project number 392050753.

# References

1. Abbaspour, R.A., Samadzadegan, F.: Time-dependent personal tour planning and scheduling in metropolises. Expert Syst. Appl. **38**(10), 12439–12452 (2011)
2. Angelelli, E., Archetti, C., Filippi, C., Vindigni, M.: The probabilistic orienteering problem. Comput. Oper. Res. **81**, 269–281 (2017)
3. Applegate, D., Cook, W., Rohe, A.: Chained Lin-Kernighan for large traveling salesman problems. INFORMS J. Comput. **15**, 82–92 (2003)
4. Archetti, C., Speranza, M.G., Vigo, D.: Vehicle routing problems with profits. In: Toth, P., Vigo, D. (eds.) Vehicle Routing, pp. 273–297 (2014)
5. Campos, V., Marti, R., Sánchez-Oro Calvo, J., Duarte, A.: Grasp with path relinking for the orienteering problem. J. Oper. Res. Soc. **65**, 1800–1813 (2014)
6. Cook, W., Applegate, D., Bixby, R., Chvátal, V.: Concorde TSP solver (2005). http://www.math.uwaterloo.ca/tsp/concorde/downloads/downloads.htm
7. Erkut, E., Zhang, J.: The maximum collection problem with time-dependent rewards. Naval Res. Logist. **43**(5), 749–763 (1996)
8. Fomin, F.V., Lingas, A.: Approximation algorithms for time-dependent orienteering. Inf. Process. Lett. **83**(2), 57–62 (2002)
9. Gavalas, D., Konstantopoulos, C., Pantziou, G.: A survey on algorithmic approaches for solving tourist trip design problems. J. Heuristics **20**(3), 291–328 (2014)
10. Golden, B., Levy, L., Vohra, R.: The orienteering problem. Naval Res. Logist. **34**, 307–318 (1987)
11. Gunawan, A., Lau, H.C., Vansteenwegen, P.: Orienteering problem: a survey of recent variants, solution approaches and applications. Eur. J. Oper. Res. **255**(2), 315–332 (2016)
12. Hansen, P., Mladenović, N., Pérez, J.M.: Variable neighbourhood search: methods and applications. Ann. Oper. Res. **175**, 367–407 (2010). https://doi.org/10.1007/s10479-009-0657-6
13. İlhan, T., Iravani, S.M.R., Daskin, M.S.: The orienteering problem with stochastic profits. IIE Trans. **40**(4), 406–421 (2008)
14. INRIX: Durchschnittliche Geschwindigkeit* im Automobilverkehr in ausgewählten deutschen Städten im Jahr 2018 (in Meilen pro Stunde). Statista (2019). https://de.statista.com/statistik/daten/studie/994676/umfrage/innerstaedtische-durchschnittsgeschwindigkeit-im-autoverkehr-in-deutschen-staedten/
15. Kara, I., Bicakci, P.S., Derya, T.: New formulations for the orienteering problem. In: 3rd Global Conference on Business, Economics, Management and Tourism, vol. 39, pp. 849–854 (2016)

16. Karlin, M., Heikkilä, J.: OSM-graph-parser (2017). https://github.com/rovaniemi/osm-graph-parser
17. Kataoka, S., Morito, S.: An algorithm for single constraint maximum collection problem. J. Oper. Res. Soc. Jpn. **31**(4), 515–560 (1988)
18. Keshtkaran, M., Ziarati, K.: An efficient evolutionary algorithm for the orienteering problem. J. Heuristics **22**, 699–726 (2016)
19. Kobeaga, G., Merino, M., Lozano, J.A.: An efficient evolutionary algorithm for the orienteering problem. Comput. Oper. Res. **90**, 42–59 (2018)
20. Kobeaga, G., Merino, M., Lozano, J.A.: OPLib: test instances for the orienteering problem (2018). https://github.com/bcamath-ds/OPLib/tree/master/instances
21. Laporte, G., Martello, S.: The selective travelling salesman problem. Discrete Appl. Math. **26**(2), 193–207 (1990)
22. Le, H.T.: DynamicOrienteeringAlgorithms (2020). https://github.com/L-HT/DynamicOrienteeringAlgorithms/
23. Mann, M., Zion, B., Rubinstein, D., Linker, R., Shmulevich, I.: The orienteering problem with time windows applied to robotic melon harvesting. J. Optim. Theory Appl. **168**, 246–267 (2015). https://doi.org/10.1007/s10957-015-0767-z
24. Marinakis, Y., Politis, M., Marinaki, M., Matsatsinis, N.: A memetic-GRASP algorithm for the solution of the orienteering problem. In: Le Thi, H.A., Pham Dinh, T., Nguyen, N.T. (eds.) Modelling, Computation and Optimization in Information Systems and Management Sciences. AISC, vol. 360, pp. 105–116. Springer, Cham (2015). https://doi.org/10.1007/978-3-319-18167-7_10
25. Nagarajan, V., Ravi, R.: The directed orienteering problem. Algorithmica **60**, 1017–1030 (2011). https://doi.org/10.1007/s00453-011-9509-2
26. Nguyen, T.T., Yang, S., Branke, J.: Evolutionary dynamic optimization: a survey of the state of the art. Swarm Evol. Comput. **6**, 1–24 (2012)
27. OpenStreetMap contributors: planet dump (2020). https://planet.osm.org. https://www.openstreetmap.org
28. Ostrowski, K., Karbowska-Chilinska, J., Koszelew, J., Zabielski, P.: Evolution-inspired local improvement algorithm solving orienteering problem. Ann. Oper. Res. **253**, 519–543 (2017). https://doi.org/10.1007/s10479-016-2278-1
29. Schilde, M., Doerner, K.F., Hartl, R.F., Kiechle, G.: Metaheuristics for the bi-objective orienteering problem. Swarm Intell. **3**, 179–201 (2009). https://doi.org/10.1007/s11721-009-0029-5
30. Schneider, W.: Bbbike.org (2020). https://download.bbbike.org/osm/
31. Tsiligiridis, T.: Heuristic methods applied to orienteering. J. Oper. Res. Soc. **35**, 797–809 (1984). https://doi.org/10.1057/jors.1984.162
32. Vansteenwegen, P., Souffriau, W., Berghe, G.V., Oudheusden, D.V.: The city trip planner: an expert system for tourists. Expert Syst. Appl. **38**(6), 6540–6546 (2011)
33. Wang, X., Golden, B.L., Wasil, E.A.: Using a genetic algorithm to solve the generalized orienteering problem. In: Golden, B., Raghavan, S., Wasil, E. (eds.) The Vehicle Routing Problem: Latest Advances and New Challenges. ORCS, vol. 43, pp. 263–274. Springer, Boston (2008). https://doi.org/10.1007/978-0-387-77778-8_12
34. Weise, T., et al.: Benchmarking optimization algorithms: an open source framework for the traveling salesman problem. IEEE Comput. Intell. Mag. **9**(3), 40–52 (2014)
35. Your Europe: Road transportation workers (2020). https://europa.eu/youreurope/business/human-resources/transport-sector-workers/road-transportation-workers
36. Zhang, S., Ohlmann, J.W., Thomas, B.W.: Dynamic orienteering on a network of queues. Transp. Sci. **52**(3), 691–706 (2018)

# Runtime Analysis of the $(\mu + 1)$-EA on the Dynamic BinVal Function

Johannes Lengler[(⊠)] and Simone Riedi

Department of Computer Science, ETH Zürich, Zurich, Switzerland
johannes.lengler@inf.ethz.ch

**Abstract.** We study evolutionary algorithms in a dynamic setting, where for each generation a different fitness function is chosen, and selection is performed with respect to the current fitness function. Specifically, we consider Dynamic BinVal, in which the fitness functions for each generation is given by the linear function BinVal, but in each generation the order of bits is randomly permuted. For the $(1+1)$-EA it was known that there is an efficiency threshold $c_0$ for the mutation parameter, at which the runtime switches from quasilinear to exponential. Previous empirical evidence suggested that for larger population size $\mu$, the threshold may increase. We prove rigorously that this is at least the case in an $\varepsilon$-neighborhood around the optimum: the threshold of the $(\mu + 1)$-EA becomes arbitrarily large if the $\mu$ is chosen large enough.

However, the most surprising result is obtained by a second order analysis for $\mu = 2$: the threshold *increases* with increasing proximity to the optimum. In particular, the hardest region for optimization is *not* around the optimum. (Extended Abstract. A full version is available on arxiv at [17].)

## 1 Introduction

An important aspect of population-based optimization heuristics like evolutionary algorithms is their incremental nature. At any point in time the population represents a set of solutions. This makes population-based optimization heuristics very flexible. For example, the heuristic can be stopped after any time budget (predefined or chosen during execution), or when some desired quality of the solutions is reached. For the same reason, population-based algorithms are naturally suited for dynamic environments, in which the optimization goal ("fitness function") may change over the course of optimization.[1] In such a setting, it is not necessary to restart the algorithm from scratch when the fitness function changes, but rather we can use the current population as starting point for the new optimization environment. If the fitness function changes slowly enough,

---

[1] By this we mean that *selection* is performed according to the current fitness function as in [8]. I.e., all individuals from parent and offspring population are compared with respect to the same fitness function. Other versions exist, e.g. [4] studies the same problem as [8] without re-evaluations.

© Springer Nature Switzerland AG 2021
C. Zarges and S. Verel (Eds.): EvoCOP 2021, LNCS 12692, pp. 84–99, 2021.
https://doi.org/10.1007/978-3-030-72904-2_6

then population-based optimization heuristics may still find the optimum, or track the optimum over time [2,7,10,11,20–22,24–26]. We refrain from giving a detailed overview over the literature since an excellent review has recently been given by Neumann, Pourhassan, and Roostapour [23]. All the settings have in common that either the fitness function changes with very low frequency, or it changes only by some small local differences, or both.

Recently, a new setting called *dynamic linear functions* was proposed by Lengler and Schaller [18].[2] A class of dynamic linear functions is determined by a distribution $\mathcal{D}$ on the positive reals $\mathbb{R}^+$. For the $k$-th generation, $n$ weights $W_1^k, \ldots, W_n^k$ are chosen independently identically distributed (i.i.d.) from $\mathcal{D}$, and the fitness function for this generation is given by $f^k : \{0, 1\}^n \to \mathbb{R}^+$; $f^k(x) = \sum_{i=1}^{n} W_i^k x_i$. So the fitness in each generation is given by a linear function with positive weights, but the weights are drawn randomly in each generation. Note that for any fitness function, a one-bit in the $i$-th position will always yield a better fitness than a zero-bit. In particular, all fitness functions share a common global maximum, which is the string $\mathrm{OPT} = (1...1)$. Hence, the fitness function may change rapidly and strongly from generation to generation, but the direction of the signal remains unchanged: one-bits are preferred over zero-bits.

Several applications are discussed in [18]. One of them is a chess engine that can switch databases for different openings ON or OFF. The databases strictly improve performance in all situations, but if the engine is trained against varying opponents, then an opening may be used more or less frequently; so the weight of the corresponding bit my be high or low. Obviously, it is desirable that an optimization heuristic manages to switch all databases ON in such a situation. However, as we will see, this is not automatically achieved by many simple optimization heuristics. Rather, it depends on the parameter settings whether the optimal configuration (all databases ON) is found.

In [18], the runtime (measured as the number of generated search points) of the well-known $(1 + 1)$-EA on dynamic linear functions was studied. The authors gave a full characterization of the optimization behavior in terms of the *mutation parameter* $c$. In the $(1 + 1)$-EA, standard bit mutation is used for generating offspring, which flips each bit independently with probability $c/n$. It was shown that there is a threshold $c^* = c^*(\mathcal{D}) \in \mathbb{R}^+ \cup \{\infty\}$ such that for $c < c^*$ the $(1 + 1)$-EA optimizes the dynamic linear function with weight distribution $\mathcal{D}$ in time $O(n \log n)$. On the other hand, for $c > c^*$, the algorithm needs exponential time to find the optimum. The threshold $c^*(\mathcal{D})$ was given by an explicit formula. For example, if $\mathcal{D}$ is an exponential distribution then $c^*(\mathcal{D}) = 2$, if it is a geometric distribution $\mathcal{D} = \mathrm{GEOM}(p)$ then $c^* = (2-p)/(1-p)$. Moreover, the authors in [18] showed that there is $c_0 \approx 1.59..$ such that $c^*(\mathcal{D}) > c_0$ for every distribution $\mathcal{D}$, but for any $\varepsilon > 0$ there is a distribution $\mathcal{D}$ with $c^*(\mathcal{D}) < c_0 + \varepsilon$. As a consequence, if $c < c_0$ then the $(1 + 1)$-EA with mutation parameter $c/n$ needs time $O(n \log n)$ to optimize any dynamic linear functions, while for $c > c_0$ there are dynamic linear functions on which it needs exponential time.

---

[2] They argued that it might either be called noisy linear functions or dynamic linear functions, but we prefer the term dynamic.

In the mostly experimental paper [15,16], Lengler and Meier defined the *dynamic binary value function* DYNBV as a limiting case of dynamic linear functions. In DYNBV, in each generation a uniformly random permutation $\pi^k :$ $\{1,\ldots,n\} \rightarrow \{1,\ldots,n\}$ of the bits is drawn, and the fitness function is then given by $f^k(x) = \sum_{i=1}^{n} 2^{n-i} x_{\pi^k(i)}$. So in each generation, DYNBV evaluates the BINVAL function with respect to a permutation of the search space. Lengler and Meier observed that the proof in [18] for the $(1+1)$-EA extends to DYNBV with threshold $c^* = c_0$, i.e., the $(1+1)$-EA needs time $O(n \log n)$ for mutation parameter $c < c_0$, and exponential time for $c > c_0$. In this sense, DYNBV is the hardest dynamic linear function.

The papers [15,16] studied the $(\mu + 1)$-EA (using only mutation) and $(\mu+1)$-GA (using randomly mutation or crossover) for $\mu \in \{1,2,3,5\}$ on DYNBV by runtime simulations and found two main results. As they increased the population size $\mu$ from 1 to 5, the efficiency threshold $c_0$ increased moderately for the $(\mu + 1)$-EA (from 1.6 to 3.4, and strongly for the $(\mu + 1)$-GA (from 1.6 to more than 20). So with larger population size, the algorithms have a larger range of feasible parameter settings, and even more so when crossover is used.

Moreover, they studied which range of the search space was hardest for the algorithms, by estimating the drift towards the optimum with Monte Carlo simulations. For the $(\mu + 1)$-GA, they found that the hardest region was around the optimum, as one would expect. Surprisingly, for the $(\mu + 1)$-EA, this did not seem to be the case. They gave empirical evidence that the hardest regime was bounded away from the optimum. I.e., there were parameters $c$ for which the $(\mu+1)$-EA had positive drift (towards the optimum) in a region around the optimum. But it had *negative* drift in an intermediate region that was further away from the optimum. This finding is remarkable since it contradicts the commonplace that optimization gets harder closer to the optimum. Notably, a very similar phenomenon was proven by Lengler and Zou [19] for the $(\mu + 1)$-EA on certain monotone functions ("HOTTOPIC"), see the discussion below. Strikingly, such an effect was neither built into the fitness environments (not for HOTTOPIC, and not for DYNBV) nor into the algorithms. Rather, it seems to originate in a complex (and detrimental!) population dynamics that unfolds only in a regime of weak selective pressure. If selective pressure is strong, then the population often degenerates into copies of the same search point. As a consequence, diversity is lost, and the $(\mu + 1)$-EA degenerates into the $(1 + 1)$-EA. In these regimes, diversity *decreases* the ability of the algorithm to make progress. For HOTTOPIC functions, these dynamics are well-understood [13,19]. For dynamic linear functions, even though we can prove this behavior in this paper for the $(2 + 1)$-EA (see below), we are still far from a real understanding of these dynamics. Most likely, they are different from the dynamics for HOTTOPIC functions.

**Our Results.** We study the degenerate population drift (see Sect. 2) for the $(\mu + 1)$-EA with mutation parameter $c > 0$ on DYNBV in an $\varepsilon$-neighbourhood of the optimum. I.e., we assume that the search points in the current population have at least $(1 - \varepsilon)n$ one-bits, for some sufficiently small constant $\varepsilon > 0$. We find that for every constant $c > 0$ there is a constant $\mu_0$ such that for $\mu \geq \mu_0$ the drift is positive (towards the optimum). This means that with high probability the algorithm needs time $O(n \log n)$ to improve from $(1 - \varepsilon)n$ one-bits to the optimum. Hence, larger population sizes are helpful, as the drift of the $(1+1)$-EA around the optimum is negative for all $c > c_0 \approx 1.59..$ (which implies exponential optimization time). So for any $c > c_0$, increasing the population size to a large constant will decrease the runtime from exponential to quasi-linear. This is consistent with the experimental findings in [15] for $\mu = \{1, 2, 3, 5\}$, and it proves that population size can compensate for arbitrarily large mutation parameters.

For the $(2 + 1)$-EA, we perform a second-order analysis of the drift (i.e., we also analyze the lower order terms) and prove that in an $\varepsilon$-neighborhood of the optimum, the drift increases with the distance from the optimum. In particular, there are some values of $c$ for which the drift is positive around the optimum, but negative in an intermediate distance. It follows from standard arguments that there are $\varepsilon, c > 0$ such that the runtime is $O(n \log n)$ if the algorithm is started in an $\varepsilon$-neighborhood of the optimum, but that it takes exponential time to reach this $\varepsilon$-neighborhood. Thus we formally prove that the hardest part of optimization is not around the optimum, as was already experimentally concluded from Monte Carlo simulations in [15].

**Related Work.** Jansen [9] introduced a pessimistic model for analyzing linear functions, later extended in [1], which is *also* a pessimistic model for dynamic linear functions and DynBV *and* for monotone functions. The similarities with monotone functions go surprisingly far. It was shown in [5] that the $(1 + 1)$-EA needs exponential time on some monotone functions if the mutation parameter $c$ is too large. The construction of hard monotone instances was simplified in [12] and later called HotTopic functions. HotTopic functions were analyzed for a large set of algorithms in [13]. For the $(1 + \lambda)$-EA, the $(1 + (\lambda, \lambda))$-GA, the $(\mu + 1)$-EA, and the $(1 + \lambda)$-fEA, thresholds for the mutation parameter $c$ or related quantities were determined such that a larger mutation rate leads to exponential runtime, and a smaller mutation rate leads to runtime $O(n \log n)$. The population size $\mu$ and offspring population size $\lambda$ of the algorithms had no impact on the threshold. Crucially, all these results were obtained for parameters of HotTopic functions in which only the behavior in an $\varepsilon$-neighborhood around the optimum mattered. This dichotomy between quasilinear and exponential runtime is very similar to the situation for DynBV. However, for the $(\mu + 1)$-EA on HotTopic functions the threshold $c_0$ was independent of $\mu$, while we show that on DynBV it becomes arbitrarily large as $\mu$ grows. Thus large population sizes help for DynBV, but not for HotTopic.

As we prove, for the $(2 + 1)$-EA the region around the optimum is not the hardest region for optimization, and there are values of $c$ for which there is a positive drift around the optimum, but a negative drift in an intermediate region. As Lengler and Zou showed [19], the same phenomenon occurs for the $(\mu + 1)$-EA on HotTopic functions. In fact, they showed that larger population size even hurts: for any $c > 0$ there is a $\mu_0$ such that the $(\mu + 1)$-EA with $\mu \geq \mu_0$

has negative drift in some intermediate region (and thus exponential runtime), even if $c$ is much smaller than one! This surprising effect is due to population dynamics in which it is not the genes of the fittest individuals who survive in the long terms. Rather, individuals which are strictly dominated by others (and substantially less fit) serve as the seeds for new generations. Importantly, the analysis of this dynamics relies on the fact that for HOTTOPIC functions, the weight of the positions stay fixed for a rather long period of time (as long as the algorithm stays in the same region/level of the search space). Thus, the results do not transfer to DYNBV functions. Nevertheless, the picture looks similar insofar as the hardest region for optimization is not around the optimum in both cases. Since our analysis for DYNBV is only for $\mu = 2$, we can't say whether the efficiency threshold in $c$ is increasing or decreasing with $\mu$. The experiments in [15,16] find increasing thresholds (so the opposite effect as for HOTTOPIC), but are only for $\mu \leq 5$.

## 2   Preliminaries

**Notation and Setup.** The general setting of a $(\mu + \lambda)$ algorithm in dynamic environments on the hypercube $\{0,1\}^n$ is as follows. A population $P^k$ of $\mu$ search points is maintained. In each generation $k$, $\lambda$ offspring are generated. Then a *selection operator* selects the next population $P^{k+1}$ from the $\mu + \lambda$ search points according to the fitness function $f^k$.

In this paper, we will study the $(\mu+1)$-Evolutionary Algorithm ($(\mu+1)$-EA) with *standard bit mutation* and *elitist selection*. So for offspring generation, a parent $x$ is chosen uniformly at random from $P^k$, and the offspring is generated by flipping each bit of $x$ independently with probability $c/n$, where $c$ is the *mutation parameter*. For selection, we simply select the $\mu$ individuals with largest $f^k$-values to form population $P^{k+1}$.

For the dynamic binary value function DYNBV, for each $k \geq 0$ a uniformly random permutation $\pi^k : \{1,\dots,n\} \rightarrow \{1,\dots,n\}$ is drawn, and the fitness function for generation $k$ is then given by $f^k(x) = \sum_{i=1}^{n} 2^{n-i} x_{\pi^k(i)}$.

Throughout the paper, we assume that the population size $\mu$ and the mutation parameter $c$ are constants, whereas $n$ tends to $\infty$. We use the expression "with high probability" or whp for events $\mathcal{E}_n$ such that $\Pr(\mathcal{E}_n) \rightarrow 1$ for $n \rightarrow \infty$. For two bit-strings $x, y \in \{0,1\}^n$, $x$ *dominates* $y$ if $x_i \geq y_i$ for all $i \in \{1..n\}$.

Our main tool will be drift theory. In order to apply this, we need to identify states that we can adequately describe by a single real value. Following the approach in [13] and [15], we call a population *degenerate* if it consists of $\mu$ copies of the same individual. If the algorithm is in a degenerate population, we will study how the *next degenerate population* looks like, so we define

$$\Phi^t := \{\# \text{ of zero-bits in an individual in the } t\text{-th degenerate population}\}. \quad (1)$$

Our main object of study will be the *degenerate population drift* (or simply *drift* if the context is clear), defined as

$$\Delta(\varepsilon) := \Delta^t(\varepsilon) := \mathbb{E}[\Phi^t - \Phi^{t+1} \mid \Phi^t = \varepsilon n]. \quad (2)$$

The expression is independent of $t$ since the considered algorithms are time-homogeneous. If we want to stress that $\Delta(\varepsilon)$ depends on the parameters $\mu$ and $c$, we also write $\Delta(\mu, c, \varepsilon)$. Note that the number of generations to reach the $(t+1)$-st degenerate population is itself a random variable. So the number of generations to go from $\Phi^t$ to $\Phi^{t+1}$ is random. As in [13], its expectation is $O(1)$ if $\mu$ and $c$ are constants, and it has an exponentially decaying tail bound, see Lemma 1 below. In particular, the probability that during the transition from one degenerate population to another the same bit is touched by two different mutations is $O(\varepsilon^2)$, and likewise the contribution of this case to the drift is $O(\varepsilon^2)$, see Lemma 2.

To compute the degenerate population drift, we will frequently need to compute the expected change of the potential provided that we visit an intermediate state $S$. Here, a state $S$ is simply given by a population of $\mu$ search points. We will call this change the *drift from state $S$*, and denote it by $\Delta(S, \varepsilon)$. Formally, if $\mathcal{E}(S, t)$ is the event that the algorithm visits state $S$ between the $t$-th and $(t + 1)$-st degenerate population,

$$\Delta(S, \varepsilon) := \mathbb{E}[\Phi^t - \Phi^{t+1} \mid \Phi^t = \varepsilon n \text{ and } \mathcal{E}(S, t)]. \tag{3}$$

This term is closely related to the *contribution to the degenerate population drift from state $S$*, which also contains the probability to reach $S$ as a factor:

$$\Delta_{\mathrm{con}}(S, \varepsilon) := \Pr[\mathcal{E}(S, t) \mid \Phi^t = \varepsilon n] \cdot \Delta(S, \varepsilon). \tag{4}$$

We will study DYNBV around the optimum, i.e., we consider any $\varepsilon = \varepsilon(n) \to 0$ for $n \to \infty$, and we compute the asymptotic expansion of $\Delta(\varepsilon)$ for $n \to \infty$. As we will see, the drift is of the form $\Delta(\varepsilon) = a\varepsilon \pm O(\varepsilon^2) \pm o(1)$ for some constant $a$.[3] Analogously to [13] and [19], if $a$ is *positive* (multiplicative drift), then the algorithm starting with at most $\varepsilon_0 n$ zero-bits for some suitable constant $\varepsilon_0$ whp needs $O(n \log n)$ generations to find the optimum. On the other hand, if $a$ is *negative* (negative drift/updrift), then whp the algorithm needs exponentially many generations to find the optimum (regardless of whether it is initialized randomly or with $\varepsilon_0 n$ zero-bits). These two cases are typical. There is no constant term in the drift since for a degenerate population $P^k$ we have $P^{k+1} = P^k$ with probability $1 - O(\varepsilon)$. This happens whenever mutation does not touch any zero-bit, since then the offspring is rejected.

We will prove that, as long as we are only interested in the first order expansion (i.e., in a results of the form $a\varepsilon \pm O(\varepsilon^2) \pm o(1)$), we may assume that between two degenerate populations, the mutation operators always flip different bits. In this case, we use the following naming convention for search points. The individuals of the $t$-th degenerate population are all called $x^0$. We call other individuals $x^{(m_1 - m_2)}$, where $m_1$ stands for the extra number of ones and $m_2$ for the extra number of zeros compared to $x^0$. Hence, if $x^0$ has $m$ zero-bits then $x^{(m_1 - m_2)}$ has $m + m_2 - m_1$ zero-bits. Following the same convention, we will denote by $X_k^z$ a set of $k$ copies of $x^z$, where the string $z$ may be 0 or $(m_1 - m_2)$. In particular, $X_\mu^0$ denotes the $t$-th degenerate population.

---

[3] The $o(1)$ term is needed since we do not make assumptions about $\varepsilon$. If we assumed that $\varepsilon = \varepsilon(n)$ goes to zero sufficiently slowly, we could swallow the $o(1)$ into $O(\varepsilon^2)$.

**Duration Between Degenerate Populations.** We formalize the above asser-
tion that the number of steps between two degenerate populations satisfies exponential tail bounds, and that it is unlikely to touch a bit by two different mutations as we transition from one degenerate population to the next. We omit the
proofs of the following two lemmas due to space restrictions. The first one holds
as any population can degenerate in $\mu$ generations by creating copies.

**Lemma 1.** *For all constant $\mu, c$ there is a constant $a > 0$ such that the following
holds for the $(\mu + 1)$-EA with mutation parameter $c$ in any population $X$ on
DYNBV. Let $K$ be the number of generations until the algorithm reaches the
next degenerate population. Then for all $k \in \mathbb{N}_0$,*

$$\Pr(K \geq k \cdot \mu) \leq e^{-a \cdot k}.$$

**Lemma 2.** *Consider the $(\mu+1)$-EA with mutation parameter $c$ on DYNBV. Let
$X^t$ and $X^{t+1}$ denote the $t$-th and $(t+1)$-st degenerate population respectively.
Let $\varepsilon > 0$, and let $X$ be a degenerate population with at most $\varepsilon n$ zero-bits.*

*(a) Let $\mathcal{E}_2$ be the event that the mutations during the transition from $X^t$ to
$X^{t+1}$ flip at least two zero-bits. Then $\Pr[\mathcal{E}_2 \mid X^t = X] = O(\varepsilon^2)$. Moreover,
the contribution of this case to the degenerate population drift $\Delta$ is*

$$\Delta^*(\varepsilon) := \Pr[\mathcal{E}_2 \mid X^t = X] \cdot \mathbb{E}[\Delta^t(\varepsilon) \mid \mathcal{E}_2 \wedge X^t = X] = O(\varepsilon^2).$$

*(b) Let $S$ be a non-degenerate state such that there is at most one position which
is a one-bit in some individuals in $S$, but a zero-bit in $X$. Let $\mathcal{E}(S,t)$ be the
event that state $S$ is visited during the transition from $X^t$ to $X^{t+1}$, and let
$\mathcal{E}_1$ be the event that a zero-bit is flipped in the transition from $S$ to $X^{t+1}$.
Then $\Pr[\mathcal{E}_1 \mid \mathcal{E}(S,t) \wedge X^t = X] = O(\varepsilon)$, and the contribution to $\Delta(S, \varepsilon)$ is*

$$\Delta^*(S, \varepsilon) := \Pr[\mathcal{E}_1 \mid \mathcal{E}(S,t) \wedge X^t = X] \cdot \mathbb{E}[\Delta^t(\varepsilon) \mid \mathcal{E}_1 \wedge \mathcal{E}(S,t) \wedge X^t = X] = O(\varepsilon). \tag{5}$$

*The contribution of the case $\mathcal{E}(S,t) \wedge \mathcal{E}_1$ to the degenerate population drift is*

$$\Delta^*_{con}(S, \varepsilon) := \Pr[\mathcal{E}(S,t)] \cdot \Delta^*(S, \varepsilon) = O(\varepsilon^2).$$

The next lemma classifies how the population can degenerate if no zero-bit
is flipped. By Lemmas 1 and 2, this assumption holds with high probability.

**Lemma 3.** *Consider the $(\mu+1)$-EA on the DYNBV problem in any population
$X$. Let $x_1, x_2, ..., x_k$ be search points in $X$ such that every search point of $X$
it dominated by one of $x_1, x_2, ..., x_k$. Then either at least one zero-bit is flipped
until the next degenerate population, or the next degenerate population consists
of copies of one of the search points $x_1, x_2, ..., x_k$.*

*Proof.* Note that the transitivity holds for the domination property, in particular, if $x$ dominates $y$ and $y$ dominates $z$, we have that $x$ dominates $z$. Assume
that, starting from X, the algorithm doesn't flip any additional zero-bits. We

start by inductively showing that for all subsequent time steps, every individual in the population is still dominated by one of the search points $x_1, x_2, ..., x_k$. Suppose, for the sake of contradiction, that eventually there are individuals which are not dominated by any of the search points in $\{x_1, x_2, ..., x_k\}$, and let $x^*$ be the first such individual. Since we assumed that the algorithm doesn't flip any additional zero-bits, $x^*$ must have been generated by mutating an individual $\bar{x}$ and only flipping one-bits. So $\bar{x}$ dominates $x^*$. On the other hand, $\bar{x}$ is dominated by one of the search points $x_1, x_2, ..., x_k$ by our choice of $x^*$. This is a contradiction since domination is transitive. Therefore, using transitivity, the algorithm will not generate any individual that is not dominated by any search point in $\{x_1, x_2, ..., x_k\}$. Furthermore, the population will never degenerate to any other individual $\tilde{x} \notin \{x_1, x_2, ..., x_k\}$. In fact, let $x_i$ be the search point in $\{x_1, x_2, ..., x_k\}$ that dominates $\tilde{x}$. We have that $f(\tilde{x}) < f(x_i)$ in all iterations and for all permutations, therefore $x_i$ will never be discarded before $\tilde{x}$, which concludes the proof.    □

## 3    Analysis of the Degenerate Population Drift

In this section, we will find a lower bound for the drift $\Delta(\varepsilon) = \Delta(\mu, c, \varepsilon)$ of the $(\mu + 1)$-EA close to the optimum, when $n \to \infty$. The main result, proven later in this section, is the following.

**Theorem 4.** *For all $c > 0$ there exist $\delta, \varepsilon_0 > 0$ such that for all $\varepsilon < \varepsilon_0$ and $\mu \geq \mu_0 := e^c + 2$, if $n$ is sufficiently large,*

$$\Delta(c, \mu, \varepsilon) \geq \delta \cdot \varepsilon.$$

Lemma 3 allows us to describe the transition from one degenerate population to the next by a relatively simple Markov chain, provided that at most one zero-bit is flipped during the transition. This zero-bit needs to be flipped in order to leave the starting state, so we assume for this chain that no zero-bit is flipped afterwards. This assumptions is justified by Lemma 2. The Markov chain is shown in Fig. 1. The states of the Markov chain do not correspond one-to-one to the generations. For example, following

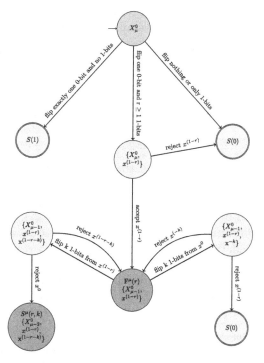

**Fig. 1.** Transition Diagram for the $(\mu + 1)$-EA

the first arrow to the left we reach a state in which one individual $x^{(1)}$ (the offspring) dominates all other individuals. By Lemma 3, such a situation must degenerate into $\mu$ copies of $x^{(1)}$, so we immediately mark this state as a degenerate state with $\Phi^{t+1} - \Phi^t = 1$.

The key step will be to give a lower bound for the contribution to the drift from state $F^\mu(r)$. Once we have a bound on this, it is straightforward to compute a bound on the degenerate population drift. Before we turn to the computations, we first introduce a bit more notation.

**Definition 5.** *Consider the $(\mu+1)$-EA in state $F^\mu(r)$ in generation $k-1$. We re-sort the $n$ positions of the search points descendingly according to the next fitness function $f^k$. So by the "first" position we refer to the position which has highest weight according to $f^k$, and the $j$-th bit of a bitstring $z$ is given by $z_i$ such that $\pi^k(i) = j$. Then we define:*

- $B_z^k :=$ *position of the first zero-bit in $z$;*
- $B_0^k :=$ *position of the first flipped bit in the $k$-th mutation;*
- $z_1^k := arg\ min\{f^k(z) \mid z \in \{X_{\mu-1}^0, X_1^{(1-r)}\}\};$
- $z_2^k := arg\ max\{f^k(z) \mid z \in \{X_{\mu-1}^0, X_1^{(1-r)}\}\}.$

*In particular, the search point to be discarded in generation $k$ is either $z_1^k$ or it is the offspring generated by the $k$-th mutation. We define $B_0^k$ to be $\infty$ if no bits are flipped in the $k$-th mutation.*

Now we are ready to bound the drift of state $F^\mu(r)$. We remark that the statement for $\mu = 2$ was also proven in [16], but the proof there was much longer, since it did not make use of the hidden symmetry of the selection process.

**Lemma 6.** *Consider the $(\mu+1)$-EA on the DYNBV function in the state $F^\mu(r)$ for some $r \geq 1$, and let $\varepsilon > 0$. Then the drift from $F^\mu(r)$ is*

$$\Delta(F^\mu(r), \varepsilon) \geq \frac{1-r}{1 + (\mu - 1) \cdot r} + \mathcal{O}(\varepsilon).$$

*For $\mu = 2$ the bound holds with equality, i.e. $\Delta(F^2(r), \varepsilon) = (1-r)/(1+r) + \mathcal{O}(\varepsilon).$*

*Proof.* Let us assume that the algorithm will not flip an additional zero-bit through mutation before it reaches the next degenerate population. In fact, the contribution to the drift in case it does flip another zero-bit can be summarized by $\mathcal{O}(\varepsilon)$ due to Eq. (5) in Lemma 2. So from now on, we assume that the algorithm doesn't flip an additional zero-bit until it reaches the next degenerate population.

The idea is to follow the Markov chain as shown in Fig. 1. We will compute the conditional probabilities of reaching different states from $F^\mu(r)$, conditional on actually leaving $F^\mu(r)$. More precisely, we will condition on the event that an offspring $\bar{x}$ is generated and accepted into the population.[4] Recall that $F^\mu(r)$

---

[4] Here we use the convention that if an offspring is identical to the parent, and they have lowest fitness in the population, then the offspring is rejected. Since the outcome of ejecting offspring or parent is the same, this convention does not change the course of the algorithm.

corresponds to the population of $\{X^0_{\mu-1}, X^{(1-r)}_1\}$, i.e., $\mu - 1$ copies of $x^0$ and one copy of $x^{(1-r)}$. So if the offspring is accepted, one of these search points must be ejected from the population. Let us first consider the case that $x^{(1-r)}$ is ejected from the population. Then the population is dominated by $x^0$ afterwards, and will degenerate into $X^0_\mu$ again by Lemma 3. The other case is that one of the $x^0$ individuals is ejected, which is described by state $S^\mu(r, k)$. It is complicated to compute the contribution of this state precisely, but by Lemma 3 we know that this population will degenerate either to copies of $x^0$ or of $x^{(1-r)}$. Hence, we can use the pessimistic bound $\Delta(S^\mu(r, k), \varepsilon) \geq (1 - r)$ for the drift of $S^\mu(r, k)$. Summarizing, once a new offspring is accepted, if a copy of $x^0$ is discarded we get a contribution of at most $1 - r$ to the drift and if $x^{(1-r)}$ is discarded we get a contribution of 0. It only remains to compute the conditional probabilities with which these cases occur.

To compute the probabilities is not straightforward, but we can use a rather surprising symmetry, using the terminology from Definition 5. Assume that the algorithm is in iteration $k$. We make the following observation: an offspring is accepted if and only if it is mutated from $z^k_2$ and $B^k_0 > B^k_{\min} :=$ $\min\{B^k_{x^0}, B^k_{x^{(1-r)}}\}$. Hence, we need to compute the probability

$$\hat{p} := \Pr\left(f^k(x^{(1-r)}) \geq f^k(x^0) \mid \{\text{mutated } z^k_2\} \wedge \{B^k_0 > B^k_{\min}\}\right),$$

since then we can bound $\Delta(F^\mu(r), \varepsilon) \geq (1 - r)\hat{p} + \mathcal{O}(\varepsilon)$ by Lemma 2.

Clearly, the events $\{f^k(x^{(1-r)}) \geq f^k(x^0)\}$ and $\{B_0 > B^k_{\min}\}$ are independent. We emphasize that this is a rather subtle symmetry of the selection process. Using conditional probability, $\hat{p}$ simplifies to:

$$\hat{p} = \frac{\Pr\left(f(x^{(1-r)}) \geq f(x^0) \wedge \{\text{mutated } z^k_2\}\right)}{\Pr\left(\{\text{mutated } z^k_2\}\right)}. \tag{6}$$

To compute the remaining probabilities, we remind the reader that $x^{(1-r)}$ has exactly $r$ more zero-bits and 1 more one-bit, than $x^0$. Hence, in order to compare them, we only need to look at the relative positions of these $r + 1$ bits in which they differ. In particular, $x^{(1-r)} = z^k_2$ holds if and only if the permutation $\pi^k$ places the one-bit from $x^{(r-1)}$ before the $r$ one-bits of $x^0$, and this happens with probability $1/(r+1)$. Moreover, recall that there are $\mu - 1$ copies of $x^0$ and only one $x^{(1-r)}$, so the probability of picking them as parents is $(\mu - 1)/\mu$ and $1/\mu$, respectively. Therefore, by using the law of total probability,

$$\Pr\left(\{\text{mutated } z^k_2\}\right) = \Pr\left(\{\text{mutated } z^k_2\} \mid x^{(1-r)} = z^k_2\right) \cdot \Pr\left(x^{(1-r)} = z^k_2\right)$$
$$+ \Pr\left(\{\text{mutated } z^k_2\} \mid x^0 = z^k_2\right) \cdot \Pr\left(x^0 = z^k_2\right)$$
$$= \frac{1}{\mu} \cdot \frac{1}{r+1} + \frac{\mu - 1}{\mu} \cdot \frac{r}{r+1}$$

Plugging this into (6) yields

$$\hat{p} = \left(\frac{1}{r+1} \cdot \frac{1}{\mu}\right) \Big/ \left(\frac{1}{\mu} \cdot \frac{1}{r+1} + \frac{\mu - 1}{\mu} \cdot \frac{r}{r+1}\right) = \frac{1}{1 + (\mu - 1)r}.$$

Together with Lemma 2 and the lower bound $\Delta(S^\mu(r,k),\varepsilon) \geq 1 - r$ on the contribution to the drift from $S^\mu(r,k)$, this concludes the proof for general $\mu$. For $\mu = 2$, the bound $1 - r$ is tight since $x^{(1-r)}$ dominates the only other search point in $S^\mu(r,k)$.                                                                    □

Now we are ready to bound the degenerate population drift and prove Theorem 4.

*Proof (of Theorem 4).* To prove this theorem, we refer to Fig. 1. By Lemma 2, the contribution of all states that involve flipping more than one zero-bit is $O(\varepsilon^2)$. If we flip no zero-bits at all, then the population degenerates to $X^0_\mu$ again, which contributes zero to the drift. So we only need to consider the case where we flip exactly one zero-bit in the transition from the $t$-th to the $(t+1)$-st degenerate population. This zero-bit needs to be flipped in the first mutation, since otherwise the population does not change. We denote by $p_r$ the probability to flip exactly one zero-bit and $r$ one-bits in $x^0$, thus obtaining $x^{(1-r)}$. If $f^k(x^{(1-r)}) > f^k(x^0)$ then $x^{(1-r)}$ is accepted into the population and we reach state $F^\mu(r)$. This happens if and only if among the $r+1$ bits in which $x^{(1-r)}$ and $x^0$ differ, the zero-bit of $x^0$ is the most relevant one. So $\Pr[f^k(x^{(1-r)}) > f^k(x^0)] = 1/(r+1)$ Finally, by Lemma 6, the drift from $F^\mu(r)$ is at least $-(r-1)/(1+(\mu-1)r)+O(\varepsilon)$. Summarizing all this into a single formula, we obtain

$$\Delta(\varepsilon) \geq O(\varepsilon^2) + p_0 + \sum_{r=1}^{(1-\varepsilon)n} p_r \cdot \left[\Pr[f^k(x^{(1-r)}) > f^k(x^0)] \cdot \Delta(F^\mu(r),\varepsilon)\right]$$

$$\geq O(\varepsilon^2) + p_0 - \sum_{r=1}^{(1-\varepsilon)n} p_r \cdot \frac{1}{r+1} \cdot \left(\frac{r-1}{1+(\mu-1)r} + O(\varepsilon)\right). \tag{7}$$

For $p_r$, we use the following standard estimate, which holds for all $r = o(n)$.

$$p_r = (1+o(1)) \cdot c^{r+1}/r! \cdot e^{-c} \cdot \varepsilon \cdot (1-\varepsilon)^r.$$

The summands for $r = \Omega(n)$ (or $r = \omega(1)$, actually) in (7) are negligible since $p_r$ decays exponentially in $r$. Plugging $p_0$ and $p_r$ into (7), we obtain

$$\Delta(\varepsilon) \geq O(\varepsilon^2) + (1+o(1))\varepsilon c e^{-c}\left[1 - \underbrace{\sum_{r=1}^{(1-\varepsilon)n} \frac{c^r}{(r+1)!} \cdot \frac{(1-\varepsilon)^r(r-1)}{(1+(\mu-1)r)}}_{=:f(r,c,\mu)}\right]. \tag{8}$$

To bound the inner sum, we use $(r-1)/(r+1) \leq 1$ and obtain

$$f(r,c,\mu) \leq \frac{c^r}{(r+1)!} \cdot \frac{r-1}{(1+(\mu-1)\cdot r)} \leq \frac{c^r}{r!} \cdot \frac{1}{1+(\mu-1)\cdot r} \leq \frac{c^r}{r!} \cdot \frac{1}{\mu-1}.$$

We plug this bound into (8). Moreover, summing to $\infty$ instead of $(1-\varepsilon)n$ only makes the expression in (8) smaller, and allows us to use the identity $\sum_{r=1}^\infty c^r/r! = e^c - 1 \leq e^c$, yielding

$$\Delta(\varepsilon) \geq O(\varepsilon^2) + (1+o(1))\varepsilon c e^{-c}\left(1 - e^c/(\mu-1)\right).$$

If $n$ is large enough and $\varepsilon$ so small that the $\mathcal{O}(\varepsilon^2)$ term can be neglected, then by picking $\mu_0 = 2 + e^c$ we get $\Delta(\varepsilon) \gtrsim \varepsilon c e^{-c}/(e^c + 1) > 0$ and therefore we can set for example $\delta = \frac{1}{2}ce^{-c}/(e^c + 1)$, which concludes the proof. □

We conclude this section by giving a formal statement on the runtime.

**Theorem 7.** *Assume that the $(\mu + 1)$-EA runs on the* DYNBV *function with constant parameters $c > 0$ and $\mu \geq e^c + 2$. Let $\varepsilon_0$ be as in Theorem 4 and let $\varepsilon < \varepsilon_0$. If the $(\mu + 1)$-EA is started with a population in which all individuals have at most $\varepsilon n$ zero-bits, then whp it finds the optimum in $\mathcal{O}(n \log n)$ steps.*

*Proof.* We only sketch the proof since the argument is mostly standard, e.g. [13]. First we note that the number of generations between two degenerate populations satisfies a exponential tail bound by Lemma 1. As an easy consequence, the total number of flipped bits between two degenerate populations also satisfies an exponential tail bound, and so does the difference $|\Phi^t - \Phi^{t+1}|$. This allows us to conclude from the *negative drift theorem* [3,14] that whp $\Phi^t < \varepsilon_0 n$ for an exponential number of steps. However, in the range $\Phi^t \in [0, \varepsilon_0 n]$, by Theorem 4 the drift is positive and multiplicative, $\mathbb{E}[\Phi^t - \Phi^{t+1} \mid \Phi^t] \geq \delta \Phi^t/n$. Therefore, by the *multiplicative drift theorem* [6,14] whp the optimum shows up among the first $O(n \log n)$ degenerate populations. Again by Lemma 1, whp this corresponds to $O(n \log n)$ generations. □

## 4  Second-Order Analysis of the Drift of the $(2 + 1)$-EA

In this section we investigate the $(2+1)$-EA. We compute a second order approximation of $\mathbb{E}[\Phi^t - \Phi^{t+1} \mid \Phi^t = \varepsilon n]$, that is we compute the drift up to $\mathcal{O}(\varepsilon^3)$ error terms. This analysis is rather similar to the proof of Theorem 4, but more involved. Due to space restrictions, we only give the final result. The derivation can be found in the full version [17]. The drift is

$$\Delta(c, \varepsilon) = \varepsilon(1 + o(1))f_0(c) + \varepsilon^2(1 + o(1))f_1(c) + \mathcal{O}(\varepsilon^3), \tag{9}$$

where

$$f_0(c) = ce^{-c} \cdot \left(1 + \sum_{r=1}^{\infty} \frac{c^r}{(r+1)!} \cdot \frac{1-r}{r+1}\right), \quad \text{and}$$

$$f_1(c) = c^2 e^{-c} + \sum_{r=1}^{\infty} \frac{(r+1)(4-2r)}{r+2} e^{-c} \frac{c^{r+2}}{(r+2)!} \tag{10}$$

$$+ \frac{e^{-2c}}{2} \sum_{r=1}^{\infty} \frac{c^{r+2}}{(r+1)!} \cdot \frac{\frac{6+6r-3r^2}{(r+1)(r+2)} + \sum_{k=0}^{\infty} \frac{c^k}{k!}(\Delta_A(k) + \Delta_B(k))}{\sum_{k=0}^{\infty} \frac{c^k}{k!} e^{-c} \frac{r+1}{r+k+1}}.$$

Here, $\Delta_A$ and $\Delta_B$ are the drifts from certain intermediate states, which we don't specify exactly for space reasons (see the full version [17] for details), and

$$\Delta_A = \mathcal{O}(\varepsilon) + \frac{r(r+2)(1-k)+k(k+2)(1-r)+2-2rk}{(r+1)(k+1)(r+k+2)},$$

$$\Delta_B = \mathcal{O}(\varepsilon) + \frac{-2r^2k - rk^2 - 3r^2 - 4rk + k^2 + 4r + k + 4}{(r+1)(k+1)(k+r+2)}.$$

As a sanity check, we note that $\varepsilon f_0(c)$ is essentially the same as (8) from the proof of Theorem 4 after setting $\mu = 2$ and ignoring all lower order terms. The formulas are complicated, but allow us to prove the following main result.

**Theorem 8.** *There are $c^* > 0$ and $\varepsilon^* > 0$ such that the $(2+1)$-EA with mutation parameter $c^*$ has positive drift $\Delta(c^*, \varepsilon) = \Omega(\varepsilon)$ for all $\varepsilon \in (0, \frac{1}{2}\varepsilon^*)$ and has negative drift $\Delta(c^*, \varepsilon) = -\Omega(1)$ for all $\varepsilon \in (\frac{3}{2}\varepsilon^*, 2\varepsilon^*)$.*

In a nutshell, Theorem 8 shows that the hardest part for optimization is not around the optimum. In other words, it shows that the range of efficient parameters settings is larger close to the optimum. We remark that we "only" state the result for one concrete parameter $c^*$, but the same argument could be extended to show that the "range of efficient parameter settings" becomes larger. Moreover, with standard arguments similar to the proof of Theorem 7, which we omit here, it would be possible to translate positive and negative drift into optimization times. I.e., one could show that whp the algorithm has optimization time $O(n \log n)$ if the algorithm is started in the range $\varepsilon \in (0, \varepsilon^*/4)$, but that the optimization time is exponential if it is started in the range $\varepsilon > 2\varepsilon^*$.

*Proof (of Theorem 8).* Inspecting $f_0$ in (10), we see that the sum goes over negative terms, if we omit the zero term for $r = 1$. Thus the factor in the bracket is strictly decreasing in $c$, ranging from 1 (for $c = 0$) to $-\infty$ (for $c \to \infty$). In particular, there is exactly one $c_0 > 0$ such that $f_0(c_0) = 0$. Numerically we find $c_0 = 2.4931\ldots$ and $f_1(c_0) = -0.4845\ldots < 0$.

In the following, we will fix some $c^* < c_0$ and set $\varepsilon^* := -f_0(c^*)/f_1(c^*)$. Note that by choosing $c^*$ sufficiently close to $c_0$ we can assume that $f_1(c^*) < 0$, since $f_1$ is a continuous function. Due to the discussion of $f_0$ above, the choice $c^* < c_0$ also implies $f_0(c^*) > 0$. Thus $\varepsilon^* > 0$. Moreover, since $f_0(c) \to 0$ for $c \to c_0$, if we choose $c^*$ close enough to $c_0$ then we can make $\varepsilon^*$ as close to zero as we wish.

To add some intuition to these definitions, note that $\Delta(c, \varepsilon) = \varepsilon(f_0(c) + \varepsilon f_1(c) + \mathcal{O}(\varepsilon^2))$, so the condition $\varepsilon = -f_0(c)/f_1(c)$ is a choice for $\varepsilon$ for which the drift is approximately zero, up to the error term. We will indeed prove that for fixed $c^*$, the sign of the drift switches around $\varepsilon \approx \varepsilon^*$. More precisely, we will show that the sign switches from positive to negative as we go from $\Delta(c^*, \varepsilon^* - \varepsilon')$ to $\Delta(c^*, \varepsilon^* + \varepsilon')$, for $\varepsilon' \in (0, \varepsilon^*)$. Actually, we need to constrict to $\varepsilon' \in (\varepsilon^*/2, \varepsilon^*)$ so that we can handle the error terms. This implies that the value $c^*$ yields positive drift close to the optimum (in the range $\varepsilon \in (0, \frac{1}{2}\varepsilon^*)$), but yields negative drift further away from the optimum (in the range $\varepsilon \in (\frac{3}{2}\varepsilon^*, 2\varepsilon^*)$). This will imply Theorem 8.

To study the sign of the drift, we define

$$\Delta^*(c, \varepsilon) := \frac{\Delta(c, \varepsilon)}{\varepsilon} = (1 + o(1)) \cdot \left(f_0(c) + \varepsilon \cdot f_1(c) + O(\varepsilon^2)\right).$$

It is slightly more convenient to consider $\Delta^*$ instead of $\Delta$, but note that both terms have the same sign. So it remains to investigate the sign of $\Delta^*(c^*, \varepsilon^* - \varepsilon')$ and $\Delta^*(c^*, \varepsilon^* + \varepsilon')$ for $\varepsilon' \in (\varepsilon^*/2, \varepsilon^*)$. We will only study the first term, the second one can be analyzed analogously. Recalling the definition of $\varepsilon^*$ and that $f_1(c^*) < 0$, we have

$$\Delta^*(c^*, \varepsilon^* - \varepsilon') = (1 + o(1)) \left( f_0(c^*) + (\varepsilon^* - \varepsilon') f_1(c^*) \right) + \mathcal{O}((\varepsilon^*)^2)$$

$$= (1 + o(1)) ( \underbrace{f_0(c^*) + \varepsilon^* f_1(c^*)}_{=0} ) - (1 + o(1)) \underbrace{\left( \varepsilon' f_1(c^*) \right)}_{< \varepsilon^* f_1(c^*)/2} + \mathcal{O}((\varepsilon^*)^2)$$

$$> -(1 + o(1)) \tfrac{1}{2} \varepsilon^* f_1(c^*) + \mathcal{O}((\varepsilon^*)^2).$$

Recall that we may choose $\varepsilon^*$ as small as we want. In particular, we can choose it so small that the above term has the same sign as the main term, which is positive due to $f_1(c^*) < 0$. Hence $\Delta^*(c^*, \varepsilon^* - \varepsilon') > 0$, as desired. The inequality $\Delta^*(c^*, \varepsilon^* + \varepsilon') < 0$ follows analogously. This concludes the proof.     □

## 5   Conclusion

We have explored the DYNBV function, and we have found that the $(\mu + 1)$-EA profits from large population size, close to the optimum. In particular, for all choices of the mutation parameter $c$, the $(\mu + 1)$-EA is efficient around the optimum if $\mu$ is large enough. However, surprisingly the region around the optimum may not be the most difficult region. For $\mu = 2$, we have proven that it is not.

This surprising result, in line with the experiments in [15], raises much more questions than it answers. Does the $(\mu + 1)$-EA with increasing $\mu$ turn efficient for a larger and larger ranges of $c$, as the behavior around the optimum suggests? Or is the opposite true, that the range of efficient $c$ shrinks to zero as the population grows, as it is the case for the $(\mu + 1)$-EA on HOTTOPIC functions? Where is the hardest region for larger $\mu$? Around the optimum or elsewhere?

For the $(\mu + 1)$-GA, the picture is even less complete. Experiments in [15] indicated that the hardest region of DYNBV for the $(\mu + 1)$-GA is around the optimum, and that the range of efficient $c$ increases with $\mu$. But the experiments were only run for $\mu \leq 5$, and formal proofs are missing. Should we expect that the discrepancy between $(\mu + 1)$-GA (hardest region around optimum) and $(\mu + 1)$-EA (hardest region elsewhere) remains if we increase the population size, and possible becomes stronger? Or does it disappear? For HOTTOPIC functions, we know that around the optimum, the range of efficient $c$ becomes arbitrarily large as $\mu$ grows (similarly as we have shown for the $(\mu + 1)$-EA on DYNBV), but we have no idea what the situation away from the optimum is.

The similarities of results between DYNBV and HOTTOPIC functions are striking, and we are pretty clueless where they come from. For example, the analysis of the $(\mu + 1)$-EA on HOTTOPIC away from the optimum in [19] clearly does not generalize to DYNBV since the very heart of the proof is that the weights do not change over long periods. In DYNBV, they change every round. Nevertheless, experiments and theoretical results indicate that the outcome is similar in both cases. Perhaps one could gain insight from "interpolating" between DYNBV and HOTTOPIC by re-drawing the weights not every round, but only every $k$-th round.

In general, the situation away from the optimum is governed by complex population dynamics, which is why the $(\mu+1)$-EA and the $(\mu+1)$-GA might behave very differently. Currently, we lack the theoretic means to understand population dynamics in which the internal population structure is complex and essential. The authors believe that developing tools for understanding such dynamics is one of the most important projects for improving our understanding of population-based search heuristics.

# References

1. Colin, S., Doerr, B., Férey, G.: Monotonic functions in EC: anything but monotone! In: Genetic and Evolutionary Computation Conference (GECCO), pp. 753–760. ACM (2014)
2. Dang-Nhu, R., Dardinier, T., Doerr, B., Izacard, G., Nogneng, D.: A new analysis method for evolutionary optimization of dynamic and noisy objective functions. In: Genetic and Evolutionary Computation Conference (GECCO), pp. 1467–1474. ACM (2018)
3. Doerr, B., Goldberg, L.A.: Adaptive drift analysis. Algorithmica **65**(1), 224–250 (2013). https://doi.org/10.1007/s00453-011-9585-3
4. Doerr, B., Hota, A., Kötzing, T.: Ants easily solve stochastic shortest path problems. In: Genetic and Evolutionary Computation Conference (GECCO), pp. 17–24. ACM (2012)
5. Doerr, B., Jansen, T., Sudholt, D., Winzen, C., Zarges, C.: Mutation rate matters even when optimizing monotonic functions. Evol. Comput. **21**(1), 1–27 (2013)
6. Doerr, B., Johannsen, D., Winzen, C.: Multiplicative drift analysis. Algorithmica **64**(4), 673–697 (2012). https://doi.org/10.1007/s00453-012-9622-x
7. Droste, S.: Analysis of the $(1+1)$ EA for a dynamically changing OneMax-variant. In: Congress on Evolutionary Computation (CEC), vol. 1, pp. 55–60. IEEE (2002)
8. Horoba, C., Sudholt, D.: Ant colony optimization for stochastic shortest path problems. In: Genetic and Evolutionary Computation Conference (GECCO), pp. 1465–1472. ACM (2010)
9. Jansen, T.: On the brittleness of evolutionary algorithms. In: Stephens, C.R., Toussaint, M., Whitley, D., Stadler, P.F. (eds.) FOGA 2007. LNCS, vol. 4436, pp. 54–69. Springer, Heidelberg (2007). https://doi.org/10.1007/978-3-540-73482-6_4
10. Kötzing, T., Lissovoi, A., Witt, C.: $(1+1)$ EA on generalized dynamic OneMax. In: Foundations of Genetic Algorithms (FOGA), pp. 40–51. Springer (2015). https://dblp.org/rec/conf/foga/KotzingLW15.html?view=bibtex
11. Kötzing, T., Molter, H.: ACO beats EA on a dynamic pseudo-Boolean function. In: Coello, C.A.C., Cutello, V., Deb, K., Forrest, S., Nicosia, G., Pavone, M. (eds.) PPSN 2012. LNCS, vol. 7491, pp. 113–122. Springer, Heidelberg (2012). https://doi.org/10.1007/978-3-642-32937-1_12
12. Lengler, J., Steger, A.: Drift analysis and evolutionary algorithms revisited. Comb. Probab. Comput. **27**(4), 643–666 (2018)
13. Lengler, J.: A general dichotomy of evolutionary algorithms on monotone functions. IEEE Trans. Evol. Comput. **24**(6), 995–1009 (2019)
14. Lengler, J.: Drift analysis. Theory of Evolutionary Computation. NCS, pp. 89–131. Springer, Cham (2020). https://doi.org/10.1007/978-3-030-29414-4_2

15. Lengler, J., Meier, J.: Large population sizes and crossover help in dynamic environments. In: Bäck, T., et al. (eds.) PPSN 2020. LNCS, vol. 12269, pp. 610–622. Springer, Cham (2020). https://doi.org/10.1007/978-3-030-58112-1_42
16. Lengler, J., Meier, J.: Large population sizes and crossover help in dynamic environments, full version. arXiv preprint. http://arxiv.org/abs/2004.09949 (2020)
17. Lengler, J., Riedi, S.: Runtime analysis of the $(\mu + 1)$-EA on the dynamic BinVal function, full version. arXiv preprint. http://arxiv.org/abs/2010.13428 (2020)
18. Lengler, J., Schaller, U.: The (1+1)-EA on noisy linear functions with random positive weights. In: Symposium Series on Computational Intelligence (SSCI), pp. 712–719. IEEE (2018)
19. Lengler, J., Zou, X.: Exponential slowdown for larger populations: the $(\mu+1)$-EA on monotone functions. In: Foundations of Genetic Algorithms (FOGA), pp. 87–101. ACM (2019)
20. Lissovoi, A., Witt, C.: MMAS versus population-based EA on a family of dynamic fitness functions. Algorithmica **75**(3), 554–576 (2016). https://doi.org/10.1007/s00453-015-9975-z
21. Lissovoi, A., Witt, C.: A runtime analysis of parallel evolutionary algorithms in dynamic optimization. Algorithmica **78**(2), 641–659 (2016). https://doi.org/10.1007/s00453-016-0262-4
22. Lissovoi, A., Witt, C.: The impact of a sparse migration topology on the runtime of island models in dynamic optimization. Algorithmica **80**(5), 1634–1657 (2018). https://doi.org/10.1007/s00453-017-0377-2
23. Neumann, F., Pourhassan, M., Roostapour, V.: Analysis of evolutionary algorithms in dynamic and stochastic environments. Theory of Evolutionary Computation. NCS, pp. 323–357. Springer, Cham (2020). https://doi.org/10.1007/978-3-030-29414-4_7
24. Neumann, F., Witt, C.: On the runtime of randomized local search and simple evolutionary algorithms for dynamic makespan scheduling. In: International Joint Conference on Artificial Intelligence (IJCAI), pp. 3742–3748. AAAI Press (2015)
25. Pourhassan, M., Gao, W., Neumann, F.: Maintaining 2-approximations for the dynamic vertex cover problem using evolutionary algorithms. In: Genetic and Evolutionary Computation Conference (GECCO), pp. 903–910. ACM (2015)
26. Shi, F., Schirneck, M., Friedrich, T., Kötzing, T., Neumann, F.: Reoptimization time analysis of evolutionary algorithms on linear functions under dynamic uniform constraints. Algorithmica **81**(2), 828–857 (2019). https://doi.org/10.1007/s00453-018-0451-4

# Tabu-Driven Quantum Neighborhood Samplers

Charles Moussa[1]([✉]), Hao Wang[1], Henri Calandra[2], Thomas Bäck[1], and Vedran Dunjko[1]

[1] LIACS, Leiden University, Niels Bohrweg 1, 2333 CA Leiden, Netherlands
c.moussa@liacs.leidenuniv.nl
[2] TOTAL SA, Courbevoie, France

**Abstract.** Combinatorial optimization is an important application targeted by quantum computing. However, near-term hardware constraints make quantum algorithms unlikely to be competitive when compared to high-performing classical heuristics on large practical problems. One option to achieve advantages with near-term devices is to use them in combination with classical heuristics. In particular, we propose using quantum methods to sample from classically intractable distributions – which is the most probable approach to attain a true provable quantum separation in the near-term – which are used to solve optimization problems faster. We numerically study this enhancement by an adaptation of Tabu Search using the Quantum Approximate Optimization Algorithm (QAOA) as a neighborhood sampler. We show that QAOA provides a flexible tool for exploration-exploitation in such hybrid settings and can provide evidence that it can help in solving problems faster by saving many tabu iterations and achieving better solutions.

**Keywords:** Quantum computing · Combinatorial optimization · Tabu search

## 1 Introduction

In the Noisy Intermediate-Scale Quantum (NISQ) era [39], hardware is limited in many aspects (e.g., the number of qubits, decoherence, etc.), which prevent the execution of fault-tolerant implementations of quantum algorithms. Therefore, hybrid quantum-classical algorithms were designed for near-term applications. Examples include algorithms for quantum chemistry problems [26,33], quantum machine learning [7] and combinatorial optimization [19]. They generally consist of one or many so-called parameterized quantum circuits (or variational quantum circuits), where the circuit architecture is fixed but the parameters of individual gates are adapted in a classical loop to achieve a computational objective.

Designed for combinatorial optimization, the Quantum Approximate Optimization Algorithm (QAOA) [19] consists of a quantum circuit of a user-specified depth $p$, involving $2p$ real parameters. To the limit of infinite depth, it converges

© Springer Nature Switzerland AG 2021
C. Zarges and S. Verel (Eds.): EvoCOP 2021, LNCS 12692, pp. 100–119, 2021.
https://doi.org/10.1007/978-3-030-72904-2_7

to the optimum. While numerous works have been studying various theoretical and empirical properties of QAOA [10,13,34,42,46], many practical challenges remain. Indeed, only small-sized problems and very limited $p$ can be run on real hardware, which severely limits the quality of solution obtained empirically [2,4,30]. Lastly, many open questions still remain, e.g., regarding the comparison of QAOA with other heuristic methods on various cases of instances which stem from particular problem domains, with different optimizers, and with varying levels of experimental (or simulated) noise. One reason of why so many uncertainties remain is that classical simulation is computationally very expensive, and quantum devices are still scarce to prevent large real world tests [2,32].

In contrast to optimization problems, quantum advantage has been demonstrated in sampling [1]. Indeed, theoretical results establish quantum advantage in producing samples according to certain distributions of constant-depth quantum circuits [11]. In this direction, it has been demonstrated that the sampling of the QAOA circuit, even at $p = 1$, cannot be efficiently simulated classically [20]. The above considerations point to a possibility of utilizing sampling features of QAOA for neighborhood explorations with the added benefit that, since the neighbourhood may be limited to fewer variables, a smaller quantum device may already lead to improved performance of a large instance.

It is interesting to delve into sampling aspects in the domain of classical local search algorithms, where we seek the optimum in the vicinity of the current solution with respect to either the original optimization problem or a subproblem thereof, using a deterministic or stochastic sampling strategy [3]. Such a sampling-based local procedure is typically realized by the combination of some parametric distribution family for drawing local trial points (e.g., the binomial or a power-law distribution [15]) and a selection method for choosing good trial points, and hence the overall outcome of this procedure results in the family of sampling distributions [14,29,44].

In this work, we propose to use QAOA circuits as local neighborhood samplers, having a malleable support in (many) good local optima but still allowing a level of exploration (which is desirable since local optima may not lead to global optima). This introduces the topics of sampling and multiobjective aspects of QAOA that allow balancing between exploration and exploitation. To this end, we study its combination with tabu search (TS), a metaheuristic that has been successfully applied in practice for combinatorial optimization by local search. Moreover, to control the trade-off between exploration and exploitation, we add this critical component in TS to the specification of the standard QAOA circuit.

**Contributions -** In this work, we construct an algorithm incorporating QAOA in TS with the usual attribute-based short-term memory structure (a.k.a the tabu list). With our approach, we kill two birds with one stone: we gain quantum enhancements, while the local properties of tabu search can make the required quantum computations naturally economic in terms of needed qubit numbers, which is vital in the near-term quantum era. We analyse and benchmark this incorporation with *small QAOA depths* against a classical TS procedure on Quadratic Unconstrained Binary Optimization (QUBO) problems of up to 500 variables. We also propose a penalized version of QAOA incorporating knowledge from a current solution. We find that QAOA is often beneficial in terms of saved iterations, and can exceptionally find shorter paths towards better solutions. The structure of the paper is as follows. Section 2 provides the necessary background on QUBOs, TS and QAOA. In Sect. 3, we detail the TS procedure incorporating a short-term memory structure with QAOA. The results of our simulations are presented in Sect. 4. We conclude our paper with a discussion in Sect. 5.

## 2   Background

The QUBO formulation can express an exceptional variety of combinatorial optimization (CO) problems such as Quadratic Assignment, Constraint Satisfaction Problems, Graph Coloring, Maximum Cut [28]. It is specified by the optimization problem $\min_{x \in \{0,1\}^n} \sum_{i \leq j} x_i Q_{ij} x_j$ where $n$ is the dimensionality of the problem and $Q \in \mathbb{R}^{n \times n}$. This formulation is connected to the task of finding so called "ground states", i.e., configurations of binary labels $\{1, -1\}$ minimising the energy of spin Hamiltonians, commonly tackled in statistical physics and quantum computing, i.e.,:

$$\min_{s \in \{-1,1\}^n} \sum_i h_i s_i + \sum_{j>i} J_{ij} s_i s_j, \tag{1}$$

where $h_i$ are the biases and $J_{ij}$ the interactions between spins.

The QAOA approach has been designed to tackle CO problems and was inspired from adiabatic quantum computing [19]. Firstly, the classical cost function is encoded in a quantum Hamiltonian defined on $N$ qubits by replacing each variable $s_i$ in Eq. (1) by the single-qubit operator $\sigma_i^z$:

$$H_C = \sum_i h_i \sigma_i^z + \sum_{j>i} J_{ij} \sigma_i^z \sigma_j^z. \tag{2}$$

Here, $H_C$ corresponds to the target Hamiltonian and the bitstring corresponding to the ground state of $H_C$ also minimizes the cost function. Secondly, a so-called *mixer Hamiltonian* $H_B = \sum_{j=1}^N \sigma_j^x$ is leveraged during the procedure. These hamiltonians are then used to build a layer of a quantum circuit with real parameters. This circuit is initialized in the $|+\rangle^{\otimes N}$ state, corresponding to all bitstrings in superposition with equal probability of being measured. Then, applying the layer $p$ times sequentially yields the following quantum state:

$$|\gamma, \beta\rangle = e^{-i\beta_p H_B} e^{-i\gamma_p H_C} \cdots e^{-i\beta_1 H_B} e^{-i\gamma_1 H_C} |+\rangle^{\otimes N},$$

defined by $2p$ real parameters $\gamma_i, \beta_i, i = 1...p$ or *QAOA angles* as they correspond to angles of parameterized quantum gates. Such output corresponds to a probability distribution over all possible bitstrings. The classical optimization challenge of QAOA is to identify the sequence of parameters $\gamma$ and $\beta$ so as to minimize the expected value of the cost function from the measurement outcome. In the limit of infinite depth, the distribution will converge to the optima.

Tabu Search (TS) [23] is a meta-heuristic that guides a local heuristic search procedure to explore the search space beyond local optimality. One of the main components of TS is its use of adaptive memory, which creates a more flexible search behavior. Such framework allows using a quantum algorithm as a local search search tool, for solving large instances with limited-sized quantum devices. Various works leveraged TS for solving QUBOs [22,24,27,36,37] using short-term and long-term strategies used during the search. We note also different hybrid settings that combine a basic TS procedure with another framework such as genetic search [31] and Path Relinking [43]. TS was also incorporated with quantum computers to tackle larger problems beyond their limitations. Indeed, finding methods to leverage smaller devices is of main importance. Many divide and conquer approaches have been designed for quantum circuits and algorithms [12,17,38,40]. In this paper, the size of the QC comes into play more naturally as a hyperparameter defining the «radius» of the search space.

With respect to the interplay between TS and quantum techniques, to our knowlege TS has only been considered from the perspective of D-Wave quantum annealers. The first approach of this kind is an algorithm called qbsolv [9]. It starts with an initial TS run on the whole QUBO. Then the problem is partitioned into several subproblems solved independently with the annealer. Subproblems are created randomly, by selecting variables. Non-selected ones have their values fixed (clamping values) from the TS solution. The subsolutions are then merged and a new TS is run as an improvement method. The second approach is an iterative solver designed in [41]. At each iteration, a subproblem is submitted to the annealer. The subproblem is obtained by clamping values from a current solution. A tabu list is used in which each element is a list of variables of length $k$. Each element is kept tabu for a user-defined number of iterations. In contrast in this work, we consider using QAOA in combination with TS.

## 3   Tabu-Driven QAOA Sampling

Inspired by the above-mentioned works, we use a simple TS procedure where QAOA is added in the neighborhood generation phase to solve QUBO problems. Note that we could also apply QAOA in more sophisticated frameworks, but a simpler approach is easier for understanding the benefits of QAOA with TS.

Local search algorithms explore a search space by generating sequences of possible solutions which are refined. At each step we generate so-called neighborhood from a current solution. In particular, if we denote the current solution $x$, a generated neighborhood corresponds to candidates $x'$ that differ by at most $k$ bits. We denote this set as $N_k(x) = \{x' \in \{0,1\}^n | \delta_H(x', x) \le k\}$, where $\delta_H$ denotes the Hamming distance. For a simple one bit-flip generation strategy, this corresponds to $k = 1$ and TS uses a modified neighborhood due to tabu conditions. Although increasing $k$ could help exploration, the neighborhood generation comes at exponential cost. But this could mean finding better solutions in fewer TS iterations, and thus also in principle overall faster if a fast good method for neighborhood exploration is devised.

This motivates the use of a quantum algorithm as a proxy for exploring $N_k(x)$. Specifically, we will use QAOA which $2p$ real parameters are tweaked in a continuous optimization scheme resulting in a probability distribution on $N_k$. Increasing $p$ (assuming the optima are found over the parameters) will improve the quality of the output (likelihood of returning an actual global optimum).

However, in the case of local search, a greedy strategy that tends to select the best point in the neighborhood would not only lead to potential stagnation, but also result in longer optimization time (unless the neighborhoods are already the size of the overall problem) [8]. Indeed, one may also consider modifications which impose (various) notions of locality, which are usually not considered in standard QAOA uses where it is used for the entire instance, with the sole goal of finding optima. To this end, we first outline the basic TS procedure generally used to solve QUBO problems [24,31,43]. Then, we show how QAOA can be combined with the latter. Finally, we propose a modification of QAOA that balances between going for the global optimum, and prioritizing local improvements relative to the current TS solution.

### 3.1  The Basic TS Algorithm

The basic TS procedure for solving a QUBO with objective function $f(x)$ is described in Algorithm 1, for $k = 1$ excluding the green-highlighted part. It uses a simple *tabu list* recording the number of iterations a variable remains tabu during the search. A variable can be set tabu for a fixed number of iterations (denoted Tabu tenure TT) but also with a random tenure. Each iteration can be considered as updating a current solution denoted $x$, exploring a modified neighborhood $N_1'$ due to the tabu considerations. Generally, $x$ is chosen greedily when evaluating the objective function over candidates $x \in N_1'$.

For large problem instances, there exists an efficient evaluation technique for QUBO solvers leveraging one-bit flip move [21]. Let $\Delta_x = f(x') - f(x)$ be a move value, that is the effect in objective of going to $x'$ from $x$. For one-bit flip moves, we denote as $\Delta_x(i)$ the move value upon flipping the $i$-th variable, which can be computed using only the QUBO coefficients. The procedure records a data structure storing those move values, which is updated after each TS iteration.

Initially, all variables can be flipped (line 5). At each iteration, the tabu solution $x$ is updated by flipping the variable that minimizes the objective over the neighborhood obtained by one-bit flip moves over non-tabu variables (lines 6–8). If the new tabu solution improves over the best recorded solution, the aspiration is activated. In this case, the tabu attribute of the flipped variable is removed. The tabu list is finally updated (lines 18–23) and iterations continue until the stopping criterion is reached. This can be either a maximum number of TS iterations, and/or a maximum number of TS iterations allowed without improvement of the best solution (*improvement cutoff*).

## 3.2   QAOA Neighborhood Sampling

In the usual TS algorithms, the neighborhood consists of candidates with Hamming distance one relative to the current tabu solution $x$. We note that sometimes considering also neighbors that are at most $k$-Hamming distance away from $x$ helps in finding better solutions. The number $k$ can be set in our case as large as the (limited) number of available qubits in a quantum hardware.

To study the exploration of such neighborhoods, a brute-force generation approach is initially tested, thereafter replaced by QAOA. As stated before, getting the optimum for subproblems in TS may lead to getting stuck during the search. QAOA, by definition, is a flexible framework as an exploration-exploitation tool. On the one hand, QAOA generates better solutions the deeper the circuit ($p$), and the better the classical optimization procedure within QAOA is. It is known it can have advantages over various standard algorithms, e.g. Simulated and Quantum Annealing [42]. To extract all advantages from the capacities of QAOA, we can further modulate the distribution over outputs it produces by limited depth or, as we present next, modifications to the QAOA objective to prioritize a more local behaviour. Such flexibility is important, not only for the exploration-exploitation trade-off as it provides interesting ways fine-tune the algorithm depending on the instance to solve.

First, the choice of variables to run QAOA on needs to be adressed. Considering the $\binom{N}{k}$ possibilities would be intractable. Variables can be chosen randomly but an approach incorporating one-bit flip move values can help in guiding towards an optimum. The $k$ variables can be chosen amongst the non-tabu ones at each step. Plus, this means QAOA is an attempt at improving over the solution one would get with the one-bit flip strategy outlined previously. One can either select the $k$ variables greedily or add randomness by using the one-flip gains as weights for defining a probability to be chosen. For simplification, we consider the greedy selection based on one-bit flip move values. If we consider the chosen variables that were flipped, the update step of the incremental evaluation strategy can be applied. Let $l \leq k$ be the number of different bits. The newly generated candidate can be considered as a result of $l$ sequential one-bit flips. Thus, $l$ calls to the above-mentioned efficient procedure are required.

A second consideration concerns the tabu strategy for updating the tabu list. We choose to set as tabu the variables amongst the $k$ chosen ones that were flipped. Choosing to flip all chosen ones could be problematic as it could lead to all variables being tabu very early during TS. An aspiration criterion can be used if the new candidate gives the best evaluation found during the search.

Finally, the question of how to run QAOA is of main importance. In our first scenario, QAOA will be run as a proxy for brute-force (with exploration properties) to optimize the subproblem defined over the $k$ chosen variables. This is done by fixing in the QUBO the non chosen ones from the current tabu solution $x$. The depth $p$ of QAOA can be user-defined. In this work, we limit $p$ to 2 to showcase sampling aspects of QAOA at small depth.

---

**Algorithm 1:** QAOA-featured Tabu Search for solving QUBO.

---

**Input**: An initial solution $x_0$, Cost function $f(x)$
**Parameter** : Tabu tenure TT, Random tabu tenure rTT, subproblem size $k$
**Output**: The best solution achieved $x^*$

1   $x^* \leftarrow x \leftarrow x_0$;
2   $Tabu(i) \leftarrow 0, \Delta_x(i) \leftarrow 0$ for $i \in [1..N]$;
3   **while** *stopping criterion not reached* **do**
4   $\quad$ $x^{\text{pre}} \leftarrow x$;
5   $\quad$ **for** $i \in [1..N] \cap Tabu(i) = 0$ **do**
6   $\quad\quad$ $x_i^{(i)} \leftarrow 1 - x_i, x^{(i)} \leftarrow x$;
7   $\quad\quad$ one bit-flip gain: $\Delta_x(i) = f(x^{(i)}) - f(x)$;
8   $\quad$ $x^{\text{1-bit}} \leftarrow x^{(j)}$, $j \leftarrow \arg\min \Delta_x(i)$;
9   $\quad$ Select greedily or randomly a subset of variables $K \subseteq I$ s.t. $|K| = k$;
10  $\quad$ Get a new QUBO by fixing the $N - k$ other variables in $x$;
11  $\quad$ Run QAOA and get the best sample $\hat{x}$ minimizing the new QUBO;
12  $\quad$ $x_i^{k\text{-bit}} \leftarrow x_i$ for $i \in I \backslash K$ and $x_i^{k\text{-bit}} \leftarrow \hat{x}_i$ for $i \in K$;
13  $\quad$ $x \leftarrow$ the better out of $x^{\text{1-bit}}$ and $x^{k\text{-bit}}$;
14  $\quad$ Update move values $\Delta_x(i)$ for $i \in [1..N] \cap x_i^{\text{pre}} \neq x_i$;
15  $\quad$ $aspiration \leftarrow$ False;
16  $\quad$ **if** $f(x) < f(x^*)$ **then**
17  $\quad\quad$ $x^* \leftarrow x$, $aspiration \leftarrow$ True;
18  $\quad$ $Tabu(i) \leftarrow Tabu(i) - 1$ for $i \in [1..N] \cap Tabu(i) > 0$;
19  $\quad$ **for** $j \in [1..N] \cap x_j^{\text{pre}} \neq x_j$ **do**
20  $\quad\quad$ **if** *aspiration* **then**
21  $\quad\quad\quad$ $Tabu(j) \leftarrow 0$;
22  $\quad\quad$ **else**
23  $\quad\quad\quad$ $Tabu(j) \leftarrow$ TT + Random(rTT);

---

Our QAOA-featured TS is outlined in Algorithm 1 where QAOA addition is indicated in green shades. It starts with the same steps as with the standard Tabu search algorithm until line 8. The QAOA part kicks in from line 9 by firstly choosing a subset of $k$ variables, and executing QAOA on the sub-QUBO problem where we optimize over the chosen $k$ variables while keeping the remaining bits the same with the current point $x$. After obtaining the best point from QAOA (lines 11–13), we select the better one from the QAOA outcome $x^{k\text{-bit}}$ and the best one-bit flip point $x^{\text{1-bit}}$ and use it to update the current search point. The move values are then updated by $l$ calls of the fast incremental method (line 14). Finally, if the best-so-far point is improved by the updated search point, we drop the accepted bit flips from the tabu list (line 21). Otherwise, we reset their tabu value to the sum of tabu tenure and a random tenure (line 23).

## 3.3   Enforcing Locality with Penalized QAOA

As a tool in local search algorithms, QAOA may be useful with modifications which impose notions of locality. We incorporate these notions in the cost hamiltonian so that they are captured during the QAOA evolution. This can be done through the cost function by adding a penalty term. A possibility is to consider the Hamming distance with a current tabu solution $x$. Hence, the objective for QAOA becomes:

$$\min_{x'} [f(x') + A\delta_H(x', x)], \tag{3}$$

where $\delta_H$ corresponds to the Hamming distance and $A$ is a constant. The right-hand side additive term of Eq. (3) aims to encourage the output of candidates that differ by few bits from the current solution if $A > 0$, and vice-versa. As a neighborhood sampler, the penalty may help in enforcing locality. However, setting the parameter $A$ is non trivial for activating the effect of the extra term.

There is also a possibility to add information about the fitness gain in differing from $x$. If switching a bit is an improving move, it would be prioritized. Our algorithm uses the one-bit flip move values $\Delta_x$ in order to select the variables to run QAOA on, from which we can construct a weighted penalty term:

$$-\tfrac{1}{2} \sum_{j=1}^{N} \Delta_x(j)(-1)^{x_j} x'_j. \tag{4}$$

For a minimization problem, $\Delta_x(j) < 0$ characterizes encouraging flipping the $j$-th variable in new candidates. For a candidate in this case, $\Delta_x(j)$ is added to the cost. Conversely, $\Delta_x(j) > 0$ would result in penalizing candidates with the $j$-th bit value flipped. One can also multiply by a positive constant A for enforcing more the locality effects. The penalty translates in an additional operator term $H_{\text{penalty}}$, following an application of a usual cost operator in QAOA, where:

$$H_{\text{penalty}} = -\tfrac{1}{2} \sum_{j=1}^{N} \Delta_x(j)(-1)^{x_j} \sigma_z^j \tag{5}$$

The corresponding quantum circuit of depth one is very simple and given by $\bigotimes_i R_Z((-1)^{x_i} \gamma \Delta_x(i)), \gamma \in \mathbb{R}$.

## 4   Simulations

We performed extensive simulations over instances of QUBO problems publicly available in the well-know OR-Library [5,6,22]. In designing them, our objectives are 1) investigating whether exploring larger neighborhood can facilitate faster convergence (in terms of TS iterations), 2) elucidating the effect of locality on QAOA output given by the introduced penalty in Eq. (5), and 3) studying the utility of QAOA as a proxy for brute-force.

### 4.1    Larger Neighborhood Exploration Benefits

To study the first objective we replaced QAOA by brute-force search in Algorithm 1. Starting from the all-zero initial solution, we first run the basic TS with different constant tabu tenures (rTT = 0). Then, we do the same with brute-force TS for different values of $k$ up to 20. We study two regimes that differ in how TS search results, namely when $k$ is not comparatively small to $N$ on instances where $N = 20$, and when it is on instances of 100, 200, and 500 variables.

**$k/N$ Relatively Large.** In the first regime, we assume that we can explore a large percentage (more than 25%) of the instance size greedily. As instances, we take the first eight instances of *bqpgka* (named 1a–8a consisting of 30–100 variables), for which we solve by TS. Then we select randomly 20 variables out of $N$ and clamp values of the non-selected ones from the solutions. We do so five times per instance, resulting in 40 instances of size 20. Then TS with different $k/N$ values for subproblems $(0.9, 0.75, 0.5, 0.25)$ were tried on this suite. When using brute-force, we set $TT = 2$ and tried many values for the basic TS.

Figure 1 shows the proportion of (run, target value) pairs aggregated over all functions for 10 targets generated by linear spacing using the benchmarking and profiling tool IOHprofiler [16] for iterative optimization heuristics. The target values were normalized by the optimum of the problems. We observed that for $k/N \geq 0.5$, these instances are straightforward to solve (in 3 iterations). The case $k = 5$ required 8 iterations on one instance but managed to achieve optimality on all of them. However, the basic TS, run for 20000 iterations, failed to solve the same instance. Letting this instance aside, 5 iterations would be required for $k = 5$, and 22 for the classical TS procedure. Hence, we clearly observe, as expected, degrading performances as $k/N$ decreases. This also enabled us to confirm numerically that a flip-gain based approach when considering subproblems is in general beneficial towards solving QUBOs.

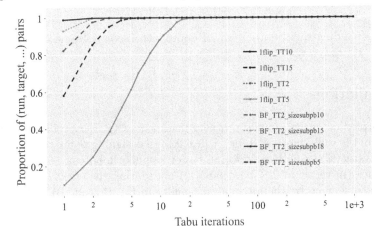

**Fig. 1.** Empirical cumulative distribution function (ECDFs) of tabu iterations for each algorithm aggregated over all 20-variable problems with 10 target values evenly spanning the range of all observed function values.

**Table 1.** Best values of tabu tenure (ranging from 2 to 10, adding 15) achieving the optimum obtained with the first TS iteration(s) to reach the corresponding maximum given in [22]. The best performances per instance are highlighted in bold. The mention All means all $TT$ values reached the same solution. The NA mention means no run returns the optimum, with the best value obtained in parenthesis.

| Algo. | 1d (6333) | 2d (6579) | 3d (9261) | 4d (10727) | 5d (11626) |
|---|---|---|---|---|---|
| Basic | 15 / 300 | All / 71 | <9 / 90 | 2 / 67 | 6 / 171 |
| $k = 10$ | All / 13 | >2 / 61 | 6 / 39 | 2 / 31 | All / 19 |
| $k = 15$ | All / 10 | 2+5 / **26** | All / 8 | 3 / 18 | 5 / 42 |
| $k = 20$ | **All / 7** | 5 / 37 | **All / 5** | **All / 11** | **2 / 18** |
| Algo | 1e (16464) | 2e (23395) | 3e (25243) | 4e (35594) | 5e (35154) |
| Basic | 9 / 419 | 6 / 493 | 10 / 190 | 15 / 170 | 6 / 238 |
| $k = 10$ | 8 / 485 | 7 / 111 | All / 21 | 6 / 55 | All / 25 |
| $k = 15$ | **All / 13** | 6+7 / **31** | All / 13 | 3 / 41 | **2 / 20** |
| $k = 20$ | 2 / 35 | NA (23370)/ 82 | 3 / 44 | **3 / 25** | 4 / 31 |
| Algo | 1f (61194) | 2f (100161) | 3f (138035) | 4f (172771) | 5f (190507) |
| Basic | 8 / 970 | 7 / 795 | 4 / 449 | NA (172734) / 1148 | NA (190502) / 647 |
| $k = 10$ | **15 / 213** | 7 / 337 | 15 / 77 | NA (172449) / 110 | NA (190502) / 126 |
| $k = 15$ | 8 / 225 | 8 / **173** | 15 / 139 | NA (172734) / 46 | NA (190502) / 383 |
| $k = 20$ | NA (61087) / 184 | NA (100158) / 101 | 4 / 57 | **4 / 57** | 9 / 150 |

**$k/N$ Relatively Small.** In this regime, we study the case when $k/N$ is relatively small (less than 20%). The simulations are carried out on 15 QUBO instances from bqpgka: 100-variable (named 1d–5d), 200-variable (1e–5e) and 500-variable problems (1f–5f). Algorithms are run until the best objective value found in [22] is reached or a maximum number of TS iterations is reached.

We run the algorithms with different $TT$ values ranging from 2 to 10 and adding 15, and for $k = 10, 15, 20$. We limit the number of iterations to 20000 for the basic version, 1000 for the brute-force approach (200 for $k = 20$ though). Table 1 shows for the different values of $TT$, which ones achieved the best performances in terms of target only. In these cases, we observe that we can find a $k$ such that better solutions are found using less iterations, especially on the most dense instances 4f and 5f. These instances, when considering the underlying graph given by the coefficients connecting different variables, have a density of respectively 0.75 and 1 (a non-zero coefficient for each pair of variables). For $k = 20$ the proposed approach achieved optimality where the basic TS failed. Again, we observe performances, in terms of target achieved, depend on setting well the tabu tenure in accordance with $k$. Intuitively, one could think that the larger $k$, the smaller $TT$ has to be to save iterations and achieve a better objective value. But we clearly observe counter-examples.

Larger $k$ exploration, in this regime, turns out to not always be beneficial. This seems conter-intuitive when considering a target objective only, on an instance to instance basis comparison. Figure 2 shows the proportion of (run, target value) pairs aggregated over all functions for 1000 targets generated by linear spacing. The target values were normalized by the optimum of the problems.

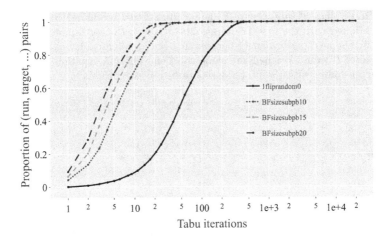

**Fig. 2.** Empirical cumulative distribution function (ECDFs) of tabu iterations for each algorithm aggregated over all dimensions, all problems, and 1000 target values evenly spanning the range of all observed function values.

Again, we observe that, with larger $k$, the proportion of successes is higher, when measured at the same number of iterations. Note that this can only be observed for $k = 20$ up to 200 iterations. The proportion for the basic TS was close to 0.9 while the brute-force approach was superior to 0.99. Hence, we can reach very good solutions with less iterations as $k$ increases.

In summary, as opposed to the previous regime, the structure of the problems becomes very important that we have to look at performances in an aggregated way to witness the benefits of exploring larger neighborhood. Having outlined some performances given by the brute-force approach on subproblems, we switch to QAOA, and study its sampling effect as a proxy.

### 4.2   QAOA as a Proxy for Brute-Force

The second part of our simulations studies the output of QAOA as a proxy for brute-force. To this end, we first study an example TS run from our previous simulations. We take the subproblem QUBOs obtained at each step (except the first one), and run QAOA at $p = 1$ and 2, and we study the distribution of the energy given by $|\gamma, \beta\rangle$, after optimization, with and without the penalty term.

Having outlined the properties of the QAOA output, we run Algorithm 1, and study its performances in comparison to the basic TS. From the optimized angles, we try different sampling strategies to generate a candidate per iteration: just sampling once, sampling 10 times and choosing a candidate greedily, and finally consider all samples (even during optimization) greedily. The latter corresponds to a quasi brute-force (BF) approach.

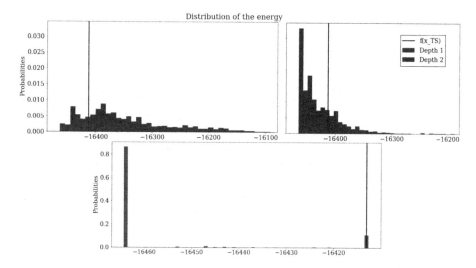

**Fig. 3.** QUBO evaluation distribution given by sampling $10^5$ times from the QAOA quantum state, and by running simulated annealing 1000 times (bottom), for the last TS iteration done on instance 1e.

**Energy Distribution of QAOA.** As a first step, we study how the QAOA output distribution looks like at small depth, with the purpose of elucidating how it can help avoiding detrimental greedy search behavior. We consider, as an example, instance 1e for which $k = 15, TT = 5$ used 13 iterations greedily. The subproblem QUBOs are kept and we run QAOA on them as previously stated. The third and last iteration are interesting when considering the penalty term. The former is a case where the optimum is obtained by flipping all bits except one and where all flip moves are favorable. The last iteration has flip moves from the current tabu solution discourage flipping all bits. Plus, very few candidates (0.2%) improve over the tabu solution. For these iterations, we look at the quantum state given by QAOA and analyse the distribution of the QUBO evaluation (or energy in an Ising context).

Figure 3 shows the distribution given by $10^5$ samples from the last iteration's quantum state at $p = 1$ and 2. At $p = 1$, we observe a homogeneous spread with two major humps on the left and right side of the tabu solution evaluation. There is a probability of 23.9% of improving from it by the quantum state. At $p = 2$, we see the distribution being shifted to lower energies, yielding an improvement probability of 75%. The average energy is 16350 for QAOA $p = 1$ and 16428 for $p = 2$. Moreover, the standard deviation of the output decreases from 79.2 to 42. This is expected as to the limit of infinite depth, QAOA converges to the optimum with less variance.

When running simulated annealing 1000 times with a temperature of 17.5 using 100 steps, we observed that unlike QAOA, the energy spread is restrained to a few points, the optimum being most present. Decreasing slightly the temperature or the number of steps would always yield the optimum. However, in terms of exploration opportunities, QAOA could allow visiting different paths that may lead to fewer iterations required towards improved solutions.

Theory shows that increasing the depth would permit QAOA to find the optimum assuming optimal parameters are found. But by limiting the depth, we can control how other good candidates are spread from the optimum. This could engender new paths to solve a problem differently, where a suboptimal solution on a subproblem leads to an easier one for QAOA towards better candidates. Note this can be done in different ways. One way would be using brute-force and perturbating or mixing the solution, which is not efficient. We could also have the same effect with simulated annealing. But in many cases, we can fail finding the optimum, and even less finding a bunch of candidates around it. Finally, due to its flexibility, QAOA permits to leverage modifications to introduce locality notions in a multiobjective scheme for local search, such as the previously mentioned penalty. In the following, we study its effect on the QAOA distribution.

**Penalty Effect.** In Sect. 3.3, we introduced a penalty term to impose notions of locality in QAOA as a local search tool. An extra operator based on the hamiltonian given in Eq. (5), translates to a circuit of depth one concatenated with a QAOA layer. We study its effect on the QAOA distribution obtained on the resulting sub-qubos at the third and last iteration.

On iteration 3, for both QAOA and its penalized version, the distribution tends to output candidates with largest Hamming distances to the current point as expected. Also, the most likely candidate is the one that completely differs from the current point, which is more favored by the penalty effect. No significant changes were observed when penalizing at $p = 2$.

For the last iteration, Fig. 4 shows that the original QAOA at $p = 1$ results in many probability peaks compared to the penalized version, which evolve to a major peak with an increased depth. The penalized version demonstrates two major peaks, from which the optimum and a close candidate to $x$ are preferred.

This characterizes the interplay between optimizing and penalizing. At $p = 1$, the penalized version has a better probability of improving $x$ (0.34 vs 0.239) and a higher probability of finding the optimum (0.1 vs 0.02). However, the unpenalized $p = 2$ version was more likely to output the optimum (respectively 0.75 and 0.26, where the penalty at $p = 2$ yields 0.62 and 0.26).

In summary, using the penalty creates a balance between the greedy approach and one-bit flip gains knowledge from the current solution. This could result in smoothening the distribution while favouring interesting candidates for both objectives. This will modify the search path taken during TS depending on the outcome. Having studied numerically the output of the quantum state one can get with QAOA, and the penalty effect, we switch to less idealized simulations where the subproblems depend on the QAOA output during the TS search.

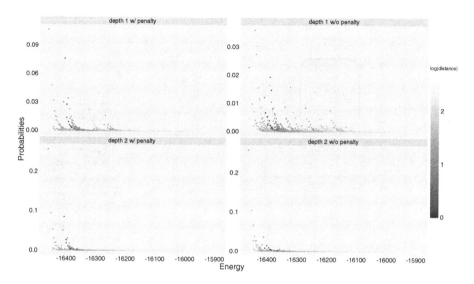

**Fig. 4.** Distribution of the evaluations (or energies) for the last iteration obtained on instance 1e given by the QAOA output state, with and without the penalty term. A colormap is given for the Hamming distance with the current tabu solution.

**QAOA Exploration Possibilities.** After looking at examples of QAOA output and outlining a few possible exploration opportunities, we carried out a few extra simulations but not considering the subproblems obtained with brute-force. Hence, QAOA (and its penalized version) was called once per iteration with BIPOP-CMAES [25,35] optimizing from one set of angles[1]. In the following, we give a few examples to illustrate the exploration possibilities.

We take instances where BF simulations required few iterations and where the basic TS was beaten in target. Namely, instance 1d for $k = 15$ and $TT = 5$ and 1e for $k = 15$ and $TT = 10$. We run Algorithm 1 for 10 times, limiting them to 20 TS iterations. Different numbers of samples are used for generating a new candidate per iteration: just once, 10 times and 1000 times. The latter could be considered as a quasi-brute-force approach in these runs. We denote as $m$ the number of samples used in the following.

Figures 6 and 5 show the median of the best normalized evaluation obtained per run by iteration. We observe from iteration 7 for 1d and 10 for 1e that $m = 1000$ is equivalent (in median) to the BF generation. In general, the more samples used, the better the solution found. However, we had at a few iterations median runs that achieved higher values than BF. For instance, this happened for the penalized $p = 2$ QAOA at the 5th iteration with $m = 10$ for 1d and $m = 1000$ at the 12th for 1e. We consider also the frequency of runs for which

---

[1] When using BIPOP-CMAES, we run circuits with 1000 measurements to estimate expectation values. The optimizer stops when it has reached 2000 evaluations. We obtained great performances in terms of averaged ratios (as the evaluations divided by the optimum of the subproblem), superior to 0.97 at the considered depths.

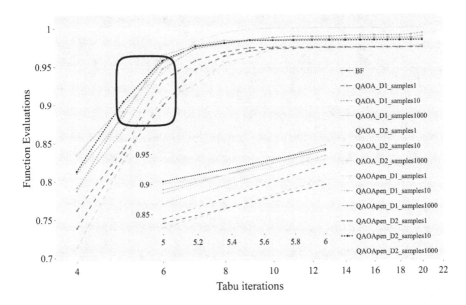

**Fig. 5.** Median of best normalized evaluations achieved over TS iterations for instance 1d. The mentions D1, D2, mean respectively $p = 1, 2$ and the penalized version is indicated by «pen». A higher curve corresponds to better solutions reached. At iteration 20, the basic TS value would be 0.38, while the lowest QAOA curve value is 0.967. The $m = 1000$ runs and BF are over 0.99 starting at the 8th iteration. At iteration 5, the penalized $p = 2, m = 10$ version is slightly better than the others, even BF (0.9049 vs 0.9007). At the 6th, the $m = 10$ versions are above BF (respectively 0.9645 and 0.9578 for unpenalized and penalized $p = 2$, 0.9555 and 0.9470 for $p = 1$, and 0.9409 for BF).

the basic TS was beaten, and the optimum was found. Table 2 summarizes our results. Increasing $m$ improves the frequency of successful runs. We observe that new paths were found, mainly with an extra iteration or two but exceptionally one run or two over 10 could save one iteration. This was the case on instance 1e with 12 iterations instead of 13, exclusively with the penalized version. These examples are numerically in favor of a greedy (or quasi) approach in solving sub-problems. However, QAOA allows, through a trade-off between exploration and exploitation, discovering new paths towards optimality that are still interesting in terms of number of iterations.

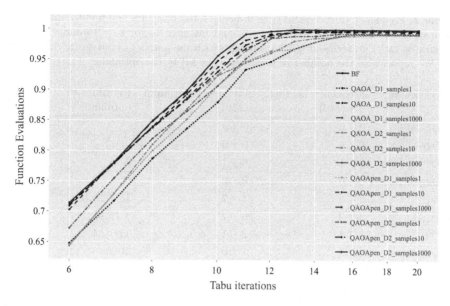

**Fig. 6.** Median of best normalized evaluations achieved over TS iteration for instance 1e. At iteration 20, the lowest QAOA curve value is 0.991 (while 0.19 for the basic TS). The $m = 1000$ runs and BF are over 0.99 starting at the 11th iteration. At iteration 7, the original $p = 2$ QAOA using 10 samples point is higher than the others, even BF (0.7827 vs 0.7808). This also happens at iteration 12, with the penalized $p = 2, m = 1000$ (0.997085 vs 0.996902).

## 5  Conclusion and Outlook

In this work, we studied sampling aspects when quantum approaches, specifically the QAOA algorithm, are considered in combinatorial optimization. We considered a practically relevant setting where a gate-based quantum algorithm, limited in the number of qubits, is utilized in a hybrid quantum-classical framework to solve large optimization instances faster. Our framework constitutes a powerful yet simple heuristic, Tabu Search, in tandem with QAOA as a local neighborhood sampler.

As a starting point, numerical experiments over open-source QUBO problems up to 500 variables validate using QAOA as a proxy to explore larger neighborhood, under the assumption that subproblems are solved optimally. Continuing, we investigated the exploration possibilities given by QAOA output at small depth. User-defined parameters such as depth and number of measurements used to generate a candidate, can be increased to favor exploitation. On our examples, solving subproblems emphasizing more on the latter gave better general performances. Yet, we found that exploration can be beneficial. Iterations can be saved with our QAOA procedure, illustrating that missing to generate the solution of a subproblem in previous iterations could yield to faster paths towards better solutions. Hence, the QAOA-based algorithm we introduce in this work becomes a very flexible tool in such hybrid quantum-classical settings.

**Table 2.** Frequency of successful runs in beating the basic TS and finding the optima for instance 1d for $k = 15$ and $TT = 5$, and instance 1e for $k = 15$ and $TT = 10$ (separated by /), in terms of QAOA settings (with the number of measurements noted $m$). We report also the number of iterations that led to the optimum.

| QAOA | $m$ | Frequency beating TS (/10) | Frequency optimum (/10) | Iterations to optimum |
|---|---|---|---|---|
| D1 | | 0 / 0 | 0 / 0 | |
| D1pen | 1 | 0 / 0 | 0 / 0 | |
| D2 | | 0 / 1 | 0 / 1 | / 14 |
| D2pen | | 0 / 0 | 0 / 0 | |
| D1 | | 7 / 2 | 1 / 2 | 18 / 15,16 |
| D1pen | 10 | 4 / 2 | 0 / 2 | / 15,15 |
| D2 | | 5 / 1 | 2 / 1 | 11 / 14 |
| D2pen | | 4 / 0 | 1 / 0 | 10 / |
| D1 | | 8 / 7 | 8 / 7 | 9,10,12 / 13 |
| D1pen | 1000 | 8 / 5 | 8 / 5 | 10 / 12,13 |
| D2 | | 9 / 9 | 9 / 9 | 10,11 / 13,14 |
| D2pen | | 10/ 7 | 10 / 7 | 9,10 / 12,13 |

We see numerous possibilities for future work. First, our model allows for many hyperparameters whose function needs to be explored, and, as is usually done in many local search methods, the exploration/exploitation trade-offs can be made online-adaptive. Second, the effect of real world limitations, most importantly noise, and hardware connectivity, calls for further investigation. Although QAOA can be run on real hardware [2,45], its output quality will improve as the quantum devices decreases in error, or through quantum error mitigation [18]. Finally, it would be interesting to propose different frameworks (e.g. [24,31,36,37]) with special emphasis on the exploration possibilities given by small-depth quantum algorithms, and cross-compare with standard techniques in future works. We believe our approach combined with these types of analyses will provide new promising ways to maximize the use of limited near-term quantum computing architectures for real world and industrial optimization problems.

**Acknowledgements.** CM, TB and VD acknowledge support from Total. This work was supported by the Dutch Research Council (NWO/OCW), as part of the Quantum Software Consortium programme (project number 024.003.037). This research is also supported by the project NEASQC funded from the European Union's Horizon 2020 research and innovation programme (grant agreement No 951821).

# References

1. Arute, F., et al.: Quantum supremacy using a programmable superconducting processor. Nature **574**(7779), 505–510 (2019). https://doi.org/10.1038/s41586-019-1666-5
2. Arute, F., et al.: Quantum approximate optimization of non-planar graph problems on a planar superconducting processor (2020)
3. Bäck, T.: Evolutionary Algorithms in Theory and Practice - Evolution Strategies, Evolutionary Programming, Genetic Algorithms. Oxford University Press, Oxford (1996)
4. Barkoutsos, P.K., Nannicini, G., Robert, A., Tavernelli, I., Woerner, S.: Improving variational quantum optimization using CVaR. Quantum **4**, 256 (2019)
5. Beasley, J.E.: OR-library: distributing test problems by electronic mail. J. Oper. Res. Soc. **41**(11), 1069–1072 (1990). http://www.jstor.org/stable/2582903
6. Beasley, J.: QUBO instances link - file bqpgka.txt. http://people.brunel.ac.uk/~mastjjb/jeb/orlib/bqpinfo.html
7. Benedetti, M., Lloyd, E., Sack, S., Fiorentini, M.: Parameterized quantum circuits as machine learning models. Quantum Sci. Technol. **4**(4), 043001 (2019). https://doi.org/10.1088/2058-9565/ab4eb5
8. Beyer, H.: The theory of evolution strategies. In: Natural Computing Series. Springer, Berlin (2001). https://doi.org/10.1007/978-3-662-04378-3
9. Booth, M., Reinhardt, S.P.: Partitioning optimization problems for hybrid classical/quantum execution technical report (2017)
10. Brandão, F.G.S.L., Broughton, M., Farhi, E., Gutmann, S., Neven, H.: For fixed control parameters the quantum approximate optimization algorithm's objective function value concentrates for typical instances arXiv:1812.04170 (2018)
11. Bravyi, S., Gosset, D., König, R.: Quantum advantage with shallow circuits. Science **362**(6412), 308–311 (2018). https://doi.org/10.1126/science.aar3106, https://science.sciencemag.org/content/362/6412/308
12. Bravyi, S., Smith, G., Smolin, J.A.: Trading classical and quantum computational resources. Phys. Rev. X **6** (2016). https://doi.org/10.1103/PhysRevX.6.021043, https://link.aps.org/doi/10.1103/PhysRevX.6.021043
13. Crooks, G.E.: Performance of the quantum approximate optimization algorithm on the maximum cut problem (2018). https://arxiv.org/abs/1811.08419
14. Doerr, B., Doerr, C.: Optimal static and self-adjusting parameter choices for the $(1+(\lambda, \lambda))$ genetic algorithm. Algorithmica **80**(5), 1658–1709 (2018). https://doi.org/10.1007/s00453-017-0354-9
15. Doerr, B., Le, H.P., Makhmara, R., Nguyen, T.D.: Fast genetic algorithms. In: Bosman, P.A.N. (ed.) Proceedings of the Genetic and Evolutionary Computation Conference, GECCO 2017, Berlin, Germany, 15–19 July 2017, pp. 777–784. ACM (2017). https://doi.org/10.1145/3071178.3071301
16. Doerr, C., Wang, H., Ye, F., van Rijn, S., Bäck, T.: IOHprofiler: a benchmarking and profiling tool for iterative optimization heuristics. arXiv e-prints:1810.05281, October 2018. https://arxiv.org/abs/1810.05281
17. Dunjko, V., Ge, Y., Cirac, J.I.: Computational speedups using small quantum devices. Phys. Rev. Lett. **121**, 250501 (2018). https://doi.org/10.1103/PhysRevLett.121.250501, https://link.aps.org/doi/10.1103/PhysRevLett.121.250501
18. Endo, S., Cai, Z., Benjamin, S.C., Yuan, X.: Hybrid quantum-classical algorithms and quantum error mitigation. J. Phys. Soc. Jpn. **90**(3), 032001 (2020)

19. Farhi, E., Goldstone, J., Gutmann, S.: A quantum approximate optimization algorithm (2014)
20. Farhi, E., Harrow, A.W.: Quantum supremacy through the quantum approximate optimization algorithm (2016)
21. Glover, F., Hao, J.K.: Efficient evaluations for solving large 0–1 unconstrained quadratic optimisation problems. Int. J. Metaheuristics $\mathbf{1}(1)$, 3–10 (2010). https://doi.org/10.1504/IJMHEUR.2010.033120
22. Glover, F., Kochenberger, G., Alidaee, B.: Adaptive memory tabu search for binary quadratic programs. Manage. Sci. $\mathbf{44}$, 336–345 (1998). https://doi.org/10.1287/mnsc.44.3.336
23. Glover, F.W.: Tabu search. In: Handbook of Combinatorial Optimization, pp. 1537–1544. Springer, US, Boston, MA (2013). https://doi.org/10.1007/978-1-4419-1153-7_1034
24. Glover, F.W., Lü, Z., Hao, J.K.: Diversification-driven tabu search for unconstrained binary quadratic problems. 4OR $\mathbf{8}$, 239–253 (2010)
25. Hansen, N.: Benchmarking a BI-population CMA-ES on the BBOB-2009 function testbed. In: ACM-GECCO Genetic and Evolutionary Computation Conference. Montreal, Canada, July 2009. https://hal.inria.fr/inria-00382093
26. Kandala, A., et al.: Hardware-efficient variational quantum eigensolver for small molecules and quantum magnets. Nature $\mathbf{549}$, 242–246 (2017). https://doi.org/10.1038/nature23879
27. Kochenberger, G., et al.: The unconstrained binary quadratic programming problem: a survey. J. Comb. Optim. $\mathbf{28}(1)$, 58–81 (2014). https://doi.org/10.1007/s10878-014-9734-0
28. Kochenberger, G.A., Glover, F.: A unified framework for modeling and solving combinatorial optimization problems: a tutorial. Multiscale Optim. Methods Appl. 101–124. Springer, US, Boston, MA (2006). https://doi.org/10.1007/0-387-29550-X_4
29. Lehre, P.K., Yao, X.: Crossover can be constructive when computing unique input-output sequences. Soft. Comput. $\mathbf{15}(9)$, 1675–1687 (2011)
30. Li, L., Fan, M., Coram, M., Riley, P., Leichenauer, S.: Quantum optimization with a novel gibbs objective function and ansatz architecture search. Phys. Rev. Res. $\mathbf{2}(2)$, 023074 (2019)
31. Lü, Z., Glover, F.W., Hao, J.K.: A hybrid metaheuristic approach to solving the UBQP problem. Eur. J. Oper. Res. $\mathbf{207}$, 1254–1262 (2010)
32. Medvidovic, M., Carleo, G.: Classical variational simulation of the quantum approximate optimization algorithm (2020)
33. Moll, N., et al.: Quantum optimization using variational algorithms on near-term quantum devices. Quantum Sci. Technol. $\mathbf{3}(3)$, 030503 (2018). https://doi.org/10.1088/2058-9565/aab822
34. Moussa, C., Calandra, H., Dunjko, V.: To quantum or not to quantum: towards algorithm selection in near-term quantum optimization. Quantum Sci. Technol. $\mathbf{5}(4)$, 044009 (2020). https://doi.org/10.1088/2058-9565/abb8e5
35. Niko, A., Yoshihikoueno, Y., Brockhoff, D., Chan, M.: ARF1: CMA-ES/pycma: r3.0.3, April 2020. https://doi.org/10.5281/zenodo.3764210
36. Palubeckis, G.: Multistart tabu search strategies for the unconstrained binary quadratic optimization problem. Ann. Oper. Res. $\mathbf{131}$, 259–282 (2004). https://doi.org/10.1023/B:ANOR.0000039522.58036.68
37. Palubeckis, G.: Iterated tabu search for the unconstrained binary quadratic optimization problem. Informatica (Vilnius) $\mathbf{17}(2)$, 279–296 (2006)

38. Peng, T., Harrow, A.W., Ozols, M., Wu, X.: Simulating large quantum circuits on a small quantum computer. Phys. Rev. Lett. **125**(15), 150504 (2020). https://doi.org/10.1103/PhysRevLett.125.150504, https://link.aps.org/doi/10.1103/PhysRevLett.125.150504
39. Preskill, J.: Quantum Computing in the NISQ era and beyond. Quantum **2**, 79 (2018). https://doi.org/10.22331/q-2018-08-06-79
40. Rennela, M., Laarman, A., Dunjko, V.: Hybrid divide-and-conquer approach for tree search algorithms (2020)
41. Rosenberg, G., Vazifeh, M., Woods, B., Haber, E.: Building an iterative heuristic solver for a quantum annealer. Comput. Optim. Appl. **65**, 845–869 (2016)
42. Streif, M., Leib, M.: Comparison of QAOA with quantum and simulated annealing, arXiv:1901.01903 (2019)
43. Wang, Y., Lü, Z., Glover, F.W., Hao, J.K.: Path relinking for unconstrained binary quadratic programming. Eur. J. Oper. Res. **223**, 595–604 (2012)
44. Watson, R.A., Jansen, T.: A building-block royal road where crossover is provably essential. In: Proceeding of Genetic and Evolutionary Computation Conference (GECCO 2007), pp. 1452–1459. ACM (2007). https://doi.org/10.1145/1276958.1277224
45. Willsch, M., Willsch, D., Jin, F., De Raedt, H., Michielsen, K.: Benchmarking the quantum approximate optimization algorithm. Quantum Inf. Process. **19**(7), 197 (2020). https://doi.org/10.1007/s11128-020-02692-8
46. Zhou, L., Wang, S.T., Choi, S., Pichler, H., Lukin, M.D.: Quantum approximate optimization algorithm: performance, mechanism, and implementation on near-term devices, arXiv:1812.01041 (2018)

# On Hybrid Heuristics for Steiner Trees on the Plane with Obstacles

Victor Parque[✉]

Department of Modern Mechanical Engineering, Waseda University, 3-4-1 Okubo, Shinjuku, Tokyo 169-8555, Japan
parque@aoni.waseda.jp

**Abstract.** Minimal-length Steiner trees in the two-dimensional Euclidean domain are of special interest to enable the efficient coordination of multi-agent and interconnected systems. We propose an approach to compute obstacle-avoiding Steiner trees by using the hybrid between hierarchical optimization of shortest routes through sequential quadratic programming over constrained two-dimensional convex domains, and the gradient-free stochastic optimization algorithms with a convex search space. Our computational experiments involving 3,000 minimal tree planning scenarios in maps with convex and non-convex obstacles show the feasibility and the efficiency of our approach. Also, our comparative study involving relevant classes of gradient-free and nature inspired heuristics has shed light on the robustness of the selective pressure and exploitation abilities of the Dividing Rectangles (DIRECT), the Rank-based Differential Evolution (RBDE) and the Differential Evolution with Successful Parent Selection (DESPS). Our approach offers the cornerstone mechanisms to further advance towards developing efficient network optimization algorithms with flexible and scalable representations.

**Keywords:** Minimal trees · Steiner trees · Optimization · Planning

## 1 Introduction

Minimal-length trees are relevant to enable the efficient coordination of multi-agent systems in interconnected and distributed environments, e.g. flocking [1], formation control [2] and leader selection [3]. With origins in the formulation of Pierre de Fermat [4], the concept of minimal trees was popularized by Vojtěch-Kössler in the 30's [5] and by Courant in the 40's [6]. One of the relevant well-known constructs is the *Steiner tree* in *graphs*. Here, for an undirected graph $G = (V, E)$ a minimal network connecting a subset of $k$ elements (nodes) of $V$ is to be found. Naturally, if the subset contains two (all) elements of $V$, the problem can be reduced to the Shortest-Path (Minimum Spanning Tree), which is computable in polynomial time. The (geometric) *Euclidean Steiner Tree* problem (EST) aims at computing minimal-length trees spanning $n$ terminal nodes while allowing the addition of extra (Steiner) points in the Euclidean domain. Here, we study the obstacle-avoiding Euclidean Steiner tree problem, requiring that no point in the tree lies in the interior of obstacles.

© Springer Nature Switzerland AG 2021
C. Zarges and S. Verel (Eds.): EvoCOP 2021, LNCS 12692, pp. 120–135, 2021.
https://doi.org/10.1007/978-3-030-72904-2_8

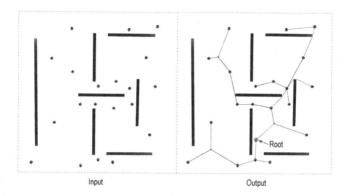

Input                               Output

**Fig. 1.** Input: a set of nodes and an arbitrary polygonal set in the Euclidean plane. Output: a collision-free Steiner tree.

## 2  Steiner Tree Problem

### 2.1  Definition

Basically, we tackle the obstacle-avoiding Euclidean Steiner tree problem in a polygonal map with obstacles. Here,

- **Inputs** are a set of nodes and a polygonal map as portrayed by Fig. 1.
- The **output** is a collision-free tree as portrayed by Fig. 1.
- The **quality** of a solution is the overall tree length.

### 2.2  Background

In the early 90's, Winter and Smith proposed an $O(n)$ algorithm to compute Euclidean Steiner trees with three terminals and one obstacle in the plane, in which $n$ is the number of extreme vertices in the obstacle [7]. In [8], Winter used visibility graphs to generate and concatenate *Steiner trees* with two, three and four terminals. In the late 90's, Zachariasen and Winter proposed the first exact algorithm for the obstacle-avoiding Euclidean Steiner tree in the plane by generating and concatenating full Steiner trees [9]. In the early 2000's, Weng and Smith studied the Steiner Trees with one obstacle and presented an $O(n^2 + n\log^2 w)$ algorithm, in which $n$ is the cardinality of terminal points in the plane, and $w$ is the number of extreme points in the convex obstacle [10]. They also generalized their result to a non-convex obstacle running with $O(n^2 + nw.\log w)$. Winter et al. devised a polynomial heuristic concatenating the exact solutions of Steiner trees with up to four terminals, in a simple polygon [11]. In 2010, Müller-Hannemann and Tazari devised a polynomial-time approximation scheme with running time $O(n \log^2 n)$, in which $n$ is number of terminals plus obstacle vertices [12]. Their approach extends the algorithm of Borradaile et al. [13] to find $(1 + \epsilon)$-approximations of Steiner trees in planar graphs whose size is $O(n \log n)$. In 2018, Cohen and Nutov studied connectivity problems in a convex metric space and proposed the improved approximation ratios for both Steiner trees and Steiner forests with minimum number of Steiner points [14]. In 2019, Chen et al. proposed the connectivity restoration of wireless networks based on obstacle-avoiding quadrilateral Steiner trees [15].

Nature-inspired and evolutionary algorithms (EAs) are general stochastic heuristics; and as such, several investigations have shown the potential of EAs in tackling various forms of minimal-length tree problems, achieving competitive results and approximation guarantees in various scenarios. For instance, a mathematical model of a Physarum polycephalum slime mold network was used to construct Steiner trees in graphs, showing the exponential convergence towards the optimum [16]. And, two variants of Physarum-inspired algorithms were shown to outperform Genetic Algorithm (GA), discrete Particle Swarm Optimization (PSO) [17]. Also, evolutionary approaches were shown to be effective in the bi-level tree formulations [18,19]. The node-based and the path-based Variable Neighborhood Search (VNS) were used to solve the Steiner tree problem in graphs, in which solutions were initialized by using the Prim's algorithm, and whose effectiveness was shown in 40 instances derived from the OR-Library [20]. Also, it was shown that the $(1 + 1)$ Evolutionary Strategy attains the $3/2$-approximation ratio in constructing Steiner trees for a class of quasi-bipartite graphs [21].

Also, minimal trees has attracted the attention in the circuit design and the pipe routing communities. Whereas circuit design is key to VLSI systems [22], pipe routing is key to offshore oil and gas industries [23]. In the octilinear Euclidean Steiner tree, sometimes referred as the X-architecture, edges are required to be oriented at $0°$, $45°$, $90°$, $135°$. Examples of the heuristics include Particle Swarm Optimization [24–26], variants of Genetic Algorithms [27–29], Ant Colony Optimization [30], and more recently Differential Evolution [31,32]. For an overview on Steiner tree algorithms for VLSI applications, refer to the survey by Chen et al. [22].

Although the above-mentioned studies have rendered minimal tree layouts being compact and free of clutter, the study of obstacle-avoiding minimal Steiner trees under arbitrary configurations of obstacles and terminal nodes by nature-inspired and gradient-free optimization algorithms has received little attention. Whereas tailored algorithms for the rectilinear and the octilinear versions of the Euclidean Steiner tree problems exist, the study of the general versions by gradient-free heuristics is potential to elucidate the feasibility to render minimal trees in the plane. In particular, we formulated and evaluated five relevant classes of gradient-free optimization algorithms for obstacle avoiding Euclidean Steiner tree problems. Compared to above-mentioned works, this paper expands the scope of evaluated gradient-free optimization heuristics by considering diverse forms of selection pressure and exploration-exploitation trade-offs. Thus, it becomes possible to elucidate the role of exploration-exploitation in the context of searching obstacle-avoiding Euclidean trees. In particular, we evaluated the following algorithms: (1) Differential Evolution with Global and Local Interpolation Vectors (DEGL), (2) Rank-Based Differential Evolution (RBDE),

(3) Differential Evolution with Successful Parent Selection (DESPS), (4) Particle Swarm Optimization with Niching Properties (NPSO), and (5) Dividing Rectangles Optimization Algorithm (DIRECT). We also present insights from rigorous computational experiments involving the comparative convergence performance of the above-described algorithms. We present the statistical comparative study over a large set of instances comprising 3,000 planning scenarios and 600,000 evaluations of obstacle-avoiding Euclidean Steiner tree instances. Our computational experiments have shown that selection pressure and exploitation play important roles to render topologically compact and minimal Euclidean Steiner tree layouts. As such, the Dividing Rectangles (DIRECT), the Rank-based Differential Evolution (RBDE) and the Differential Evolution with Successful Parent Selection (DESPS) are potential to tackle the Euclidean Steiner tree problem in arbitrary configuration of polygonal obstacles.

## 3   Proposed Approach

In this section we describe our proposed approach, and the following sections describe the key components and dynamics.

### 3.1   Basic Algorithm

The basic outline of our algorithm for building a minimal tree is depicted by Fig. 2 (route bundling). Given the coordinates of a *root*, our goal is to construct a tree minimizing the total length, in which the tree layout represents an obstacle-avoiding network topology over the polygonal map. Here, our approach describes the case of a fixed *root* location; and in the next section, we describe the notion of flexible location of *roots* being optimized by gradient-free sampling with various forms of selection pressure.

- **Inputs.** A set of nodes and a polygonal map in the Euclidean plane as portrayed by Fig. 2-(1).
- **Shortest Paths.** The shortest paths from the *root* towards nodes are computed, as exemplified by Fig. 2-(2).
- **Hierarchical Clustering.** The shortest paths are clustered by the *hierarchical (agglomerative)* approach in which the distance between two shortest paths is computed as

$$d(A, B) = \left( \sum_{k=1}^{K} ||A^k - B^k||^2 \right) \left( \cos^{-1} \left( \frac{\boldsymbol{a}.\boldsymbol{b}}{|\boldsymbol{a}||\boldsymbol{b}|} \right) \right) \tag{1}$$

$$\boldsymbol{a} = A^{end} - A^{init}, \tag{2}$$

where $A^k$ is the $k$-th sampled point along the shortest path $A$, and $K$ is the number of 1-D equally-distanced interpolated points along the path $A$ and $B$. The vector $\boldsymbol{a}$ is defined as such $A^{init}$ and $A^{end}$ is the *origin* and *destination* coordinate of the shortest path $A$, respectively. The above distance metric

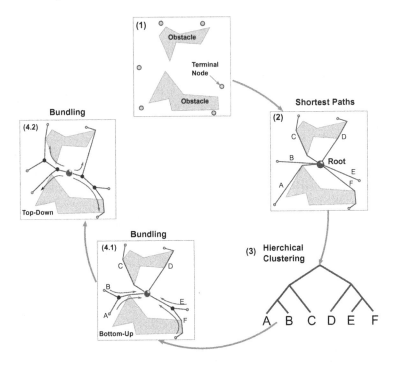

**Fig. 2.** Basic concept of the route bundling approach.

combines *piecewise gaps* and *orientation gaps*, thus two shortest paths are similar if they overlap each other and share the same direction vector. For simplicity and without loss of generality, we use the product rule between *piecewise gaps* and *orientation gaps*, our future work will study other possible algebraic combinations. The hierarchical clustering of shortest paths renders a *dendrogram* **Z**, as exemplified by Fig. 2-(3).

- **Hierarchical Bundling.** The minimal tree is computed by using the hierarchical ordering structure of the *dendrogram* **Z**, and by using a nature-inspired approach which considers the *merging*, the *expansion* and the *shrinkage* of tree structures. Basically, the location of intermediate nodes (Steiner Points) are computed hierarchically in a *bottom-up* approach (from the terminal nodes towards the root), complemented by a *top-down* approach (from root to terminal nodes), as exemplified by Fig. 2(4)–(5) and Fig. 3. Both the *bottom-up* and the *top-down* approaches aim at co-adaptation of the global hierarchy when either roots or leaves change in local topology. During bundling, Steiner points are inserted within the convex hull of the shortest paths by searching for locations that minimize the tree length over the local region (convex hull) of the shortest paths, as exemplified by Fig. 3. The search procedure is realized by the trust region method using Sequential Quadratic Programming (SQP) [33]. For a region involving two shortest paths $A$ and $B$, the location of the Steiner Points are encoded by the tuple $r = (\tau, r_1, r_2)$ where $\tau \in [N_{A,B}]$

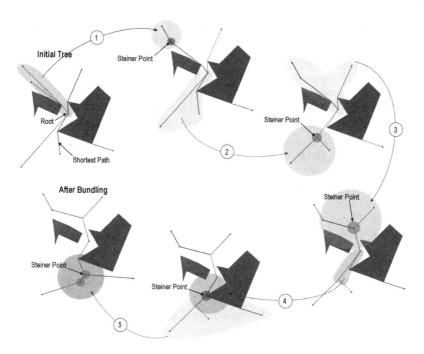

**Fig. 3.** Basic concept of the stages during tree bundling. The shortest paths are described in the top-left tree as 'Initial Tree'. The regions in which bundling occurs are highlighted in colored regions. (Color figure online)

is the index of the $\tau$-th triangle derived from the set of *Delaunay Triangulation* of the *free space* of the polygonal environment in the lócal search space, $N_{A,B}$ is the number of Delaunay triangles involved in the free region of the shortest paths $A$ and $B$, and $r_1, r_2 \in [0, 1]$. The above encoding ensures that the Steiner points are located inside the free space, and avoids the complexity of validating whether points are outside the obstacles, thus enabling to build a convex search space of potential Steiner points. Also, the equivalent 2-dimensional cartesian coordinate $(r_x, r_y)$ of the root $r$ can be computed by $(r_x, r_y) = (1 - r_1)a_\tau + \sqrt{r_1}(1 - r_2)b_\tau + \sqrt{r_1}r_2c_\tau$, where $a_\tau$, $b_\tau$, $c_\tau$ are the 2-dimensional coordinates of the vertices of the $\tau$-th triangle derived from the *Delaunay Triangulation* of the *free space* [34].

### 3.2   Root Optimization

The coordinate of the root of the Steiner tree is determined by solving

$$\min_{x} f(\boldsymbol{x}) = L(T(\boldsymbol{x})) = \sum_{s \in \lambda(T)} \ell(s)$$

$$\text{s.t. } \boldsymbol{x} \in \mathcal{F} \tag{3}$$

where $T(\boldsymbol{x})$ is the Steiner tree with root at $\boldsymbol{x} \in \mathcal{F}$, $\mathcal{F}$ is the free space in the polygonal map, $\lambda(T)$ is the set of edges of $T(\boldsymbol{x})$, and $\ell(s)$ is the shortest Euclidean distance connecting two nodes in the edge $s \in T$. Basically, the cost function $L(T(\boldsymbol{x}))$ aims at evaluating the length of the Steiner tree. Thus, by solving the above-mentioned problem one seeks to find the root at which the obstacle-avoiding Euclidean Steiner tree attains its minimal length. The root of the Steiner tree is encoded by the tuple $r = (\tau, r_1, r_2)$, thus the free space $\mathcal{F}$ is equivalent to $\mathbb{N}^{N_{\mathcal{F}}} \times \mathbb{R}^{[0,1]} \times \mathbb{R}^{[0,1]}$, where $N_{\mathcal{F}}$ is the number of Delaunay triangles involved in the free space. Therefore, due to the observations mentioned above, the search space $\mathcal{F}$ is basically of convex nature, in which the lower bound is $(1, 0, 0)$, and the upper bound is $(N_{\mathcal{F}}, 1, 1)$. This feature enhances the representation used in [31] and [35] by enabling the convex search space of Steiner points given arbitrary obstacles in the plane [19].

Also, due to the non-linear landscape of the above-mentioned cost function, and the multimodal nature of the shortest paths when dealing with non-convex obstacles, we use the class of gradient-free optimization algorithms considering features of multimodality and balance between exploration and exploitation. In this paper, we use *Differential Evolution (DE)* [36], Particle Swarm Optimization and DIRECT due to the flexibility to realize diversity of exploration and exploitation during search. In line of the above, we used five relevant classes of gradient-free optimization heuristics, each of which denotes distinct modes of selection pressure and balance between *exploration* and *exploitation*. Considering other nature-inspired heuristics is straightforward, yet including a large number of such heuristics is out of the scope of this paper. In the following sections we describe the variants used in our study.

**Differential Evolution with Global and Local Interpolation Vectors (DEGL).** Basically, in *Differential Evolution* (DE)

$$\boldsymbol{x}_{t+1} = \begin{cases} \boldsymbol{u}_t & f(\boldsymbol{u}_t) \leq f(\boldsymbol{x}_t) \\ \boldsymbol{x}_t & \text{otherwise} \end{cases} \tag{4}$$

$$\boldsymbol{u}_t = \boldsymbol{x}_t + \boldsymbol{m}_t \circ (\boldsymbol{v}_t - \boldsymbol{x}_t), \tag{5}$$

where $\boldsymbol{x}_t$ is the root of the Euclidean Steiner tree at iteration $t$, $\boldsymbol{u}_t$ is the *trial* root at iteration $t$, $\circ$ is the *Hadamard* product (element-wise), $\boldsymbol{v}_t$ is the *mutant* vector at iteration $t$, $\boldsymbol{m}_t$ is a binary vector defined by the crossover probability $CR$. In Differential Evolution with Global and Local Interpolation Vectors (DEGL) [37], the mutant vector $\boldsymbol{v}_t$ is computed by:

$$\boldsymbol{v}_t = w^{\boldsymbol{x}_t} . \boldsymbol{g}_t + (1 - w^{\boldsymbol{x}_t}) . \boldsymbol{l}_t, \tag{6}$$

where $w^{\boldsymbol{x}_t}$ is the weight associated to vector $\boldsymbol{x}_t$, $\boldsymbol{g}_t$ is the global interpolation vector, $\boldsymbol{l}_t$ is the local interpolation vector. Here, the weight $w^{\boldsymbol{x}_t}$ is computed adaptively in the *neighborhood* with ring topology and radius $\eta$. DEGL's sampling scheme focuses on the self-balance of both exploration and exploitation by considering not only the entire population, but also the local neighborhood organized in a ring structure.

**Rank-Based Differential Evolution (RBDE).** Here, the mutant vector is

$$v_t = x_t^1 + F(x_t^2 - x_t^3), \qquad (7)$$

where $F$ is the mutation factor, and the parent individuals $x_t^1$, $x_t^2$ and $x_t^3$ are selected from the Whitley Distribution [38] with $x_t^i = P_j^S$ for $i = \{1, 2, 3\}$, with

$$j = \left\lfloor \frac{|P|}{2(\beta - 1)} \left( \beta - \sqrt{\beta^2 - 4(\beta - 1)\sigma} \right) \right\rfloor, \qquad (8)$$

where $P$ is the population in Differential Evolution, $|P|$ is the population size, $P_j^S$ is the $j$-th element of the *population sorted* by fitness, $\beta$ is a user-defined bias term, and $\sigma$ is a random number in $U[0, 1]$. RBDE's sampling scheme focuses on the selective pressure to allow *exploitative power* through a rank-based ordering of the population, in which smaller values of $\beta \in (1, 3]$ are found to be favorable to speed-up the learning convergence in non-separable problems [38].

**Differential Evolution with Successful Parent Selection (DESPS).** Here, the mutant vector $v_t$ is computed by the DE/best/1/bin strategy and selection follows the successful parent selection principle [39]:

$$x_t^{best}, \ x_t^1, \ x_t^2 \in \begin{cases} P_T, \text{ if } q_t \leq Q \\ SP_t, \text{ otherwise} \end{cases}, \qquad (9)$$

where $P_T$ is the population at iteration $t$, $Q$ is a user-defined parameter to count *stagnation* during the convergence, and $SP_t$ is an archive (set) of successful solutions at iteration $t$, in which the initial $SP$ is a copy of the initial population for $t = 0$ (initialization), and subsequent successful solutions replace the oldest by selective pressure.

**Particle Swarm Optimization with Niching Properties (NPSO).** Here, the global best is based on the Fitness Euclidean Ratio [40]

$$gbest_x = \underset{x}{\text{argmax}} \ \text{FER}(x, y) \qquad (10)$$

$$\text{FER}(x, y) = \gamma \frac{f(p_y) - f(p_x)}{||p_y - p_x||}, \qquad (11)$$

where $p_x$ is the personal best of particle $x$, $p_g$ is the global best in the population, $p_w$ is the global worst in the population, $\gamma$ is a normalization factor, $L_d, U_d$ are the lower and upper bounds in the dimension $d$. In the above, the normalization factor ensures that the global best of each particle is set to the particles in the population that are close and offer fitness improvement. Thus, the above scheme allows NPSO to let solutions move towards the *best* and *closest* particles in the neighborhood, thus allowing the self-formation of communities in the search space.

**Dividing Rectangles Optimization Algorithm (DIRECT).** DIRECT is a gradient-free global optimization algorithm suitable for nonlinear, nonsmooth and multimodal problems. DIRECT uses the DIviding RECTangles concept to interpolate vectors through the mid of hypercubes, in which potential hypercubes are selected by iterative subdivisions [41]. We use the DIRECT algorithm due to its highly selective pressure allowing the balance of exploration and exploitation during search. Basically, DIRECT balances the exploration-exploitation trade-off by selecting not one but several divisions of the search space by using all possible relative weightings of local versus global (subdivisions) hypercubes of the search space. Whereas the purely exploitative (explorative) version of a sampling algorithm would select the best (largest) subdivisions (neighborhoods) in the search space, DIRECT selects a plural number of potential solutions and hypercubes considering the Pareto frontier that maximizes fitness performance and search space coverage, leading to an algorithm with the efficient trade-offs between explorative and exploitative search.

## 4 Computational Experiments

We evaluated our algorithms in Matlab 2020a by using an Intel i7-4930K @ 3.4 GHz and by using 3 types of polygonal maps with arbitrary configuration of obstacle geometry and (input) terminal nodes in the plane. On maps of type 1, obstacles are represented by arbitrary non-convex polygons. In maps of type 2, obstacles are represented by squared obstacles. In maps of type 3, obstacles are represented by rectangles resembling the configuration of walls. Our main motivation of the above is due to our focus on scenarios being reminiscent of indoor map configurations, in which the reasonable number of nodes are able to co-exist in the same environment.

As for parameters in the above-mentioned classes of gradient-free optimization algorithms, we used probability of crossover $CR = 0.5$, scaling factor $F = 0.7$, population size $|P| = 5$ individuals, neighborhood ratio $\eta = 0.1$, the bias term $\beta = 3$, and the termination criterion is set as 200 function evaluations. Also, for each configuration, 20 independent runs were performed due to the stochastic nature of Differential Evolution (DE) and Particle Swarm Optimization (PSO). Other parameters were used following the respective references. The key motivation of using the above parameters is to evaluate the (competitive) performance of the gradient-free optimization algorithms under small evaluation budgets, in which evaluation of the cost function is expensive.

As a result of considering 3 types of maps, 10 independent terminal node configurations, 20 independent runs per each case and 5 algorithms, we evaluated $3,000$ planning scenarios of minimal trees and $600,000$ evaluations of the cost function (Eq. 3). The fine-tuning of the above-mentioned gradient-free optimization algorithms is out of the scope of this study. Figure 4, 5 and 6 show examples of obtained minimal trees by (a) DEGL, (b) RBDE, (c) DESPS, (d) NPSO and (e) DIRECT after 200 function evaluations and under the same initial conditions and settings over independent runs. In these figures, the location of the suggested root is highlighted in yellow color. By looking at these tree topologies, we can observe differences in the tree structure by each heuristic. Also, Fig. 7 shows an example inspired by pipe routing in a real-world building environment. Also, to evaluate the convergence of the gradient-free optimization algorithms, Fig. 8, 9 and Fig. 10 describe the *convergence behaviour* of the tree length as a function of the number of evaluations, and Fig. 11 show the Wilcoxon test on the converged cost function of each gradient-free optimization algorithm over independent runs.

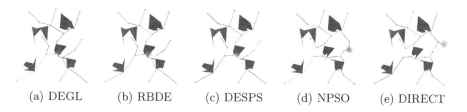

   (a) DEGL       (b) RBDE      (c) DESPS     (d) NPSO     (e) DIRECT

**Fig. 4.** Example of minimal tree in maps of **type 1**. (Color figure online)

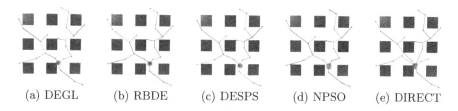

   (a) DEGL       (b) RBDE      (c) DESPS     (d) NPSO     (e) DIRECT

**Fig. 5.** Example of minimal tree in maps of **type 2**. (Color figure online)

By observing the above-described results, we note the following facts: (1) DIRECT algorithm shows a competitive convergence in all maps (as depicted by green color in Fig. 8, 9 and Fig. 10) when compared to the classes of Differential Evolution (DE) and Particle Swarm Optimization (PSO). (2) Fast convergence is attained by DIRECT algorithm, within [50, 100] function evaluations.

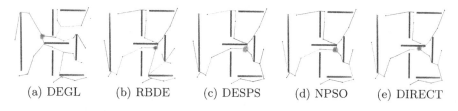

   (a) DEGL       (b) RBDE      (c) DESPS     (d) NPSO     (e) DIRECT

**Fig. 6.** Example of minimal tree in maps of **type 3**. (Color figure online)

(a) View 1    (b) View 2

**Fig. 7.** Views of the topology of a minimal tree in an environment layout. The red (green) node (s) denote the root (leaves) of the tree. (Color figure online)

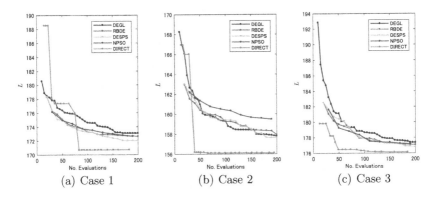

(a) Case 1    (b) Case 2    (c) Case 3

**Fig. 8.** Mean of convergence of the tree length over 20 independent runs in **Map 1**. (Color figure online)

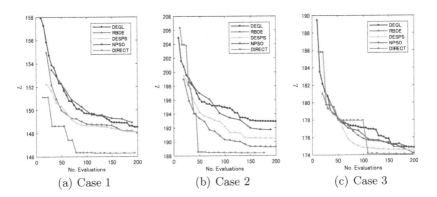

(a) Case 1    (b) Case 2    (c) Case 3

**Fig. 9.** Mean of convergence of the tree length over 20 independent runs in **Map 2**. (Color figure online)

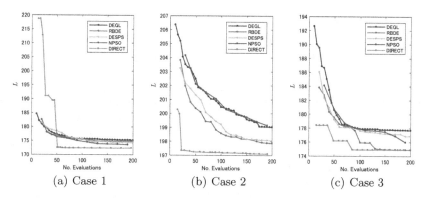

Fig. 10. Mean of convergence of the tree length over 20 independent runs in **Map 3**. (Color figure online)

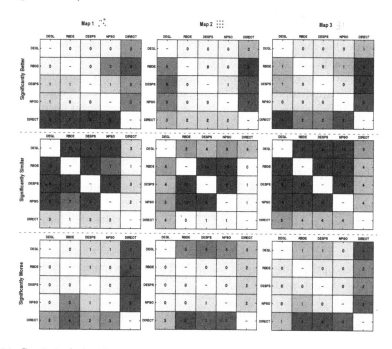

Fig. 11. Statistical significance test using Wilcoxon at 5% significance level. Numbers represent planning instances where an algorithm in the row is significantly better, similar and worse to an algorithm in the column (higher values in dark).

The above results offer relevant observations that regard the suitability of the class of Differential Evolution, Particle Swarm Optimization and Dividing Rectangles approaches. The evaluated scenarios allows to elucidate the nature of the search space and the fitness landscape of the Euclidean Steiner tree problem in arbitrary configurations of the polygonal domain: due to the convexity of the encoding and our formulated search space at Eq. (3), the selective pressure and exploitation abilities of DIRECT becomes beneficial for faster convergence in polygonal environments with irregular and non-convex obstacle geometry. On the other hand, evolutionary computing algorithms with strong foci on exploitation such as the Rank-based Differential Evolution (RBDE) and Differential Evolution with Successful Parent Selection (DESPS) become beneficial in polygonal environments with regular and convex obstacle geometry. Thus, among the above-studied scenarios, selective pressure and exploitation schemes are advantageous to render shorter Steiner trees. Conversely, DEGL and NPSO offer less selective pressure due to the additional perturbation from the population neighborhood. Studying whether different sampling mechanisms or canonical representations of Steiner tree structures has a positive effect on the convergence rate is left in future agenda.

We believe the competitive performance of DIRECT is due to the amenability of the division of rectangles concept to the convex representation of the feasible search space (by using the Delaunay triangulation). Since the convex representation ensures that coordinates of the root are located within the free-space of the polygonal map, the division of hypercubes in DIRECT enables to sample the feasible search space efficiently. Since our proposed hierarchical bundling computes the optimal topology of the Steiner tree using roots suggested by the gradient-free optimization algorithm, the sampling of potential locations of the root of the Steiner tree in few number of evaluations of the cost function becomes essential to attain competitive convergence. DIRECT allows to sample roots of the Steiner trees efficiently in less number of function evaluations due to the Delaunay-based representation: potential roots located in proximal triangles of the search space will suggest solutions within the proximity of such potential roots. Likewise, roots being far from potential locations will be neglected by the hierarchical division of hypercubes in DIRECT. Our results offer building blocks to further advance towards developing global network optimization algorithms with flexible and scalable representations. In future work, we aim at studying Steiner tree optimization using enumerative representations [42–46] and its applications to design modular folds [47].

## 5    Conclusions

In this paper we have proposed a hybrid approach to compute obstacle-avoiding minimal trees in the plane. Our approach is based on (1) the hierarchical bundling of the shortest paths, (2) the formulation of a convex search space of Steiner points and roots, and (3) the gradient-free stochastic optimization algorithms able to realize diverse forms of selection pressure and exploration-exploitation trade-offs. Our computational experiments confirmed the feasibility and the efficiency of our proposed approach, showing the faster convergence of the Dividing Rectangles (DIRECT) approach, as well as the positive benefits of selection pressure of RBDE and DESPS.

**Acknowledgment.** This research was supported by JSPS KAKENHI Grant Number 20K11998.

# References

1. Olfati-Saber, R.: Flocking for multi-agent dynamic systems: algorithms and theory. IEEE Trans. Autom. Control **51**, 401–420 (2006)
2. Li, A., Wang, L., Pierpaoli, P., Egerstedt, M.: Formally correct composition of coordinated behaviors using control barrier certificates. In: 2018 IEEE/RSJ International Conference on Intelligent Robots and Systems (IROS), pp. 3723–3729 (2018)
3. Luo, W., Khatib, S.S., Nagavalli, S., Chakraborty, N., Sycara, K.: Distributed knowledge leader selection for multi-robot environmental sampling under bandwidth constraints. In: 2016 IEEE/RSJ International Conference on Intelligent Robots and Systems (IROS), pp. 5751–5757 (2016)
4. de Fermat, P.: Method for determining maxima and minima and tangents to curved lines. Oeuvres **1**, 135 (1643)
5. Vojtěch, J., Kössler, M.: On minimal graphs containing n given points. Časopis pro pěstování matematiky a fysiky **63**, 223–235 (1934). (in Czech). Zbl 0009.13106
6. Robbins, H., Courant, R.: What is Mathematics? Oxford University Press, New York (1941)
7. Winter, P., MacGregor Smith, J.: Steiner minimal trees for three points with one convex polygonal obstacle. Ann. Oper. Res. **33**, 577–599 (1991). https://doi.org/10.1007/BF02067243
8. Winter, P.: Euclidean Steiner minimal trees with obstacles and Steiner visibility graphs. Discret. Appl. Math. **47**, 187–206 (1993)
9. Zachariasen, M., Winter, P.: Obstacle-avoiding Euclidean Steiner trees in the plane: an exact algorithm. In: Goodrich, M.T., McGeoch, C.C. (eds.) ALENEX 1999. LNCS, vol. 1619, pp. 286–299. Springer, Heidelberg (1999). https://doi.org/10.1007/3-540-48518-X_17
10. Weng, J.F., MacGregor Smith, J.: Steiner minimal trees with one polygonal obstacle. Algorithmica **29**, 638–648 (2001). https://doi.org/10.1007/s00453-001-0002-1
11. Winter, P., Zachariasen, M., Nielsen, J.: Short trees in polygons. Discret. Appl. Math. **118**, 55–72 (2002)
12. Müller-Hannemann, M., Tazari, S.: A near linear time approximation scheme for Steiner tree among obstacles in the plane. Comput. Geom. Theory Appl. **43**, 395–409 (2010)
13. Borradaile, G., Kenyon-Mathieu, C., Klein, P.: A polynomial-time approximation scheme for Steiner tree in planar graphs. In: ACM-SIAM Symposium on Discrete Algorithms, pp. 1285–1294 (2007)
14. Cohen, N., Nutov, Z.: Approximating Steiner trees and forests with minimum number of Steiner points. J. Comput. Syst. Sci. **98**, 53–64 (2018)
15. Chen, B., Chen, H., Wu, C.: Obstacle-avoiding connectivity restoration based on quadrilateral Steiner tree in disjoint wireless sensor networks. IEEE Access **7**, 124116–124127 (2019)
16. Caleffi, M., Akyildiz, I.F., Paura, L.: On the solution of the Steiner tree np-hard problem via Physarum bionetwork. IEEE/ACM Trans. Network. **23**, 1092–1106 (2015)

17. Sun, Y., Halgamuge, S.: Fast algorithms inspired by Physarum polycephalum for node weighted Steiner tree problem with multiple terminals. In: IEEE Congress on Evolutionary Computation, pp. 3254–3260 (2016)

18. Camacho-Vallejo, J.F., Garcia-Reyes, C.: Co-evolutionary algorithms to solve hierarchized Steiner tree problems in telecommunication networks. Appl. Soft Comput. **84**, 105718 (2019)

19. Parque, V., Miyashita, T.: Obstacle-avoiding Euclidean Steiner trees by n-star bundles. In: IEEE 30th International Conference on Tools with Artificial Intelligence, pp. 315–319 (2018)

20. Chuong, T.V., Nam, H.H.: A variable neighborhood search algorithm for solving the Steiner minimal tree problem. In: Cong Vinh, P., Alagar, V. (eds.) ICCASA/ICTCC -2018. LNICST, vol. 266, pp. 218–225. Springer, Cham (2019). https://doi.org/10.1007/978-3-030-06152-4_19

21. Lai, X., Zhou, Y., Xia, X., Zhang, Q.: Performance analysis of evolutionary algorithms for Steiner tree problems. Evol. Comput. **25**, 707–723 (2017)

22. Chen, X., Liu, G., Xiong, N., Su, Y., Chen, G.: A survey of swarm intelligence techniques in VLSI routing problems. IEEE Access **8**, 26266–26292 (2020)

23. Tan, W.C., Chen, I., Pantazis, D., Pan, S.J.: Transfer learning with PipNet: for automated visual analysis of piping design. In: 2018 IEEE 14th International Conference on Automation Science and Engineering (CASE), pp. 1296–1301 (2018)

24. Liu, Q., Wang, C.: Multi-terminal pipe routing by Steiner minimal tree and particle swarm optimisation. Enterp. Inf. Syst. **6**, 315–327 (2012)

25. Liu, G., Guo, W., Niu, Y., Chen, G., Huang, X.: A PSO-based timing-driven octilinear Steiner tree algorithm for VLSI routing considering bend reduction. Soft. Comput. **19**, 1153–1169 (2015)

26. Huang, X., Liu, G., Guo, W., Niu, Y., Chen, G.: Obstacle-avoiding algorithm in X-architecture based on discrete particle swarm optimization for VLSI design. ACM Trans. Des. Autom. Electron. Syst. **20**, 1–28 (2015)

27. Sui, H., Niu, W.: Branch-pipe-routing approach for ships using improved genetic algorithm. Front. Mech. Eng. **11**, 316–323 (2016). https://doi.org/10.1007/s11465-016-0384-z

28. Niu, W., Sui, H., Niu, Y., Cai, K., Gao, W.: Ship pipe routing design using NSGA-II and coevolutionary algorithm. Math. Probl. Eng. **2016**, 1–21 (2016)

29. Liu, L., Liu, Q.: Multi-objective routing of multi-terminal rectilinear pipe in 3D space by MOEA/D and RSMT. In: 2018 3rd International Conference on Advanced Robotics and Mechatronics (ICARM), pp. 462–467 (2018)

30. Jiang, W.Y., Lin, Y., Chen, M., Yu, Y.Y.: A co-evolutionary improved multi-ant colony optimization for ship multiple and branch pipe route design. Ocean Eng. **102**, 63–70 (2015)

31. Ztopuoianu, A.C., et al.: Multi-objective optimal design of obstacle-avoiding two-dimensional Steiner trees with application to ascent assembly engineering. J. Mech. Des. **140**, 061401-1–061401-11 (2018)

32. Wu, H., Xu, S., Zhuang, Z., Liu, G.: X-architecture Steiner minimal tree construction based on discrete differential evolution. In: Liu, Y., Wang, L., Zhao, L., Yu, Z. (eds.) ICNC-FSKD 2019. AISC, vol. 1074, pp. 433–442. Springer, Cham (2020). https://doi.org/10.1007/978-3-030-32456-8_47

33. Byrd, R., Gilbert, J., Nocedal, J.: A trust region method based on interior point techniques for nonlinear programming. Math. Program. **89**(1), 149–185 (2000). https://doi.org/10.1007/PL00011391

34. Parque, V., Miyashita, T.: Bundling n-Stars in polygonal maps. In: 29th IEEE International Conference on Tools with Artificial Intelligence, ICTAI 2017, Boston, MA, USA, 6–8 November 2017, pp. 358–365 (2017)

35. Zăvoianu, A.-C., et al.: On the optimization of 2D path network layouts in engineering designs via evolutionary computation techniques. In: Andrés-Pérez, E., González, L.M., Periaux, J., Gauger, N., Quagliarella, D., Giannakoglou, K. (eds.) Evolutionary and Deterministic Methods for Design Optimization and Control With Applications to Industrial and Societal Problems. CMAS, vol. 49, pp. 307–322. Springer, Cham (2019). https://doi.org/10.1007/978-3-319-89890-2_20

36. Storn, R., Price, K.: Differential evolution - a simple and efficient heuristic for global optimization over continuous spaces. J. Glob. Optim. **11**, 341–359 (1997). https://doi.org/10.1023/A:1008202821328

37. Das, S., Abraham, A., Chakraborty, U.K., Konar, A.: Differential evolution using a neighborhood-based mutation operator. IEEE Trans. Evol. Comput. **13**, 526–553 (2009)

38. Sutton, A.M., Lunacek, M., Whitley, L.D.: Differential evolution and non-separability: using selective pressure to focus search. In: The Genetic and Evolutionary Computation Conference (GECCO), pp. 1428–1435 (2007)

39. Guo, S., Yang, C., Hsu, P., Tsai, J.: Improving differential evolution with a successful-parent-selecting framework. IEEE Trans. Evol. Comput. **19**(5), 717–730 (2015)

40. Qu, B., Liang, J., Suganthan, P.: Niching particle swarm optimization with local search for multi-modal optimization. Inf. Sci. **197**, 131–143 (2012)

41. Jones, D.R.: Direct global optimization algorithm. In: Floudas, C.A., Pardalos, P.M. (eds.) Encyclopedia of Optimization, pp. 431–440. Springer, Boston (2001). https://doi.org/10.1007/0-306-48332-7_93

42. Parque, V., Miura, S., Miyashita, T.: Computing path bundles in bipartite networks. In: Proceedings of the 7th International Conference on Simulation and Modeling Methodologies, Technologies and Applications, pp. 422–427 (2017)

43. Parque, V., Miyashita, T.: Numerical representation of modular graphs. In: IEEE 42nd Annual Computer Software and Applications Conference, pp. 819–820 (2018)

44. Parque, V., Miyashita, T.: On the numerical representation of labeled graphs with self-loops. In: 29th IEEE International Conference on Tools with Artificial Intelligence, pp. 342–349 (2017)

45. Parque, V., Miyashita, T.: On succinct representation of directed graphs. In: IEEE International Conference on Big Data and Smart Computing, pp. 199–205 (2017)

46. Parque, V., Miyashita, T.: On graph representation with smallest numerical encoding. In: IEEE 42nd Annual Computer Software and Applications Conference, pp. 817–818 (2018)

47. Parque, V., Suzaki, W., Miura, S., Torisaka, A., Miyashita, T., Natori, M.: Packaging of thick membranes using a multi-spiral folding approach: flat and curved surfaces. Adv. Space Res. (2020, in press). https://doi.org/10.1016/j.asr.2020.09.040

# Flowshop NEH-Based Heuristic Recommendation

Lucas Marcondes Pavelski[1]([⊠]) [ID], Marie-Éléonore Kessaci[2] [ID],
and Myriam Delgado[3] [ID]

[1] Federal University of Technology - Paraná, Av. Sete de Setembro
- 3165 - 80230-901, Curitiba, Brazil
`lpavelski@alunos.utfpr.edu.br`
[2] Univ. Lille, CNRS, Centrale Lille, UMR 9189 CRIStAL, 59000 Lille, France
`mkessaci@univ-lille.fr`
[3] Federal University of Technology - Paraná, Curitiba, Brazil
`myriamdelg@utfpr.edu.br`

**Abstract.** Flowshop problems (FSPs) have many variants and a broad
set of heuristics proposed to solve them. Choosing the best heuristic and
its parameters for a given FSP instance can be very challenging for practitioners. Per-instance Algorithm Configuration (PIAC) approaches aim at
recommending the best algorithm configuration for a particular instance
problem. This paper presents a PIAC methodology for building models
to automatically configure the Nawaz, Encore, and Ham (NEH) algorithm which proved to be a good choice in most FSP variants (especially
when they are used to provide initial solutions). We use irace to build
the performance dataset (problem features ↔ algorithm configuration),
while training Decision Tree and Random Forest models to recommend
NEH configurations on unseen problems of the test set. Results show that
the recommended heuristics have good performance, especially those by
random forest models considering parameter dependencies.

**Keywords:** Flowshop · Heuristics · Automatic algorithm
configuration · Parameter dependencies · Machine learning

## 1 Introduction

There are different frameworks proposed in the literature to automate producing
state-of-the-art strategies for solving a broad set of problems, ranging from combinatorial optimization to satisfiability and AI planning [28]. Some well-studied
approaches include Automatic Algorithm Configuration (AAC). Practitioners
can use an AAC to search in the algorithm parameter space for the configuration that provides the best performance on a set of instances of the given problem. Per-instance algorithm/parameter selection problem has attracted more
and more interest over the past 15 years. A step further on AAC strategies would
be therefore recommending the best parameters for each target instance using

C. Zarges and S. Verel (Eds.): EvoCOP 2021, LNCS 12692, pp. 136–151, 2021.
https://doi.org/10.1007/978-3-030-72904-2_9

problem features [43], possibly generalizing for unseen instances and gaining knowledge about the problem and algorithms [36].

Flowshop Problems (FSPs) are an example of NP-hard problems that have many variants in the literature [3,45]. A general FSP formulation considers a production line processing a set of jobs sequentially. There are many variants of this problem using different objectives, like the time to process all jobs (makespan), and constraints, like no machine waiting times. Consequently, there are many different algorithms specialized in solving different FSPs. The Nawaz, Encore, and Ham (NEH) algorithm [35] is an example of a generic framework that performs well on makespan objective. Different proposals adapt NEH varying components like insertion order and tie-breaking mechanisms for different FSP variants and objectives [18,19,45].

There are proposals in the literature for automating FSP heuristic generation/recommendation by configuring its components. [20] presents a comprehensive numerical study on different NEH initial orders. It concludes that different initializations improve the state-of-the-art for makespan, idle-time, and total flowtime objectives. [14] uses irace to combine the two-phase local search with Pareto local search components and design heuristics for bi-objective (makespan and total flowtime) FSPs. [5,6] and [32] build a grammar for automatic configuration based on irace, which generates complex iterative algorithms with different NEH variants and stochastic local search components. The differences between [5] and [6] rely mainly on the type of the considered objective with [5] addressing the total completion time and [6] minimizing the makespan. [39] proposes and tests on permutation FSP, a framework that recommends stochastic local search algorithms and their parameters. [1] describes a methodology for using irace in the automatic design of FSP optimization algorithms with multiple parallel machines for makespan, total flowtime, and total weighted earliness and tardiness objectives.

Analysis of flowshop features effects on metaheuristics performance are present in works based on Fitness Landscape Analysis (FLA) [10,22,30,31,41]. Many concepts of these works contribute to our proposal, but they focus on studying the main effects of different instance features, solution representation, and neighborhood operator on flowshop search landscapes. Another closely related topic is hyper-heuristics, where search information is used to select or generate heuristics on the run [8]. Many works apply it to scheduling problems [7,44,52].

This paper expands previous works by addressing different FSPs and using Machine Learning (ML) techniques to recommend various NEH components. In the present work, NEH is chosen not only because it performs well on most of FSP variants, but also because it can provide suitable initial solutions to be further used for other (meta)heuristics, including more complex state-of-the-art algorithms). Most of the literature considers stochastic local search approaches on a limited number of FSPs formulations. Our work proposes an ML-based Per-instance Algorithm Configuration (PIAC), which considers different variants of FSPs and a wide variety of NEH components (over 1.5 million combinations).

Aiming to encompass a relatively wide range of problems, we consider various sets of FSP instances, each with different characteristics: objectives (makespan and total flowtime), variants (permutation, no-idle, and no-wait). We also consider difficult instances from the literature (VRF [49]) and generate others whose processing times better resemble real-world problems. The irace configurator adopted as AAC builds the dataset used to train the ML models. Finally, the recommended heuristics are tested on unseen problems and later compared with the traditional NEH configuration, an irace-tuned NEH with all instances, and a randomly chosen configuration. With that, we intend to contribute to the general understating of FSPs and the NEH algorithm with an empirical study of problems features and how they relate to NEH parameters. Also, by building parameter recommendation models considering parameter dependencies, this work can contribute to the growing interest in using ML for optimization [36].

The paper is organized as follows, Sect. 2 briefly reviews some basic concepts necessary to understand the proposal. Section 3 describes the proposed methodology for NEH heuristic recommendation for solving FSPs. Section 4 presents the results obtained from the proposal. Finally, Sect. 5 highlights some conclusions obtained from the experiments and presents future directions.

## 2    Background

The usual permutation FSP formulation has the following conditions [3]: (1) a set of $J$ unrelated, multiple-operation jobs is available for processing at time zero; (2) each job requires $M$ operations, and each one requires a different machine; (3) setup times for the operations are sequence-independent and included in processing times; (4) job descriptors are available in advance; (5) all machines are continuously available; (6) once an operation begins, it proceeds without interruption. Besides, a FSP instance is given by the processing times, $p_{j,m}$, for all $j = 1, \ldots, J$ and $m = 1, \ldots, M$, the amount of time required to process each job $j$ in each machine $m$.

A sequence is given by a permutation of $(1, 2, \ldots, J)$. As mentioned, each machine can process an available job immediately. In other words, machine 1 starts processing the first job in the sequence at time zero, and machine $m$ can start processing the $j$-th job as soon as job $j - 1$-th is completed on machine $m$ and $j$-th job is completed on machine $m - 1$.

Given the completion times $C_j$ for all jobs $j = 1, \ldots, J$, it is possible to define the makespan objective function as the time to complete all jobs, i.e., $C_{max} = \max\{C_j, j = 1, \ldots J\}$. Another common objective is the total flowtime given by $F = \sum_{j=1}^{J} C_j$.

Some FSPs formulations add new constraints to model real-world scenarios better. The *no-wait* FSP variant includes the requirement that there are no stores between operations, i.e., a process never waits for the next machine. Also, for the *no-idle* FSP variant, the machines have no idle time between job exchanges, which is common in steel production [3].

A pioneer work proposes an exact method for solving two-machine FSP ($M = 2$) [24]. The method iteratively removes the job with the lowest processing time and puts it in the first or the last position, depending on which machine it belongs to (the first or the last machine, respectively). Various methods use the idea behind this algorithm, known as Johnson's rule: a job $i$ precedes job $j$ in an optimal sequence if $\min\{p_{1,i}, p_{2,j}\} \leq \min\{p_{1,j}, p_{2,i}\}$.

Although exact methods are optimal, they may not be computationally efficient in larger instances. The general FSP problem with makespan minimization for $M = 3$ or more machines is NP-Hard [21]. Therefore, there is no efficient way to solve it (if P $\neq$ NP). Consequently, it is common to use heuristics that yield near-optimal solutions for large instances.

There are two basic categories of FSP heuristics: constructive and improvement heuristics. Constructive heuristics iteratively build the final sequence by adding jobs to partial permutations. Otherwise, improvement heuristics start with a full permutation and iteratively improve it, usually with swap and shift operators [15]. In [37], a constructive heuristic sums the jobs processing times on each machine, weighted by a slope order index (i.e., machine $m$ has index $M - (2m - 1)$). After that, it sorts the jobs by a non-increasing order of the resulting sum. Therefore, the heuristic tries to reduce the makespan added by the first and last jobs in the sequence.

Another well known constructive heuristic is CDS [9]. CDS idea is to divide the $M$ machines into two groups and apply the $M = 2$ machines Johnson's algorithm. It does so by considering *virtual machines* formed by the sum of processing times for multiple machines. Usually, CDS evaluates every pair of disjoint sets of machines ($M - 1$ in total). The Rapid Access (RA) heuristic [11] combines Palmer's and CDS heuristics. RA creates two virtual machines using two slope order index weights to aggregate processing times (the virtual machines processing times are $p_{j,1'} = \sum_{m=1}^{M}(M - m + 1)p_{j,m}$ and $p_{j,2'} = \sum_{m=1}^{M} mp_{j,m}$). It then uses Johnson's algorithm to generate the final permutation.

The NEH heuristic [35] is a well known constructive heuristic that works by inserting jobs in partial sequences. The following steps describe the NEH algorithm:

1. **Insertion order:** for each job, calculate the sum of its processing times ($\sum_{m=1}^{M} p_{j,m}$), and sort them by the non-increasing order of that sum;
2. Initialize the partial sequence with the first job and, for $j = 2, \ldots, J$ do:
   (a) **Insertion:** Insert $j$-th job in every position in the partial sequence;
   (b) Select the best partial sequence and continue.

This procedure has a complexity of $O(MJ^3)$, later improved to $O(MJ^2)$ for the makespan objective because, given the previous partial sequences completion times, it is easy to recalculate the makespan from each insertion [47]. Other accelerations are available for no-wait FSP with makespan objective in $O(J)$ [38] and no-idle FSP with makespan and total flowtime in $O(MJ^2)$ [16]. With extensive tests, many proposals found NEH to be one of the best constructive heuristics for FSPs [45]. Thus, many recent heuristics and metaheuristics proposals internally use NEH or some variant of NEH.

Some variants of NEH alter its initialization phase by using some ordering heuristic [20], for example, including the mean and standard deviation to the sum of the processing times on each machine [13]. In some cases, randomness can generate better initial permutations [42]. The Nagano-Mocellin (NM) initialization penalizes jobs in the initial sequence according to a lower bound for its waiting time (time between the end of the operation on machine $j$ and the beginning of the next operation on machine $j + 1$) [33]. Also, inverting the original insertion order yields slightly better results in no-idle FSPs [38]. KK1 heuristic [25] calculates a slope index for each job ($a_j = \sum_{M=1}^{M}((M-1)*(M-2)/2+M-m)p_{j,m}$ and $b_j = \sum_{m=1}^{M}((M-1)*(M-2)/2+m-1)p_{j,m}$) and builds the initial permutation according to the non-increasing order of $min(a_j, b_j)$. KK2 heuristic [26] adds the skewness to the job processing times sum. Some order heuristics for sequence-dependent FSPs draw inspiration from Traveling Salesperson Problem construction procedures [3,46]. Other methods apply known FSP lower-bounds as metrics to construct schedules iteratively [40]. Finally, the Liu-Reeves (LR) heuristic is used mainly for FSPs with the flowtime objective and uses a weighted sum of idle, completion, and artificial job times as an indicator to sort the jobs in a schedule [29,34].

Another aspect of the NEH heuristic explored in the literature is the insertion phase. If different positions have the same fitness, a tie-breaking strategy decides the best position. In KK1 and KK2 [26], a tie-breaking mechanism is added to the insertion phase, using principles of Johnson's rule and Palmer's weighted sums of job times. An insertion strategy based on balancing the utilization among all machines is proposed in [13]. Recently, in [18] a NEH variant with tie-breaking for makespan objective uses estimated idle times.

## 3   Methodology

### 3.1   Problem and Feature Space

As described in Sect. 2, FSPs have different formulations depending on the instance, objective and variant. To build the recommendation database, we need a big enough sample of the problem space while maintaining a reasonable computation time to collect the performance data. In the experiments, we consider our own generated instances: (i) exponential and Erlang instances: with exponential (1/50 rate) and Erlang (with shape 4 and 4/50 rate) distributions, known to better represent practical problems [2,19]; (ii) generated structured instances: with uniform distributions such that the processing times have a correlation of 0.9 between jobs (job-correlated) or machines (machine-correlated), where less complex strategies have better performance [51]. Also, instances selected from the literature: VRF [49] instances, widely used in the permutation FSP, originally generated to be hard for state-of-the-art heuristics.

The instance sizes (job, machines) use ranges from small instances, all combinations of $J = \{10, 20, 30, 40, 50, 60\}$ jobs and $M = \{5, 10, 15, 20\}$ machines, and large instances, with all combinations of $J = \{100, 200, 300, 400, 500\}$ jobs and

$M = \{20, 40, 60\}$ machines. For each instance set and size there are 10 instances samples using different random seeds.

We considered all combinations of the above problem features to build the performance dataset. In total there are (2 objectives × 3 variants × 5 processing time distributions × 39 instance sizes) = 1,170 FSP formulations with 10 instances each.

We use different types of problem features as inputs for the recommendation models. Some simple features come directly from the problem definition, like the number of jobs $(J)$, the number of machines $(M)$, FSP variants (PERM, NOWAIT, NOIDLE), and objectives (MAKESPAN, FLOWTIME). Since, during the recommendation, the practitioner might not know the processing times distribution, we include some calculated statistics, such as standard deviations, skewness and kurtosis per job (machine).

Other advanced instance features consider Fitness Landscape Analysis (FLA) [10]. A fitness landscape consists of a solution representation, a neighborhood connecting them, and the fitness function. In this paper, we compute FLA features for each problem over the space of permutations connected by the insertion neighborhood. The literature proposes FLA metrics mainly to measure problem difficulty and gain insights about neighborhoods and operators. For example, a low correlation of fitness values between neighbor solutions means that the fitness landscape is rugged.

The paper adopts FLA based on solution types and random walk samples. For solution-based FLAs we sample 100 solutions for all problems with $J \leq 100$, 50 solutions when $100 < J \leq 300$ and 30 solutions when $300 < J \leq 500$. By computing the neighborhood, the estimation of different metrics is possible like percentage of strict local minimum (maximum), local minimum (maximum), plateau, ledge, and slopes, as well as the proportion of up, down and side edges [23].

For random walk samples, we collect the objective values from a random walk of $T = 10,000$ steps on each problem fitness landscape. From each sequence of fitness $(f_1, \ldots f_T)$, we can estimate its autocorrelation. Considering a pair $pq$ of fitness in the sequence as up, down and side movements we calculate features such as entropy, partial information, information stability, and density basin [50].

Regarding their computational costs, problem and processing time based features are inexpensive to compute, and the random-walks involve 10,000 fitness evaluations. A neighborhood sweep computation is most costly on permutation flowshop and is proportional to $J^3$ [48]. The NEH insertion phase also involves a neighborhood sweep of the same cost. Therefore the features computed are usually less expensive than testing thousands of NEH configurations.

As shown in Table 1, in total, there are 27 features used as inputs for the proposed recommendation models (4 problem features + 8 features based on processing times + 10 solution statistics + 5 random walk statistics).

## 3.2 Algorithm Space

There are many FSP heuristics available in the literature. Some of them are general, while others are specific for different variants and objectives. Despite that, common heuristics consist of one or more phases like index-development, solution construction, and solution improvement [19].

**Table 1.** Input features for the recommendation models.

| Scope | Feature | Value |
|---|---|---|
| Problem | Number of jobs | $J$ |
| | Number of machines | $M$ |
| | FSP variant | {PERM, NOWAIT, NOIDLE} |
| | Objective | {MAKESPAN, FLOWTIME} |
| Proc. times | Jobs per machines ratio | $J/M$ |
| | Standard deviation | $\sigma[p_{jm}]$ |
| | Standard deviation per job | $E_{j=1\ldots J}[\sigma[p_{jm}]]$ |
| | Standard deviation per machine | $E_{m=1\ldots M}[\sigma[p_{jm}]]$ |
| | Skewness per job | $E_{j=1\ldots J}[((p_{jm}-E[p_{jm}])/\sigma[p_{jm}])^3]]$ |
| | Skewness per machine | $E_{m=1\ldots M}[((p_{jm}-E[p_{jm}])/\sigma[p_{jm}])^3]]$ |
| | Kurtosis per job | $E_{j=1\ldots J}[((p_{jm}-E[p_{jm}])/\sigma[p_{jm}])^4]]$ |
| | Kurtosis per machine | $E_{m=1\ldots M}[((p_{jm}-E[p_{jm}])/\sigma[p_{jm}])^4]]$ |
| Solution statistics | Strict local minimum | $\|\{s:f(s)<f(s')\wedge s'\in N(s)\}\|$ |
| | Local minimum | $\|\{s:f(s)\leq f(s')\wedge s'\in N(s)\}\|$ |
| | Strict local maximum | $\|\{s:f(s)>f(s')\wedge s'\in N(s)\}\|$ |
| | Local maximum | $\|\{s:f(s)\geq f(s')\wedge s'\in N(s)\}\|$ |
| | Plateau | $\|\{s:f(s)=f(s')\wedge s'\in N(s)\}\|$ |
| | Slopes | $\|\{s:f(s)\neq f(s')\wedge s'\in N(s)\}\|$ |
| | Ledge | $\|\{s:s\text{ is none of the above}\}\|$ |
| | Up edges | $\|\{(s,s'):f(s)<f(s')\wedge s'\in N(s)\}\|$ |
| | Down edges | $\|\{(s,s'):f(s)>f(s')\wedge s'\in N(s)\}\|$ |
| | Side edges | $\|\{(s,s'):f(s)=f(s')\wedge s'\in N(s)\}\|$ |
| Random walk statistics | Auto-correlation | $Cov[f_t,f_{t+1}]/\sqrt{Var[f_t]Var[f_{t+1}]}$ |
| | Entropy | $-\sum_{p\neq q}P(pq)\log_6 P(pq)$ |
| | Partial information | $(T-\|\{f_t=f_{t+1}:t=1\ldots T-1\}\|)/T$ |
| | Information stability | $\max_{t=1\ldots T-1}\|f_t-f_{t+1}\|$ |
| | Density basin | $-\sum_{p=q}P(pq)\log_3 P(pq)$ |

Index-development addresses different job-ordering algorithms, from a simple decreasing sum of total processing time to heuristics like LR [29], which involves computing idle-times and artificial jobs processing times. The NEH algorithm [35], using different insertion orders and tie-breaking mechanisms, supports the construction phase. Finally, solution improvement adopts an iterative local search algorithm like Iterated Local Search (ILS) or Iterated Greedy (IG) [23].

This work focuses o the first two phases of index development and solution construction by NEH. These phases might be further complemented/improved using powerful heuristics like ILS and IG. Aiming to produce both job-ordering algorithms and NEH-based constructive heuristics, we propose the following algorithm framework (also shown in Fig. 1):

1. Order $\lfloor J \times IOR \rfloor$ jobs using a given order heuristic;
2. Order $\lceil J \times (1 - IOR) \rceil$ jobs using NEH insertion order;
3. Starting from the ordered jobs, iteratively insert each job from NEH insertion order using the given tie-breaking strategy.

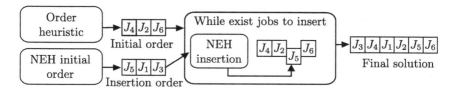

**Fig. 1.** The proposed heuristic framework (an example with $IOR = 0.5$).

Using the initial order ratio $IOR$, it is possible to generate different NEH heuristics that combine index development and solution construction phases. For example, the original NEH uses $IOR = 0$ with a decreasing sum of processing times as NEH initial order and first best insertion strategy. Also, using $IOR = 0.75$, completion time on the last machine as the initial order, sum of processing times as NEH initial order would resemble LR-NEH [34], proposed for minimizing flowtime on no-idle FSPs.

The orderings used in steps 1 and 2 in the framework can consider different indicators based on the problem data (see Table 2 for the list addressed in this paper). All indicators are computed as an aggregation of terms of the jobs processing times on each machine. A weighted version of each indicator (IOW and NOW parameters) is defined by multiplying each aggregated term by $(M - m + 1)$, where $m = 1, \ldots, M$ is the term's machine index, similar to Palmer's and RA heuristics. Different sorting options of each indicator can also be applied (IOS and NOS parameters), as shown in Fig. 2.

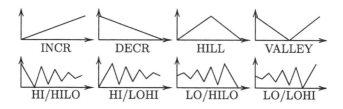

**Fig. 2.** Order indicator sorting orders (based on [20]).

Finally, we consider five different NEH tie-breaking (NTB) mechanisms in NEH insertion phase: first best, last best, NM [33], KK1 and KK2 [26]. In summary, Table 3 describes all parameters of the proposed heuristic. In total there are $1,695,456$ possible configurations for the NEH variants addressed in this work. Regarding irace configuration, the IOR parameter is ordinal ranging from $[0,1]$ with 0.25 steps and the remaining are categorical. When $IOR = 0$ the initial order parameters (IOW, IOS, and IOI) are not sampled by irace. Also, when $IOR = 1$ NEH insertion parameters (NOI, NOS, NOW, NTB) are not sampled.

**Table 2.** Order heuristic and NEH insertion order indicators.

| Name | Ref |
|---|---|
| Sum of job processing times | [35] |
| Sum of the processing times standard deviation | [13] |
| Sum processing times standard deviation, average and skewness | [13] |
| Absolute difference of processing times | [20] |
| Absolute residuals sum | [20,46] |
| Square residuals sum | [20,46] |
| Sum of the negative residuals with negative residual carry-over | [20,46] |
| Sum of the absolute residuals with negative residual carry-over | [20,46] |
| Sum of the squared residuals with negative residual carry-over | [20,46] |
| Sum of the absolute residuals with double negative residuals, no carryover | [20,46] |
| Sum of the lower bounds on the completion times | [20,40] |
| Sum of possible idle times | [20,40] |
| Sum of possible waiting time of jobs and idletime of machines | [20,40] |
| Iterative idle, completion and artificial total flowtime (LR) | [29] |
| Iterative idle, completion times (LR variant) | [29] |
| Iterative idle times (LR variant) | [29] |
| Iterative artificial total flowtime (LR variant) | [29] |
| Iterative completion times (LR variant) | [29,34] |
| Sum of job processing times minus waiting time lower bound (NM1) | [27,33] |
| NEHKK1 | [27] |
| NEHKK2 | [25] |

**Table 3.** Parameters of the proposed approach.

| Parameter | Symbol | Possible values |
|---|---|---|
| Initial order ratio | IOR | 0, 0.25, 0.5, 0.75 or 1.0 |
| Initial order weighted by machine | IOW | Yes or no |
| Initial order indicator sorting criteria | IOS | Figure 2 |
| Initial order indicator | IOI | Table 2 |
| NEH order weighted by machine | NOW | Yes or no |
| NEH order indicator sorting criteria | NOS | Figure 2 |
| NEH order indicator | NOI | Table 2 |
| NEH tie-breaking | NTB | First best, last best, NM, KK1 or KK2 |

### 3.3   Performance Data and Recommendation

To generate performance data, irace (used as AAC) finds the best parameters for each FSP problem, using ten instances generated with different random seeds. Irace runs with its default configuration, in deterministic mode with a maximum budget of 5, 000 evaluations of candidate configurations.

After that, the proposed approach calculates the problem features listed in Sect. 3.1. A pre-processing step removes some features with zero variance and high correlation with others (absolute value above 0.95). Then, for each parameter, a training and test set is built using problem features as inputs and the best parameter values as outputs.

The training phase uses Decision Trees (DT) and Random Forests (RF) [4] to learn how to recommend parameters. The models implementation are taken from the R packages `rpart` v4.1--15 and `ranger` v0.12.1. While the RF number of trees is fixed as 2, 000, this phase considers a simple parameter grid with 10-fold cross-validation to tune different values for DT maximum tree depth (1 to 5), RF number of predictors sampled for each split (1 or 10) and RF number of data points sampled for each split (2 or 40).

The DT and RF models are trained and tested associated with two recommendation strategies. The first strategy considers only the problem features as inputs and tries to predict the best parameter value (no dependencies). The second approach recommends the parameters sequentially, and the output of one recommendation is fed as input to the next parameters (with dependencies). The second strategy aims to model implicit dependencies among the parameters. For example, the NEH tie-breaking mechanism might depend on its initial sequence. The order of recommendations is the one shown in Table 3, starting from *Initial order ratio*, ending with *NEH tie-breaking*.

## 4   Results

We divided the results into two different groups (i) those obtained from a machine learning perspective when we analyze models' predictive qualities; and (ii) those obtained from the optimization perspective when we focus on the best objective function achieved by each recommended strategies. The results from both sections consider the 25% test split of the problems dataset.

### 4.1   Machine-Learning Performance

Figure 3 shows the proposed model results (DT and RF, with and without parameters dependencies) in terms of accuracy (Fig. 3(a)) and F-score (Fig. 3(b)) metrics. Both metrics agree in most cases. DTs are worse on parameters like IOI and NTB but have a good performance on IOS and IOW. RFs and parameter dependency models are slightly better on most problems.

The high performance on the IOR recommendation task is because it is highly unbalanced. In most cases (93%), a ratio of zero is recommended (only NEH phases with no initial sequence). The initial order parameters are only considered for the training set when IOR > 0, and, in this case, the recommendation models have good performance. NEH parameters are more difficult to recommend, like NOI, possibly because the 8 order indicators have similar or negligible effect.

## 4.2   Recommended Heuristics Performance

The quality of recommendation models from the optimization's perspective is measured using the relative performance given by:

$$RP = 100 * (Fitness - BestFitness)/BestFitness$$

averaged for all problems in the test set, where $BestFitness$ is the fitness of the best configuration found by irace to build the dataset.

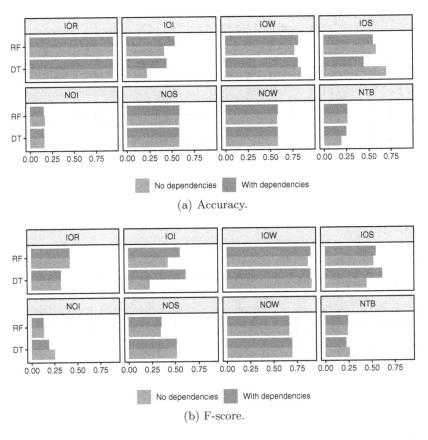

**Fig. 3.** Prediction performance of the recommendation models (higher values indicate better performance).

Figure 4 shows the average relative performance ($ARP$) statistics for the following strategies: standard NEH with original configuration, i.e., a decreasing sum of processing times order and first best insertion (Standard NEH); global best NEH configured by irace with all problems instances and budget of $5,000|P|$ evaluations of candidate configurations (Global best NEH[1]), where $|P|$ is the number of problems; proposed recommendation models (DT and RF) trained and tested with and without parameter dependencies. A random parameter choice (Random) is also compared with these strategies, but its performance is very poor (on average 4.5 times worse than irace reference configurations).

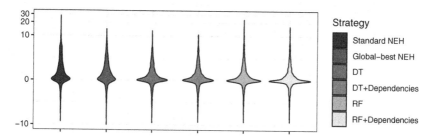

**Fig. 4.** Average relative performance for all strategies compared to the baseline irace best configuration.

From the optimization perspective, it is clear that the random parameter choice is the worst. Secondly, we can observe that choosing a single configuration for multiple types of problems like the standard NEH and global-best NEH is better than a random choice. Nevertheless, in most cases fixing parameters for a set of instances is worst than instance-based recommendations. RF models show the best performances, with more than 25% of configurations that improve upon the best found by irace. This result indicates that the model can produce highly efficient heuristics even on unseen problems.

According to the all-pairs Friedman test [12], RF models with parameter dependencies are statistically better than all of the other strategies. Also, DT models are not improved by parameter dependency are equivalent to RF models without parameter dependency. The source code and extra results regarding DT models interpretability are left available as supplementary material[2].

## 5 Conclusion

This paper has presented a methodology for recommending the Nawaz, Encore, and Ham (NEH) heuristics in the optimization of Flowshop problems (FSP).

---

[1] The global best NEH found by irace uses the hill-sorted absolute difference of processing times with NM tie-breaking strategy.

[2] Source code and supplementary material link: https://github.com/lucasmpavelski/flowshop-neh-based-heuristic-recommendation.

Moreover, by recommending NEH components sequentially, we considered modeling some parameter dependencies. We used irace as an Automatic Algorithm Configuration (AAC) tool to build a training set for a variety of FSP problems. The problems included no-idle and no-wait variants, different processing times distributions, and hard instances used in the literature. After that, we trained decision trees and random forest models using problem features as inputs.

The empirical results show the good performance of the proposed models, mainly from the optimization perspective. Like NEH initial order, some parameters proved to be hard to predict given the addressed problem features. Finally, the random forest's optimization performance with parameter dependencies was comparable to irace best configurations, even on problems unseen during the training phase.

In future works, we intend to improve the methodology using ablation [17], to find robust parameter values for the training dataset and possibly leverage the models' prediction performance. Also, investigate features' importance given by the Random Forest models. Moreover, comprehensive approaches could further incorporate stochastic local search [23] recommendation models to improve the recommended NEH heuristics's initial solutions. Other types of FSP problems, different NEH components, fitness landscape metrics, and machine learning models could also improve the empirical results.

**Acknowledgments.** M. Delgado acknowledges CNPq (a Brazilian research-funding agency) for her partial financial support, grants 309935/2017-2 and 439226/2018-0.

# References

1. Alfaro-Fernández, P., Ruiz, R., Pagnozzi, F., Stützle, T.: Automatic algorithm design for hybrid flowshop scheduling problems. Eur. J. Oper. Res. **282**(3), 835–845 (2020). https://doi.org/10.1016/j.ejor.2019.10.004

2. Baker, K.R., Trietsch, D.: Appendix A: practical processing time distributions. Principles of Sequencing and Scheduling, pp. 445–458. John Wiley & Sons Ltd., Hoboken (2009). https://doi.org/10.1002/9780470451793.app1

3. Baker, K.R., Trietsch, D.: Principles of Sequencing and Scheduling. Wiley Publishing, New Jersey (2009)

4. Breiman, L.: Random forests. Mach. Learn. **45**(1), 5–32 (2001). https://doi.org/10.1023/A:1010933404324

5. Brum, A., Ritt, M.: Automatic algorithm configuration for the permutation flow shop scheduling problem minimizing total completion time. In: Liefooghe, A., López-Ibáñez, M. (eds.) EvoCOP 2018. LNCS, vol. 10782, pp. 85–100. Springer, Cham (2018). https://doi.org/10.1007/978-3-319-77449-7_6

6. Brum, A., Ritt, M.: Automatic design of heuristics for minimizing the makespan in permutation flow shops. In: 2018 IEEE Congress on Evolutionary Computation (CEC), pp. 1–8, July 2018. https://doi.org/10.1109/CEC.2018.8477787

7. Burcin Ozsoydan, F., Sağir, M.: Iterated greedy algorithms enhanced by hyper-heuristic based learning for hybrid flexible flowshop scheduling problem with sequence dependent setup times: a case study at a manufacturing plant. Comput. Oper. Res. **125**, 105044 (2021). https://doi.org/10.1016/j.cor.2020.105044

8. Burke, E.K., Hyde, M.R., Kendall, G., Ochoa, G., Özcan, E., Woodward, J.R.: A classification of hyper-heuristic approaches: revisited. In: Gendreau, M., Potvin, J.-Y. (eds.) Handbook of Metaheuristics. ISORMS, vol. 272, pp. 453–477. Springer, Cham (2019). https://doi.org/10.1007/978-3-319-91086-4_14

9. Campbell, H.G., Dudek, R.A., Smith, M.L.: A heuristic algorithm for the n job, m machine sequencing problem. Manage. Sci. **16**(10), B630–B637 (1970)

10. Czogalla, J., Fink, A.: Fitness landscape analysis for the no-wait flow-shop scheduling problem. J. Heuristics **18**(1), 25–51 (2012). https://doi.org/10.1007/s10732-010-9155-x

11. Dannenbring, D.G.: An evaluation of flow shop sequencing heuristics. Manage. Sci. **23**(11), 1174–1182 (1977). https://doi.org/10.1287/mnsc.23.11.1174

12. Demšar, J.: Statistical comparisons of classifiers over multiple data sets. J. Mach. Learn. Res. **7**(1), 1–30 (2006)

13. Dong, X., Huang, H., Chen, P.: An improved NEH-based heuristic for the permutation flowshop problem. Comput. Oper. Res. **35**(12), 3962–3968 (2008). https://doi.org/10.1016/j.cor.2007.05.005

14. Dubois-Lacoste, J., López-Ibáñez, M., Stützle, T.: Automatic configuration of state-of-the-art multi-objective optimizers using the TP+PLS framework. In: Proceedings of the 13th annual conference on Genetic and evolutionary computation. GECCO 2011, pp. 2019–2026. Association for Computing Machinery, New York, NY, USA, July 2011. https://doi.org/10.1145/2001576.2001847

15. Emmons, H., Vairaktarakis, G.: Theoretical results, algorithms, and applications. In: Flow Shop Scheduling. International Series in Operations Research & Management Science, vol. 182, 11th edn. Springer, New York (2013). https://doi.org/10.1007/978-1-4614-5152-5

16. Fatih Tasgetiren, M., Pan, Q.K., Suganthan, P.N., Buyukdagli, O.: A variable iterated greedy algorithm with differential evolution for the no-idle permutation flowshop scheduling problem. Comput. Oper. Res. **40**(7), 1729–1743 (2013). https://doi.org/10.1016/j.cor.2013.01.005

17. Fawcett, C., Hoos, H.H.: Analysing differences between algorithm configurations through ablation. J. Heuristics **22**(4), 431–458 (2016). https://doi.org/10.1007/s10732-014-9275-9

18. Fernandez-Viagas, V., Framinan, J.M.: On insertion tie-breaking rules in heuristics for the permutation flowshop scheduling problem. Comput. Oper. Res. **45**, 60–67 (2014). https://doi.org/10.1016/j.cor.2013.12.012

19. Framinan, J.M., Gupta, J.N.D., Leisten, R.: A review and classification of heuristics for permutation flow-shop scheduling with makespan objective. J. Oper. Res. Soc. **55**(12), 1243–1255 (2004). https://doi.org/10.1057/palgrave.jors.2601784

20. Framinan, J.M., Leisten, R., Rajendran, C.: Different initial sequences for the heuristic of Nawaz, Enscore and Ham to minimize makespan, idletime or flowtime in the static permutation flowshop sequencing problem. Int. J. Prod. Res. **41**(1), 121–148 (2003). https://doi.org/10.1080/00207540210161650

21. Garey, M.R., Johnson, D.S., Sethi, R.: The complexity of flowshop and jobshop scheduling. Math. Oper. Res. **1**(2), 117–129 (1976)

22. Hernando, L., Daolio, F., Veerapen, N., Ochoa, G.: Local optima networks of the permutation flowshop scheduling problem: makespan vs. total flow time. In: 2017 IEEE Congress on Evolutionary Computation (CEC), pp. 1964–1971. IEEE, San Sebastian, Spain, June 2017. https://doi.org/10.1109/CEC.2017.7969541

23. Hoos, H.H., Stützle, T.: Stochastic Local Search: Foundations and Applications. Elsevier, San Francisco, USA (2004)

24. Johnson, S.M.: Optimal two- and three-stage production schedules with setup times included. Naval Res. Logistics Q. **1**(1), 61–68 (1954). https://doi.org/10.1002/nav.3800010110

25. Kalczynski, P.J., Kamburowski, J.: An improved NEH heuristic to minimize makespan in permutation flow shops. Comput. Oper. Res. **35**(9), 3001–3008 (2008). https://doi.org/10.1016/j.cor.2007.01.020

26. Kalczynski, P.J., Kamburowski, J.: An empirical analysis of the optimality rate of flow shop heuristics. Eur. J. Oper. Res. **198**(1), 93–101 (2009). https://doi.org/10.1016/j.ejor.2008.08.021

27. Kalczynski, P.J., Kamburowski, J.: On the NEH heuristic for minimizing the makespan in permutation flow shops. Omega **35**(1), 53–60 (2007). https://doi.org/10.1016/j.omega.2005.03.003

28. Kerschke, P., Hoos, H.H., Neumann, F., Trautmann, H.: Automated algorithm selection: survey and perspectives. Evol. Comput. **27**(1), 3–45 (2019). https://doi.org/10.1162/evco_a_00242

29. Liu, J., Reeves, C.R.: Constructive and composite heuristic solutions to the $P//\sum C_i$ scheduling problem. Eur. J. Oper. Res. **132**(2), 439–452 (2001). https://doi.org/10.1016/S0377-2217(00)00137-5

30. Marmion, M.-E., Dhaenens, C., Jourdan, L., Liefooghe, A., Verel, S.: On the neutrality of flowshop scheduling fitness landscapes. In: Coello, C.A.C. (ed.) LION 2011. LNCS, vol. 6683, pp. 238–252. Springer, Heidelberg (2011). https://doi.org/10.1007/978-3-642-25566-3_18

31. Marmion, M.-E., Regnier-Coudert, O.: Fitness landscape of the factoradic representation on the permutation flowshop scheduling problem. In: Dhaenens, C., Jourdan, L., Marmion, M.-E. (eds.) LION 2015. LNCS, vol. 8994, pp. 151–164. Springer, Cham (2015). https://doi.org/10.1007/978-3-319-19084-6_14

32. Mascia, F., López-Ibáñez, M., Dubois-Lacoste, J., Stützle, T.: Grammar-based generation of stochastic local search heuristics through automatic algorithm configuration tools. Comput. Oper. Res. **51**, 190–199 (2014). https://doi.org/10.1016/j.cor.2014.05.020

33. Nagano, M.S., Moccellin, J.V.: A high quality solution constructive heuristic for flow shop sequencing. J. Oper. Res. Soc. **53**(12), 1374–1379 (2002)

34. Nagano, M.S., Rossi, F.L., Martarelli, N.J.: High-performing heuristics to minimize flowtime in no-idle permutation flowshop. Eng. Optim. **51**(2), 185–198 (2019). https://doi.org/10.1080/0305215X.2018.1444163

35. Nawaz, M., Enscore, E.E., Ham, I.: A heuristic algorithm for the m-machine, n-job flow-shop sequencing problem. Omega **11**(1), 91–95 (1983). https://doi.org/10.1016/0305-0483(83)90088-9

36. Ochoa, G., Herrmann, S.: Perturbation strength and the global structure of QAP fitness landscapes. In: Auger, A., Fonseca, C.M., Lourenço, N., Machado, P., Paquete, L., Whitley, D. (eds.) PPSN 2018. LNCS, vol. 11102, pp. 245–256. Springer, Cham (2018). https://doi.org/10.1007/978-3-319-99259-4_20

37. Palmer, D.S.: Sequencing jobs through a multi-stage process in the minimum total time–a quick method of obtaining a near optimum. J. Oper. Res. Soc. **16**(1), 101–107 (1965). https://doi.org/10.1057/jors.1965.8

38. Pan, Q.K., Wang, L., Zhao, B.H.: An improved iterated greedy algorithm for the no-wait flow shop scheduling problem with makespan criterion. Int. J. Adv. Manuf. Technol. **38**(7), 778–786 (2008). https://doi.org/10.1007/s00170-007-1120-y

39. Pavelski, L.M., Delgado, M.R., Kessaci, M.É.: Meta-learning on flowshop using fitness landscape analysis. In: Proceedings of the Genetic and Evolutionary Computation Conference. GECCO 2019, pp. 925–933. ACM, New York, NY, USA (2019). https://doi.org/10.1145/3321707.3321846
40. Rajendran, C.: Heuristic algorithm for scheduling in a flowshop to minimize total flowtime. Int. J. Prod. Econ. **29**(1), 65–73 (1993). https://doi.org/10.1016/0925-5273(93)90024-F
41. Reeves, C.: Landscapes, operators and heuristic search. Ann. Oper. Res. **86**, 473–490 (1999). https://doi.org/10.1023/A:1018983524911
42. Ribas, I., Companys, R., Tort-Martorell, X.: Comparing three-step heuristics for the permutation flow shop problem. Comput. Oper. Res. **37**(12), 2062–2070 (2010). https://doi.org/10.1016/j.cor.2010.02.006
43. Rice, J.R.: The algorithm selection problem. In: Rubinoff, M., Yovits, M.C. (eds.) Advances in Computers, Advances in Computers, vol. 15, pp. 65–118. Elsevier, Washington, DC, USA (1976). https://doi.org/10.1016/S0065-2458(08)60520-3, iSSN: 0065-2458
44. Rodriguez, J.A.V., Petrovic, S., Salhi, A.: A combined meta-heuristic with hyper-heuristic approach to the scheduling of the hybrid flow shop with sequence dependent setup times and uniform machines. In: Baptiste, P., Kendall, G., Munier-Kordon, A., Sourd, F. (eds.) In proceedings of the 3rd Multidisciplinary International Conference on Scheduling : Theory and Applications (MISTA 2007), pp. 506–513. Paris, France (2007), issue: 0
45. Ruiz, R., Maroto, C.: A comprehensive review and evaluation of permutation flow-shop heuristics. Eur. J. Oper. Res. **165**(2), 479–494 (2005). https://doi.org/10.1016/j.ejor.2004.04.017
46. Stinson, J.P., Smith, A.W.: A heuristic programming procedure for sequencing the static flowshop. Int. J. Prod. Res. **20**(6), 753–764 (1982). https://doi.org/10.1080/00207548208947802
47. Taillard, É.: Some efficient heuristic methods for the flow shop sequencing problem. Eur. J. Oper. Res. **47**(1), 65–74 (1990). https://doi.org/10.1016/0377-2217(90)90090-X
48. Taillard, É.: Benchmarks for basic scheduling problems. Eur. J. Oper. Res. **64**(2), 278–285 (1993). https://doi.org/10.1016/0377-2217(93)90182-M
49. Vallada, E., Ruiz, R., Framinan, J.M.: New hard benchmark for flowshop scheduling problems minimising makespan. Eur. J. Oper. Res. **240**(3), 666–677 (2015). https://doi.org/10.1016/j.ejor.2014.07.033
50. Vassilev, V.K., Fogarty, T.C., Miller, J.F.: Information characteristics and the structure of landscapes. Evol. Comput. **8**(1), 31–60 (2000)
51. Watson, J.P., Barbulescu, L., Howe, A.E., Whitley, L.D.: Algorithm performance and problem structure for flow-shop scheduling. In: AAAI/IAAI, pp. 688–695. American Association for Artificial Intelligence, Menlo Park, CA, USA (1999)
52. Yahyaoui, H., Krichen, S., Derbel, B., Talbi, E.G.: A hybrid ILS-VND based hyper-heuristic for permutation flowshop scheduling problem. Procedia Comput. Sci. **60**, 632–641 (2015). https://doi.org/10.1016/j.procs.2015.08.199

# Stagnation Detection with Randomized Local Search

Amirhossein Rajabi$^{(\boxtimes)}$ and Carsten Witt

Technical University of Denmark, Kgs. Lyngby, Denmark
{amraj,cawi}@dtu.dk

**Abstract.** Recently a mechanism called stagnation detection was proposed that automatically adjusts the mutation rate of evolutionary algorithms when they encounter local optima. The so-called SD-(1+1) EA introduced by Rajabi and Witt (GECCO 2020) adds stagnation detection to the classical (1+1) EA with standard bit mutation, which flips each bit independently with some mutation rate, and raises the mutation rate when the algorithm is likely to have encountered local optima.

In this paper, we investigate stagnation detection in the context of the $k$-bit flip operator of randomized local search that flips $k$ bits chosen uniformly at random and let stagnation detection adjust the parameter $k$. We obtain improved runtime results compared to the SD-(1+1) EA amounting to a speed-up of up to $e = 2.71\ldots$ Moreover, we propose additional schemes that prevent infinite optimization times even if the algorithm misses a working choice of $k$ due to unlucky events. Finally, we present an example where standard bit mutation still outperforms the local $k$-bit flip with stagnation detection.

**Keywords:** Randomized search heuristics · Local search · Self-adjusting algorithms · Multimodal functions · Runtime analysis

## 1 Introduction

Evolutionary Algorithms (EAs) are parameterized algorithms, so it has been ongoing research to discover how to choose their parameters best. Static parameter settings are not efficient for a wide range of problems. Also, given a specific problem, there might be different scenarios during the optimization, which results in inefficiency of one static parameter configuration for the whole run. Self-adjusting mechanisms address this issue as a non-static parameter control framework that can learn acceptable or even near-optimal parameter settings on the fly. See also the survey article [6] for a detailed coverage of static and non-static parameter control.

Many studies have been conducted on frameworks which adjust the mutation rate of different mutation operators, in particular in the standard bit mutation

Supported by a grant from the Danish Council for Independent Research (DFF-FNU 8021-00260B).

C. Zarges and S. Verel (Eds.): EvoCOP 2021, LNCS 12692, pp. 152–168, 2021.
https://doi.org/10.1007/978-3-030-72904-2_10

for the search space of bit strings $\{0, 1\}^n$ to make the rate efficient on unimodal functions. For example, the $(1 + (\lambda, \lambda))$ GA using the 1/5-rule can adjust its mutation strength (and also its crossover rate) on ONEMAX [5], resulting in asymptotic speed-ups compared to static settings. Likewise, the self-adjusting mechanism in the $(1+\lambda)$ EA with two rates proposed in [9] performs on unimodal functions as efficiently as the best $\lambda$-parallel unary unbiased black-box algorithm.

The self-adjusting frameworks mentioned above are mainly designed to optimize unimodal functions. Generally, they are not able to suggest an efficient parameter setting where algorithms get stuck in a local optimum since they mainly work based on the number of successes, so there is no signal in such a situation. On multimodal functions, where some specific numbers of bits have to flip to make progress, *Stagnation Detection* (SD) introduced in [17] can overcome local optima in efficient time. This module can be added to most of the existing algorithms to leave local optima without any significant increase of the optimization time of unimodal (sub)problems. To our knowledge, no study has put forward other runtime analyses of self-adjusting mechanisms on multimodal functions. However, in a broader context of mutation-based randomized search heuristics, the heavy-tailed mutation presented in [11] has been able to leave a local optimum in a much more efficient time than the standard bit mutation does. Moreover, in the context of artificial immune systems [2] and hyperheuristics [13], there are proofs that specific search operators and selection of low-level heuristics can speed up multimodal optimization compared to the classical mutation operators.

Recent theoretical research on evolutionary algorithms in discrete search spaces mainly considers global mutations which can create all possible points in one iteration. These mutations have been functional in optimization scenarios where information about the difficulties of the local optima is not available. For example, the standard bit mutation which flips each bit independently with a non-zero probability can produce any point in the search space. However, local mutations can only create a fixed set of offspring points. The 1-bit flip mutation that often can be found in the Randomized Local Search algorithm (RLS) can only reach a limited number of search points, which results in being stuck in a local optimum with the elitist selection. Nevertheless, local mutations may outperform global mutations on unimodal functions and multimodal functions with known gap sizes. It is of special interest to use advantages of local mutations on unimodal (sub)functions additionally to overcome local optima efficiently.

This paper investigates $k$-bit flip mutation as a local mutation in the context of the above-mentioned stagnation detection mechanism. This mechanism detects when the algorithm is stuck in a local optimum and gradually increases mutation strength (i. e., the number of flipped bits) to a value the algorithm needs to leave the local optimum. Similarly, we aim to show that the algorithms using $k$-bit flip can use stagnation detection to tune the parameter $k$. One of the key benefits of such algorithms is using the efficiency of RLS, which performs very well on unimodal (sub)problems without fear of infinite running time in local optima. An additional advantage of using $k$-bit flip mutation accompanied by stagnation detection is that it overcomes local optima more efficiently than global mutations. Moreover, the outcome points out the advantages and practicability of our self-adjusting approach that makes local-mutation algorithms able to optimize functions that have been intractable to solve so far.

We propose two algorithms combining stagnation detection with local mutations. The first algorithm called SD-RLS gradually increases the mutation strength when the current strength has been unsuccessful in finding improvements for a significantly long time. In the most extreme case, the strength ends at $n$, i. e., mutations flipping all bits. With high probability, SD-RLS has a runtime that is by a factor of $\left(\frac{ne}{m}\right)^m / \binom{n}{m}$ (up to lower-order terms) smaller on functions with Hamming gaps of size $m$ than the SD-$(1+1)$ EA previously considered in [17]. This improvement is especially strong for small $m$ and amounts to a factor of $e$ on unimodal functions. Although it is unlikely that the algorithm fails to find an improvement when the current strength allows this, there is a risk that this algorithm misses the "right" strength and therefore it can have infinite expected runtime. To address this, we propose a second algorithm called SD-RLS* that repeatedly loops over all smaller strengths than the last attempted one when it fails to find an improvement. This results in expected finite optimization time on all problems and only increases the typical runtime by lower-order terms compared to SD-RLS. We also observe that the algorithms we obtain can still follow the same search trajectory as the classical RLS when one-bit flips are sufficient to make improvements. In those cases, well-established techniques for the analysis of RLS like the fitness-level method carry over to our variant enhanced with stagnation detection. This is not necessarily the case in related approaches like variable neighborhood search [12] and quasirandom evolutionary algorithms [8] both of which employ more determinism and do not generally follow the trajectory of RLS.

We shall investigate the two suggested algorithms on unimodal functions and functions with local optima of different so-called gap sizes, corresponding to the number of bits that need to be flipped to escape from the optima. Many results are obtained following the analysis of the SD-$(1+1)$ EA [17] which uses a global operator with self-adjusted mutation strength. In fact, often the general proof structure could be taken over almost literally but with improved overall bounds. In conclusion, the self-adjusting local mutation seems to be the preferred alternative to the SD-$(1+1)$ EA with global mutation. However, we will also investigate carefully chosen scenarios where global mutations are superior.

This paper is structured as follows: in Sect. 2, we state the classical RLS algorithm and introduce our self-adjusting variants with stagnation detection; moreover, we collect important mathematical tools. Section 3 shows runtime results for the simpler variant SD-RLS, concentrating on the probability of leaving local optima, while Sect. 4 gives a more detailed analysis of the variant SD-RLS* on benchmark functions like ONEMAX and JUMP. Section 5 analyzes an example function which the standard $(1+1)$ EA with standard bit mutation can solve in polynomial time with high probability whereas the $k$-bit flip mutation with stagnation detection needs exponential time. Through improved upper bounds, we give in Sect. 6 indications for that our approach may also be superior to static settings on instances of the minimum spanning tree problem. This problem and other scenarios are investigated experimentally in Sect. 7 before we finally conclude the paper. Due to space restrictions, several proofs had to be omitted from this paper and have been replaced by proof sketches, but note that these proofs can be found in the preprint [18].

## 2 Preliminaries

### 2.1 Algorithms

In this paper, we consider pseudo-boolean functions $f \colon \{0,1\}^n \to \mathbb{R}$ that w. l. o. g. are to be maximized. One of the first randomized search heuristics studied in the literature is *randomized local search* (RLS) [4] displayed in Algorithm 1. This heuristic starts with a random search point and then repeats mutating the point by flipping $s$ uniformly chosen bits (without replacement) and replacing it with the offspring if it is not worse than the parent.

---

**Algorithm 1.** RLS with static strength $s$

---
Select $x$ uniformly at random from $\{0,1\}^n$
**for** $t \leftarrow 1, 2, \ldots$ **do**
    Create $y$ by flipping $s$ bit(s) in a copy of $x$.
    **if** $f(y) \geq f(x)$ **then**
        $x \leftarrow y$.

---

The *runtime* or the *optimization time* of a heuristic on a function $f$ is the first point time $t$ where a search point of maximal fitness has been created; often the expected runtime, i. e., the expected value of this time, is analyzed.

Theoretical research on evolutionary algorithms mainly studies algorithms on simple unimodal well-known benchmark problems like

$$\text{ONEMAX}(x_1, \ldots, x_n) := |x|_1,$$

but also on the multimodal $\text{JUMP}_m$ function with gap size $m$ defined as follows:

$$\text{JUMP}_m(x_1, \ldots, x_n) = \begin{cases} m + |x|_1 & \text{if } |x|_1 \leq n - m \text{ or } |x|_1 = n \\ n - |x|_1 & \text{otherwise} \end{cases}$$

The mutation used in RLS is a local mutation as it only produces a limited number of offspring. This mutation, which we call $s$-flip in the following (in the introduction, we used the classical name $k$-bit flip), flips exactly $s$ bits randomly chosen from the bit string of length $n$, so for any point $x \in \{0,1\}^n$, RLS can just sample from $\binom{n}{s}$ possible points. As a result, $s$-flip is often more efficient compared to global mutations when we know the difficulty of making progress since the algorithm just looks at a certain part of the search space. To be more precise, we recall the so-called *gap* of the point $x \in \{0,1\}^n$ defined in [17] as the minimum Hamming distance to points with the strictly larger fitness function value. Formally,

$$\text{gap}(x) := \min\{H(x,y) : f(y) > f(x), y \in \{0,1\}^n\}.$$

It is not possible to make progress by flipping less than $\text{gap}(x)$ bits of the current search point $x$. However, if the algorithm uses the $s$-flip with $s = \text{gap}(x)$, it can make progress with a positive probability. In addition, on unimodal functions where the gap of all points in the search space (except for global optima) is one, the algorithm makes progress with strength $s = 1$.

Nevertheless, understanding the difficulty of a local optimum has not generally been possible so far, and benefiting from domain knowledge to use it to determine the strength is not always feasible in the perspective of black-box optimization. Therefore, despite the advantages of $s$-flip, global mutations, e. g. standard bit mutation, which can produce any point in the search space, have been used in the literature frequently. For example, the (1+1) EA that uses a similar approach to Algorithm 1 benefits from standard bit mutation that implicitly uses the binomial distribution to determine how many bits must flip. Consequently, even if the algorithm uses strength 1 (e.g. mutation rate $1/n$), with a positive probability, the algorithm can escape from any local optimum.

We study the search and success probability of Algorithm 1 and its relation to stagnation detection more closely. With similar arguments as presented in [17], if the gap of the current search point is 1 then the algorithm makes an improvement with probability $1/R$ at strength 1, and the probability of not finding it in $n \ln R$ steps is at most $(1 - 1/n)^{n \ln R} \leq 1/R$ (where $R$ is a parameter to be discussed). Similarly, the probability of not finding an improvement for a point with gap of $k$ within $\binom{n}{k} \ln R$ steps is at most

$$\left(1 - \frac{1}{\binom{n}{k}}\right)^{\binom{n}{k} \ln R} \leq \frac{1}{R}.$$

Hence, after $\binom{n}{k} \ln R$ steps without improvement there is a probability of at least $1 - 1/R$ that no improvement at Hamming distance $k$ exists, so for enough large $R$ the probability of failing is small.

We consider this idea to develop the first algorithm. We add the stagnation detection mechanism to RLS to manage the strength $s$. As shown in Algorithm 2, hereinafter called SD-RLS, the initial strength is 1. Also, there is a counter $u$ for counting the number of unsuccessful steps to find the next after the last success. When the counter exceeds the threshold of $\binom{n}{s} \ln R$, strength $s$ is increased by one, and when the algorithm makes progress, the counter and strength are reset to their initial values. In the case that the algorithm is failed to have a success where the strength is equal to the gap of the current search point, the algorithm misses the chance of making progress. Therefore, with probability $1/R$, the optimization time would be infinite. Choosing a large enough $R$ to have an overwhelming large probability of making progress could be a solution to this problem. However, we propose another algorithm that resolves this issue, although the running time is not always as efficient as with Algorithm 2.

In Algorithm 3, hereinafter called SD-RLS*, we introduce a new variable $r$ called radius. This parameter determines the largest Hamming distance from the current search point that algorithm must investigate. In details, when the radius becomes $r$, the algorithm starts with strength $r$ (i.e., $s = r$) and when the threshold is exceeded, it decreases the strength by one as long as the strength is greater than 1. This results in a more robust behavior. In the case that the threshold exceeds and the current strength is 1, the radius is increased by one to cover a more expanded space. Also, when the radius exceeds $n/2$, the algorithm increases the radius to $n$, which means that the algorithm covers all possible strengths between 1 and $n$. We note that the strategy of repeatedly returning to lower strengths remotely resembles the 1/5-rule with rollbacks proposed in [1].

---

**Algorithm 2.** RLS with stagnation detection (SD-RLS)

---

Select $x$ uniformly at random from $\{0,1\}^n$ and set $s_1 \leftarrow 1$.
$u \leftarrow 0$.
**for** $t \leftarrow 1, 2, \ldots$ **do**
    Create $y$ by flipping $s_t$ bits in a copy of $x$ uniformly.
    $u \leftarrow u + 1$.
    **if** $f(y) > f(x)$ **then**
        $x \leftarrow y$.
        $s_{t+1} \leftarrow 1$.
        $u \leftarrow 0$.
    **else if** $f(y) = f(x)$ **and** $s_t = 1$ **then**
        $x \leftarrow y$.
    **if** $u > \binom{n}{s} \ln R$ **then**
        $s_{t+1} \leftarrow \min\{s_t + 1, n\}$.
        $u \leftarrow 0$.
    **else**
        $s_{t+1} \leftarrow s_t$.

---

The parameter $R$ represents the probability of failing to find an improvement at the "right" strength. More precisely, as we will see in Theorem 1 and Lemma 2 (for SD-RLS and SD-RLS*, respectively), the probability of not finding an improvement where there is a potential of making progress is at most $1/R$. We recommend $R \geq |\text{Im } f|$ for SD-RLS (where $\text{Im } f$ is the image set of $f$), and for a constant $\epsilon$, $R \geq n^{3+\epsilon} \cdot |\text{Im } f|$ for SD-RLS*, resulting in that the probability of ever missing an improvement at the right strength is sufficiently small throughout the run.

---

**Algorithm 3.** RLS with robust stagnation detection (SD-RLS*)

---

Select $x$ uniformly at random from $\{0,1\}^n$ and set $r_1 \leftarrow 1$ and $s_1 \leftarrow 1$.
$u \leftarrow 0$.
**for** $t \leftarrow 1, 2, \ldots$ **do**
    Create $y$ by flipping $s_t$ bits in a copy of $x$ uniformly.
    $u \leftarrow u + 1$.
    **if** $f(y) > f(x)$ **then**
        $x \leftarrow y$.
        $s_{t+1} \leftarrow 1$.
        $r_{t+1} \leftarrow 1$.
        $u \leftarrow 0$.
    **else if** $f(y) = f(x)$ **and** $r_t = 1$ **then**
        $x \leftarrow y$.
    **if** $u > \binom{n}{s_t} \ln R$ **then**
        **if** $s_t = 1$ **then**
            **if** $r_t < n/2$ **then** $r_{t+1} \leftarrow r_t + 1$ **else** $r_{t+1} \leftarrow n$
            $s_{t+1} \leftarrow r_{t+1}$.
        **else**
            $r_{t+1} \leftarrow r_t$.
            $s_{t+1} \leftarrow s_t - 1$.
        $u \leftarrow 0$.
    **else**
        $s_{t+1} \leftarrow s_t$.
        $r_{t+1} \leftarrow r_t$.

---

## 2.2   Mathematical Tools

The following lemma containing some combinatorial inequalities will be used in the analyses of the algorithms. The first part of the lemma seems to be well known and has already been proved in [14] and is also a consequence of Lemma 1.10.38 in [3]. The second part follows from elementary manipulations.

**Lemma 1.** *For any integer $m \leq n/2$, we have*

*(a)* $\sum_{i=1}^{m} \binom{n}{i} \leq \frac{n-(m-1)}{n-(2m-1)} \binom{n}{m}$,
*(b)* $\binom{n}{M} \leq \binom{n}{m} \left(\frac{n-m}{m}\right)^{M-m}$ *for $m < M < n/2$.*

## 3   Analysis of the Algorithm SD-RLS

In this section, we study the first algorithm called SD-RLS, see Algorithm 2. In the beginning of the section, we show upper and lower bounds on the time for escaping from local optima. Then, in Theorem 2, we show the important result that on unimodal functions, SD-RLS with probability $1 - |\text{Im } f|/R$ behaves in the same way as RLS with strength 1, including the same asymptotic bound on the expected optimization time.

The following theorem shows the time SD-RLS takes with probability $1 - 1/R$ to make progress of search point $x$ with a gap of $m$.

**Theorem 1.** *Let $x \in \{0,1\}^n$ be the current search point of SD-RLS on a pseudo-boolean function $f\colon \{0,1\}^n \to \mathbb{R}$. Define $T_x$ as the time to create a strict improvement if $gap(x) = m$. Let $U$ be the event of finding an improvement at Hamming distance $m$. Then, we have*

$$\mathrm{E}\,(T_x \mid U) \leq \begin{cases} \binom{n}{m}(1 + O(\frac{m \ln R}{n})) & \text{if } m = o(n), \\ O\left(\binom{n}{m} \ln R\right) & \text{if } m = \Theta(n) \wedge m < n/2, \\ O(2^n \ln R) & \text{if } m \geq n/2. \end{cases}$$

*Moreover, $\Pr(U) \geq 1 - 1/R$.*

Compared to the corresponding theorems in [17], the bounds in Theorem 1 are by a factor of $(\frac{ne}{m})^m/\binom{n}{m}$ (up to lower-order terms) smaller. This speedup is roughly $e$ for $m = 1$, i.e., unimodal functions (like ONEMAX) but becomes less pronounced for larger $m$ since, intuitively, the number of flipped bits in a standard bit mutation will become more and more concentrated and start resembling the $m$-bit flip mutation.

*Proof.* The algorithm SD-RLS can make an improvement only where the current strength $s$ is equal to $m$ and the probability of not finding an improvement during this phase is

$$\left(1 - \binom{n}{m}^{-1}\right)^{\binom{n}{m} \ln R} \leq \frac{1}{R}.$$

If the improvement event happens, the running time of the algorithm to escape from this local optimum is

$$\mathrm{E}\,(T_x \mid U) < \underbrace{\sum_{i=1}^{m-1} \binom{n}{i} \ln R}_{=:S_1} + \underbrace{\binom{n}{m}}_{=:S_2},$$

where $S_1$ is the number of iterations for $s < m$ and $S_2$ is the expected number of iterations needed to make an improvement where $s = m$.

By using Lemma 1 for $m < n/2$, we have

$$\mathrm{E}\,(T_x \mid U) < \sum_{i=1}^{m-1} \binom{n}{i} \ln R + \binom{n}{m} < \frac{n-m+2}{n-2m+3}\binom{n}{m-1} \ln R + \binom{n}{m}$$

$$= \frac{n-m+2}{n-2m+3} \cdot \frac{m}{n-m+1}\binom{n}{m} \ln R + \binom{n}{m}$$

$$= \binom{n}{m}\left(\frac{n-m+2}{n-2m+3} \cdot \frac{m}{n-m+1} \ln R + 1\right),$$

and for $m \geq n/2$, we know that $\sum_{i=1}^{n} \binom{n}{i} < 2^n$, so we can compute

$$\mathrm{E}\,(T_x \mid U) = \sum_{i=1}^{m-1} \binom{n}{i} \ln R + \binom{n}{m} \leq O(2^n \ln R)$$

Altogether we achieve

$$
\mathrm{E}\left(T_x \mid U\right) \leq \begin{cases} \binom{n}{m}(1 + O(\frac{m \ln R}{n})) & \text{if } m = o(n), \\ O\left(\binom{n}{m} \ln R\right) & \text{if } m = \Theta(n) \wedge m < n/2, \\ O(2^n \ln R) & \text{if } m \geq n/2. \end{cases}
$$

$\square$

Using the previous lemma, we obtain the following result that allows us to reuse existing results for RLS on unimodal functions.

**Theorem 2.** *Let* $f \colon \{0,1\}^n \to \mathbb{R}$ *be a unimodal function and consider SD-RLS with* $R \geq |\mathrm{Im}\, f|$. *Then, with probability at least* $1 - \frac{|\mathrm{Im}\, f|}{R}$, *the SD-RLS never increases the radius and behaves stochastically like RLS before finding an optimum of* $f$.

With these two general results, we conclude the analysis of SD-RLS and turn to the variant SD-RLS* that always has finite expected optimization time. In fact, we will present similar results in general optimization scenarios and supplement them by analyses on specific benchmark functions. It is possible to analyze the simpler SD-RLS on these benchmark functions as well, but we do not feel that this gives additional insights.

# 4 Analysis of the Algorithm SD-RLS*

In this section, we turn to the algorithm SD-RLS* that iteratively returns to lower strengths to avoid missing the "right" strength. We recall $T_x$ as the number of steps SD-RLS* takes to find an improvement point from the current search point $x$. Let phase $r$ consists of all points of time where radius $r$ is used in the algorithm. When the algorithm enters phase $r$, it starts with strength $r$, but when the counter exceeds the threshold, the strength decreases by one as long as it is greater than 1. In the case of strength 1, the radius $r$ is increased to $r+1$ (or to $n$ if $r+1$ is at least $n/2$), so the algorithm enters phase $r+1$ (or phase $n$).

Let $E_r$ be the event of **not** finding the optimum within phase $r$, and $U_i^j$ for $j > i$ be the event of not finding the optimum during phases $i$ to $j-1$ and finding it in phase $j$. In other words, $U_i^j = E_i \cap \cdots \cap E_{j-1} \cap \overline{E_j}$. For $i = j$, we define $U_i^i = \overline{E_i}$. We obtain the following result on the failure probability which follows from the fact that the algorithm tries to find an improvement for $\binom{n}{m} \ln R$ iterations with a probability of success of $\binom{n}{m}^{-1}$ when the radius is at least $m$.

**Lemma 2.** *Let* $x \in \{0,1\}^n$ *be the current search point of SD-RLS* on a pseudo-boolean fitness function* $f \colon \{0,1\}^n \to \mathbb{R}$ *and let* $m = gap(x)$. *Then*

$$
\Pr(E_r) \leq \begin{cases} \frac{1}{R} & \text{if } m \leq r < \frac{n}{2} \\ 0 & \text{if } r = n. \end{cases}
$$

The following lemma bounds the time to leave a local optimum conditional on that the "right" strength was missed.

**Lemma 3.** *Let* $x \in \{0,1\}^n$ *with* $m = gap(x) < n/2$ *be the current search point of SD-RLS\* with* $R \geq n^{3+\epsilon} \cdot |\mathrm{Im}\, f|$ *for an arbitrary constant* $\epsilon > 0$ *on a pseudo-boolean function* $f \colon \{0,1\}^n \to \mathbb{R}$ *and* $T_x$ *be the time to create a strict improvement. Then, we have*

$$\mathrm{E}\left(T_x \mid E_m\right) = o\left(\frac{R}{|\mathrm{Im}\, f|} \binom{n}{m}\right),$$

*where* $E_m$ *is the event of not finding an optimum when the radius* $r$ *equals* $m$.

The reason behind the factor $R/|\mathrm{Im}\, f|$ in Lemma 3 is that for proving a running time of SD-RLS\* on a function like $f$, the event $E_m$ happens with probability $1/R$ for each point in $\mathrm{Im}\, f$, so in the worst case, during the run, there are expected $|\mathrm{Im}\, f|/R$ search points where the counter exceeds the threshold, resulting in an expected number of at most $|\mathrm{Im}\, f|/R \cdot o(R/|\mathrm{Im}\, f|\binom{n}{m})$ extra iterations for the whole run in the case of exceeding the thresholds. Also, note that we always have $R/|\mathrm{Im}\, f| = \Omega(1)$ since according to the assumption, $R > |\mathrm{Im}\, f|$.

The following theorem and its proof are similar to Theorem 1 but require a more careful analysis to cover the repeated use of smaller strengths. We note that the bounds differ from Theorem 1 only in lower-order terms unless $m$ is very big.

**Theorem 3.** *Let* $x \in \{0,1\}^n$ *be the current search point of SD-RLS\* with* $R \geq n^{3+\epsilon} \cdot |\mathrm{Im}\, f|$ *for an arbitrary constant* $\epsilon > 0$ *on a pseudo-boolean function* $f \colon \{0,1\}^n \to \mathbb{R}$. *Define* $T_x$ *as the time to create a strict improvement if* $gap(x) = m$. *Then, we have*

$$\mathrm{E}\left(T_x\right) \leq \begin{cases} \binom{n}{m}\left(1 + O\left(\frac{m^2}{n-2m} \ln R\right)\right) & \text{if } m < n/2 \\ 2^n n \ln R & \text{if } m \geq n/2 \end{cases},$$

*and* $\mathrm{E}\left(T_x\right) \geq \binom{n}{m}/W$, *where* $W$ *is the number of strictly better search points at Hamming distance* $m$.

Similarly to Lemma 2, we obtain a relation to RLS on unimodal functions and can re-use existing upper bounds based on the fitness-level method [20].

**Lemma 4.** *Let* $f \colon \{0,1\}^n \to \mathbb{R}$ *be a unimodal function and consider SD-RLS\* with* $R \geq n^{3+\epsilon} \cdot |\mathrm{Im}\, f|$ *for an arbitrary constant* $\epsilon > 0$. *Then, with probability at least* $1 - \frac{|\mathrm{Im}\, f|}{R}$, *SD-RLS\* never increases the radius and behaves stochastically like RLS before finding an optimum of* $f$.

*Denote by* $T$ *the runtime of SD-RLS\* on* $f$. *Let* $f_i$ *be the* $i$-*th fitness value of an increasing order of all fitness values in* $f$ *and* $s_i$ *be a lower bound on the probability that RLS finds an improvement from search points with fitness value* $f_i$, *then*

$$\mathrm{E}\left(T\right) \leq \sum_{i=1}^{|\mathrm{Im}\, f|} \frac{1}{s_i} + o(n).$$

Finally, we use the results developed so far to prove a bound on the JUMP function which seems to be the best available for mutation-based hillclimbers.

**Theorem 4.** *Let $n \in \mathbb{N}$. For all $2 \leq m$, the expected runtime $E(T)$ of SD-RLS\* with $R \geq n^{4+\epsilon}$ for an arbitrary constant $\epsilon > 0$ on JUMP$_m$ satisfies*

$$E(T) \leq \begin{cases} \binom{n}{m} \left(1 + O\left(\frac{m^2}{n-2m} \ln n\right)\right) & \text{if } m < n/2, \\ O(2^n n \ln n) & \text{otherwise.} \end{cases}$$

## 5    An Example Where Global Mutations Are Necessary

While our $s$-flip mutation along with stagnation detection can outperform the (1+1) EA on JUMP functions, it is clear that its different search behavior may be disadvantageous on other examples. Concretely, we will present a function that has a unimodal path to a local optimum with a large Hamming distance to the global optimum. SD-RLS will with high probability follow this path and incur exponential optimization time. However, the function has a second gradient that requires two-bit flips to make progress. The classical (1+1) EA will be able to follow this gradient and to arrive at the global optimum before one-bit flips have reached the end of the path to the local optimum.

In a broader context, our function illustrates an advantage of global mutation operators. By a simple swap of local and global optimum, it immediately turns into the direct opposite, i.e., an example where using global instead of local mutations is highly detrimental and increases the runtime from polynomial to exponential with overwhelming probability. An example of such a function was previously presented in [10]; however, both the underlying construction and the proof of exponential runtime for the (1+1) EA seem much more complicated than our example.

We will in the following define the example function called NEEDGLOBAL-MUT and give proofs for the behavior of SD-RLS and (1+1) EA. In fact, NEED-GLOBALMUT is obtained from the function NEEDHIGHMUT defined in [17] to show disadvantages of stagnation detection adjusting the rate of a global mutation operator. The only change is to adjust the length of the suffix part of the function, which rather elegantly allows us to re-use the previous technique of construction and a major part of the analysis. We also encourage the reader to read the corresponding section in [17] for further insights into the construction.

In the following, we will imagine any bit string $x$ of length $n$ as being split into a prefix $a := a(x)$ of length $n-m$ and a suffix $b := b(x)$ of length $m$, where $m$ is defined below. Hence, $x = a(x) \circ b(x)$, where $\circ$ denotes the concatenation. The prefix $a(x)$ is called *valid* if it is of the form $1^i 0^{n-m-i}$, i.e., $i$ leading ones and $n-m-i$ trailing zeros. The prefix fitness PRE$(x)$ of a string $x \in \{0,1\}^n$ with valid prefix $a(x) = 1^i 0^{n-m-i}$ equals $i$, the number of leading ones. The suffix consists of $\lceil \frac{1}{3}\sqrt{n} \rceil$ consecutive blocks of $\lceil n^{1/4} \rceil$ bits each, altogether $m \leq \frac{1}{3}n^{3/4} = o(n)$ bits. Such a block is called *valid* if it contains either 0 or 2 one-bits; moreover, it is called *active* if it contains 2 and *inactive* if it contains 0 one-bits. A suffix where all blocks are valid and where all blocks following first inactive block are also inactive is called valid itself, and the suffix fitness $\text{SUFF}(x)$ of a string $x$ with valid suffix $b(x)$ is the number of leading active blocks before the first inactive one. Finally, we call $x \in \{0, 1\}^n$ valid if both its prefix and suffix are valid.

The final fitness function is a weighted combination of $\text{PRE}(x)$ and $\text{SUFF}(x)$. We define for $x \in \{0, 1\}^n$, where $x = a \circ b$ with the above-introduced $a$ and $b$,

$$\text{NEEDGLOBALMUT}(x) :=$$

$$\begin{cases} n^2\text{SUFF}(x) + \text{PRE}(x) & \text{if } \text{PRE}(x) \leq \frac{9(n-m)}{10} \wedge x \text{ valid} \\ n^2 m + \text{PRE}(x) + \text{SUFF}(x) - n - 1 & \text{if } \text{PRE}(x) > \frac{9(n-m)}{10} \wedge x \text{ valid} \\ -\text{ONEMAX}(x) & \text{otherwise.} \end{cases}$$

The function $\text{NEEDGLOBALMUT}$ equals $\text{NEEDHIGHMUT}_\xi$ from [17] for the setting $\xi = 1/2$ (ignoring that $\xi < 1$ was disallowed there for technical reasons). We note that all search points in the second case have a fitness of at least $n^2 m - n - 1$, which is bigger than $n^2(m-1) + n$, an upper bound on the fitness of search points that fall into the first case without having $m$ leading active blocks in the suffix. Hence, search points $x$ where $\text{PRE}(x) = n - m$ and $\text{SUFF}(x) = \lceil \frac{1}{3}\sqrt{n} \rceil$ represent local optima of second-best overall fitness. The set of global optima equals the points where $\text{PRE}(x) = 9(n - m)/10$ and $\text{SUFF}(x) = \lceil \frac{1}{3}\sqrt{n} \rceil$, which implies that $(n - m)/10 = \Omega(n)$ bits have to be flipped simultaneously to escape from the local toward the global optimum.

**Theorem 5.** *With probability $1 - o(1)$, SD-RLS with $R \geq n$ needs $2^{\Omega(n)}$ steps to optimize $\text{NEEDGLOBALMUT}$. The (1+1) EA optimizes this function in time $O(n^2)$ with probability $1 - 2^{-\Omega(n^{1/3})}$.*

## 6   Minimum Spanning Trees

Our self-adjusting $s$-flip mutation operator can also have advantages on classical combinatorial optimization problems. We reconsider the minimum spanning tree (MST) problem on which EAs and RLS were analyzed before [15]. The known bounds for the globally searching (1+1) EA are not tight. More precisely, they depend on $\log(w_{\max})$, the logarithm of the largest edge weight. This is different with RLS variants that flip only one or two bits due to an equivalence first formulated in [16]: if only up to two bits flip in each step, then the MST instance becomes indistinguishable from the MST instance formed by replacing all edge weights with their rank in their increasingly sorted sequence. This results in a tight upper bound of $O(m^2 \ln n)$, where $m$ is the number of edges, for $\text{RLS}^{1,2}$, an algorithm that uniformly at random decides to flip either one or two uniformly chosen bits [21]. Although not spelt out in the paper, it is easy to see that the leading term in the polynomial $O(m^2 \ln m)$. This 2 stems from the logarithm of sum of the weight ranks, which can be in the order of $m^2$. We will see that the first factor of 2 can, in some sense, be avoided in our SD-RLS*.

The following theorem bounds the optimization time of SD-RLS* in the case that the algorithm has reached a spanning tree and the fitness function only allows spanning trees to be accepted. It is well known that with the fitness functions from [15], the expected time to find the first spanning tree is $O(m \log m)$, which also transfers to SD-RLS*; hence we do not consider this lower-order term further. However, our bound comes with an additional term related to the number of strict improvements. We will discuss this term after the theorem.

**Theorem 6.** *The expected optimization time of SD-RLS\* with $R = m^4$ on the MST problem with $m$ edges, starting with an arbitrary spanning tree, is at most*

$$(1 + o(1))\big((m^2/2)(1 + \ln(r_1 + \cdots + r_m)) + (4m \ln m)\mathrm{E}\,(S)\big)$$
$$= (1 + o(1))\big(m^2 \ln m + (4m \ln m)\mathrm{E}\,(S)\big),$$

*where $r_i$ is the rank of the ith edge in the sequence sorted by increasing edge weights and $\mathrm{E}\,(S)$ is the expected number of strict improvements that the algorithm makes conditioned on that the strength never exceeds 2.*

The term $\mathrm{E}\,(S)$ appearing in the previous theorem is not easy to bound. If $\mathrm{E}\,(S) = o(m)$, the upper bound suggests that SD-RLS may be more efficient than the classical RLS[1,2] algorithm; with the caveat that we are talking about upper bounds only. However, it is not difficult to find examples where $\mathrm{E}\,(S) = \Omega(m)$, e.g., on the worst-case graph used for the lower-bound proof in [15], which we will study below experimentally, and we cannot generally rule out that $\mathrm{E}\,(S)$ is asymptotically bigger than $m$ on certain instances. However, empirically SD-RLS* can be faster than RLS[1,2] and the (1+1) EA on MST instances, as we will see in Sect. 7. In any case, although the algorithm can search globally, the bound in Theorem 6 does not suffer from the $\log(w_{\max})$ factor appearing in the analysis of the (1+1) EA.

We also considered variants of SD-RLS* that do not reset the strength to 1 after each strict improvement and would therefore, be able to work with strength 2 for a long while on the MST problem. However, such an approach is risky in scenarios where, e.g., both one-bit flips and two-bit flips are possible and one-bit flips should be exploited for the sake of efficiency. Instead, we think that a combination of stagnation detection and selection hyperheuristics [19] based on the $s$-flip operator or the learning mechanism from [7], which performs very well on the MST, would be more promising here.

## 7     Experiments

In this section, we present the results of the experiments conducted to see the performance of the proposed algorithms for small problem dimensions. This experimental design was employed because our theoretical results are asymptotic.

In the first experiment, we ran an implementation of Algorithm 3 (SD-RLS*) on the JUMP fitness function with jump size $m = 4$ and $n$ varying from 80 to 160. We compared our algorithm against the (1+1) EA with standard mutation rate $1/n$, the (1+1) EA with mutation probability $m/n$, Algorithm (1+1) FEA$_\beta$ from [11] with three different $\beta = \{1.5, 2, 4\}$, and the SD-(1+1) EA presented in [17]. In Fig. 1, we observe that SD-RLS* outperforms the rest of the algorithms.

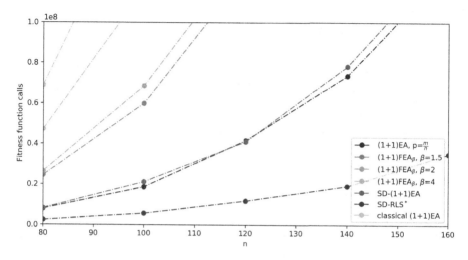

**Fig. 1.** Average number of fitness calls (over 1000 runs) the mentioned algorithms took to optimize JUMP$_4$.

In the second experiment, we ran an implementation of four algorithms SD-RLS*, (1+1) FEA$_\beta$ with $\beta = 1.5$ from [11], the standard (1+1) EA and RLS$^{1,2}$ from [15] on the MST problem with the fitness function from [15] for two types of graphs called TG and Erdős–Rényi.

The graph TG with $n$ vertices and $m = 3n/4 + \binom{n/2}{2}$ edges contains a sequence of $p = n/4$ triangles which are connected to each other, and the last triangle is connected to a complete graph of size $q = n/2$. Regarding the weights, the edges of the complete graph have the weight 1, and we set the weights of edges in triangle to $2a$ and $3a$ for the side edges and the main edge, respectively. In this paper, we consider $a = n^2$. The graph TG is used for estimating lower bounds on the expected runtime of the (1+1) EA and RLS in the literature [15]. In this experiment, we use $n = \{24, 36, 48, 60\}$. As can be seen in Fig. 2b, (1+1) FEA$_\beta$ is faster than the rest of the algorithms, but SD-RLS* outperforms the standard (1+1) EA and RLS$^{1,2}$.

Regarding the graphs Erdős–Rényi, we produced some random Erdős–Rényi graphs with $p = (2 \ln n)/n$ and assigned each edge an integer weight in the range $[1, n^2]$ uniformly at random. We also checked that the graphs certainly had a spanning tree. Then, we ran the implementation on MST of these graphs. The obtained results can be seen in Fig. 2a. As we discussed in Sect. 6, SD-RLS* does not outperform the (1+1) EA and RLS$^{1,2}$ on MST with graphs where the number of strict improvements in SD-RLS* is large.

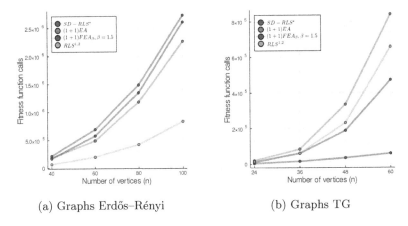

(a) Graphs Erdős–Rényi                    (b) Graphs TG

**Fig. 2.** Average number of fitness calls (over 400 runs) the mentioned algorithms took to optimize the fitness function MST of the graphs.

For statistical tests, we ran the implementation of the algorithms on the graphs TG and Erdős–Rényi 400 times, and all p-values obtained from a Mann-Whitney U-test between the algorithms, with respect to the null hypothesis of identical behavior, are less than $10^{-4}$ except for the results regarding the graph TG with $n = 24$.

## Conclusions

We have transferred stagnation detection, previously proposed for EAs with standard bit mutation, to the operator flipping exactly $s$ uniformly randomly chosen bits as typically encountered in randomized local search. Through both theoretical runtime analyses and experimental studies we have shown that this combination of stagnation detection and local search efficiently leaves local optimal and often outperforms the previously considered variants with global mutation. We have also introduced techniques that make the algorithm robust if it, due to its randomized nature, misses the right number of bits flipped, and analyzed scenarios where global mutations are still preferable. In the future, we would like to investigate stagnation detection more thoroughly on instances of classical combinatorial optimization problem like the minimum spanning tree problem, for which the present paper only gives preliminary but promising results.

# References

1. Bassin, A., Buzdalov, M.: The 1/5-th rule with rollbacks: on self-adjustment of the population size in the $(1+(\lambda, \lambda))$ GA. In: Proceedings of GECCO 2019 (Companion), pp. 277–278. ACM Press (2019)

2. Corus, D., Oliveto, P.S., Yazdani, D.: Fast artificial immune systems. In: Auger, A., Fonseca, C.M., Lourenço, N., Machado, P., Paquete, L., Whitley, D. (eds.) PPSN 2018. LNCS, vol. 11102, pp. 67–78. Springer, Cham (2018). https://doi.org/10.1007/978-3-319-99259-4_6

3. Doerr, B.: Probabilistic tools for the analysis of randomized optimization heuristics. In: Doerr, B., Neumann, F. (eds.) Theory of Evolutionary Computation. NCS, pp. 1–87. Springer, Cham (2020). https://doi.org/10.1007/978-3-030-29414-4_1

4. Doerr, B., Doerr, C.: The impact of random initialization on the runtime of randomized search heuristics. Algorithmica **75**(3), 529–553 (2016)

5. Doerr, B., Doerr, C.: Optimal static and self-adjusting parameter choices for the $(1+(\lambda, \lambda))$ genetic algorithm. Algorithmica **80**(5), 1658–1709 (2018)

6. Doerr, B., Doerr, C.: Theory of parameter control for discrete black-box optimization: provable performance gains through dynamic parameter choices. In: Doerr, B., Neumann, F. (eds.) Theory of Evolutionary Computation. NCS, pp. 271–321. Springer, Cham (2020). https://doi.org/10.1007/978-3-030-29414-4_6

7. Doerr, B., Doerr, C., Yang, J.: $k$-bit mutation with self-adjusting $k$ outperforms standard bit mutation. In: Handl, J., Hart, E., Lewis, P.R., López-Ibáñez, M., Ochoa, G., Paechter, B. (eds.) PPSN 2016. LNCS, vol. 9921, pp. 824–834. Springer, Cham (2016). https://doi.org/10.1007/978-3-319-45823-6_77

8. Doerr, B., Fouz, M., Witt, C.: Quasirandom evolutionary algorithms. In: Proceedings of GECCO 2010, pp. 1457–1464. ACM (2010)

9. Doerr, B., Gießen, C., Witt, C., Yang, J.: The $(1 + \lambda)$ evolutionary algorithm with self-adjusting mutation rate. Algorithmica **81**(2), 593–631 (2019)

10. Doerr, B., Jansen, T., Klein, C.: Comparing global and local mutations on bit strings. In: Ryan, C., Keijzer, M. (eds.) Proceedings of GECCO 2008, pp. 929–936. ACM Press (2008)

11. Doerr, B., Le, H.P., Makhmara, R., Nguyen, T.D.: Fast genetic algorithms. In: Proceedings of GECCO 2017, pp. 777–784. ACM Press (2017)

12. Hansen, P., Mladenović, N., Brimberg, J., Pérez, J.A.M.: Variable neighborhood search. In: Gendreau, M., Potvin, J.-Y. (eds.) Handbook of Metaheuristics. ISORMS, vol. 272, pp. 57–97. Springer, Cham (2019). https://doi.org/10.1007/978-3-319-91086-4_3

13. Lissovoi, A., Oliveto, P.S., Warwicker, J.A.: On the time complexity of algorithm selection hyper-heuristics for multimodal optimisation. In: Proceedings of AAAI 2019, pp. 2322–2329. AAAI Press (2019)

14. Lugo, M.: Sum of "the first $k$" binomial coefficients for fixed $n$. MathOverflow (2017). https://mathoverflow.net/q/17236. Accessed 15 Mar 2021

15. Neumann, F., Wegener, I.: Randomized local search, evolutionary algorithms, and the minimum spanning tree problem. Theoretical Comput. Sci. **378**, 32–40 (2007)

16. Raidl, G.R., Koller, G., Julstrom, B.A.: Biased mutation operators for subgraph-selection problems. IEEE Trans. Evol. Comput. **10**(2), 145–156 (2006)

17. Rajabi, A., Witt, C.: Self-adjusting evolutionary algorithms for multimodal optimization. In: Proceedings of GECCO 2020, pp. 1314–1322. ACM Press (2020)

18. Rajabi, A., Witt, C.: Stagnation detection with randomized local search (2021). CoRR abs/2101.12054. http://arxiv.org/abs/2101.12054

19. Warwicker, J.A.: On the runtime analysis of selection hyper-heuristics for pseudo-Boolean optimisation. Ph.D. thesis, University of Sheffield, UK (2019). http://ethos.bl.uk/OrderDetails.do?uin=uk.bl.ethos.786561
20. Wegener, I.: Methods for the analysis of evolutionary algorithms on pseudo-Boolean functions. In: Sarker, R., Mohammadian, M., Yao, X. (eds.) Evolutionary Optimization. Kluwer Academic Publishers, New York (2001)
21. Witt, C.: Revised analysis of the (1+1) EA for the minimum spanning tree problem. In: Proceedings of GECCO 2014, pp. 509–516. ACM Press (2014)

# An Artificial Immune System for Black Box Test Case Selection

Lukas Rosenbauer[1][(✉)], Anthony Stein[2], and Jörg Hähner[3]

[1] BSH Hausgeräte GmbH, Im Gewerbepark B35, 93059 Regensburg, Germany
lukas.rosenbauer@bshg.com
[2] University of Hohenheim, Garbenstr. 9, 70599 Stuttgart, Germany
anthony.stein@uni-hohenheim.de
[3] University of Augsburg, Eichleitner Str. 30, 86159 Augsburg, Germany
joerg.haehner@informatik.uni-augsburg.de

**Abstract.** Testing is a crucial part of the development of a new product. For software validation a transformation from manual to automated tests can be observed which enables companies to implement large numbers of test cases. However, during testing situations may occur where it is not feasible to run all tests due to time constraints. Hence a set of critical test cases must be compiled which usually fulfills several criteria. Within this work we focus on criteria that are feasible for black box testing such as system tests. We adapt an existing artificial immune system for our use case and evaluate our method in a series of experiments using industrial datasets. We compare our approach with several other test selection methods where our algorithm shows superior performance.

**Keywords:** Software validation · Test automation · Artificial immune system · Bio-inspired computing

## 1 Introduction

Testing is major part in modern product development and aims at revealing errors as quickly as possible. Thus quality can be ensured. Testing becomes more and more automated which enables companies to carry out large numbers of tests. These tests may have a variable runtime and capabilities. This has led to need to steadily optimize the test cases at hands [22,31].

Bioinspired computation has led to major improvements in several ways. For example test data or even entire test cases can be generated using *genetic algorithms* (GA) [8,23]. *Mutation testing* focuses on changing parts of the software to test in order to evaluate how effective the underlying tests are [10].

The aforementioned techniques focus on use cases where the underlying source code is available to testers which is coined *white box testing*. However, in certain industries such as the automotive sector that is not always the case since software is bought from other companies and only machine code is integrated. Hence there is only a limited or no knowledge of these components available. Software validation in such an environment is called *black box testing* [3].

© Springer Nature Switzerland AG 2021
C. Zarges and S. Verel (Eds.): EvoCOP 2021, LNCS 12692, pp. 169–184, 2021.
https://doi.org/10.1007/978-3-030-72904-2_11

Within this work we focus on the selection of appropriate test cases during black box testing. This is necessary during some situations of software validation such as *smoke testing* [6]. The aforementioned testing approach aims at deciding if a software version should be rejected or should be investigated more deeply. Thus prolonged testing times or expensive manual checks may be avoided. If only a few or minor errors are detected, then other seemingly working features can be evaluated more deeply whilst the development team tries to fix the already recognized faults.

A critical test suite should be chosen according to several criteria such as its execution time or fault revealing capabilities [22]. Hence the test case selection can be modelled as a multi-objective optimization problem [15]. Within this work we adapt a novel class of optimizers called *germinal center artificial immune system* (GCAIS) [12] to the task. This has lead to the following contributions:

- We introduce a new population initialization method to GCAIS which leads to a diverse starting solutions.
- We move with GCAIS beyond set covering use cases [12] and knapsack problems [13] and show its feasibility for the use case.
- Within our experiments we exclusively rely on industrial datasets in order to verify our method in a real world setting.
- We compare our improved GCAIS variant with a variety of other algorithms including a *nondominated sorting genetic algorithm II* (NSGA-II) based method specialized for black box test selection [15]. Statistical tests indicate the superiority of our approach.

In Sect. 2 we discuss related work. This is followed by a problem description (Sect. 3). Afterwards we introduce GCAIS and how we adapted it (Sect. 4). Then we switch to an evaluation of our approach in Sect. 5. This is followed by a short discussion of future work and a conclusion in Sect. 6.

## 2   Related Work

Choosing an appropriate test suite is a difficult task in many subbranches of testing. This includes the *test suite minimisation problem*, the *test case selection problem* and the *test case prioritisation problem* [31].

Test case prioritisation approaches try to order given test cases according to some quality criterion, for example their likelihood to fail [27]. A current trend is to apply *reinforcement learning* to train such an ordering based on historical test data [21,30]. However, test case prioritisation methods are not limited to artificial intelligence approaches. There are also several approaches outside the field as documented by Marijan et al. [17]. These methods mostly focus on failure revealing capabilities.

Test suite minimisation usually deals at specification level and only has information about the available test cases and which requirements they cover [7,31]. However, there are also variations where branch coverage [9], call stack coverage [18] or model transitions [28] are considered instead.

Test case selection occurs with the scenario mentioned in Sect. 1. Corresponding test suites can be selected according to single or multiple objectives [22,31]. If coverage metrics are used, then the corresponding problem is NP-hard since these are instances of the weighted set covering problem [31]. Arrieta et al. [1] employ a NSGA-II for a black box test case selection specialized for simulations of cyber physical systems and corresponding test objectives. Lachmann et al. [15] also developed an NSGA-II based approach that optimizes several criteria next to requirement coverage for generic black box test case selection. Other approaches are limited to white box testing since they focus for example on code coverage [19,26,29].

From an immune computation point of view we have been influenced by the works of Joshi et al. [12,14] who translated new insights on germinal centres into a generic search heuristic. The algorithm is related to other metaheuristics such as *Simple evolutionary algorithms with isolated population* (SEIP) [32] or *global simple evolutionary multi-objective optimiser* [20] since all three rely on negative selection and maintain populations of so called non-dominated solutions. However, immune computation is not limited to negative selection but also offers other unique properties combined such as memory, fault tolerance and robustness [2]. We are also not the first to use artificial immune systems for a practical use case. They have already been applied successfully in security, optimization and machine learning [16]. Especially GCAIS has already been applied to the aforementioned test suite minimisation problem (using requirement coverage as a decision basis) [25]. A succeeding work by Rosenbauer et al. [24] applied GCAIS to the test case selection problem but there they also focused solely on requirements coverage. However, several surveys [22,31] underline that a selection only based on requirement coverage is insufficient as empirical results indicate that coverage is not necessarily linked to fault revealing capabilities. Hence we move to a multi-objective version of the test case selection problem, which also includes fault revealing capabilities, to overcome this issue and furthermore change the algorithmic structure of GCAIS to fit this use case.

## 3   Problem Description

For black box testing certain objectives that are feasible for white box testing are technically not possible. For example optimizing code coverage. Hence we focus on the following objectives which are applicable in black box testing:

- **Requirement coverage:** A test case $tc$ covers at least one requirement. If a set of test cases covers a large number of requirements this indicates that most of the functionality is tested. This can be formalized as follows:

$$Cov(S) = \frac{|\bigcup_{tc \in S} r(tc)|}{R} \in [0, 1] \tag{1}$$

where $S$ is a set of selected tests and $r(tc)$ is the set of requirements that a test case covers. Further $R$ denotes the total number of requirements. We name the objective $Cov(S)$. It is worth mentioning that the coverage objective alone is NP-hard and not even in the *approximable* (APX) computational class since it corresponds to the *weighted minimum set cover problem* [4].

- **Failure probability:** Throughout the life time of a project, a test is executed several times. We use its execution history to estimate a test's probability to fail as:

$$P(tc \text{ fails}) = \frac{f_{tc}}{n_{tc}} \tag{2}$$

where $f_{tc}$ is the number of executions where $tc$ failed and $n_{tc}$ is the total number of executions of the test. In our experiments we maximize the average failure probability of the set of selected tests:

$$FP(S) = \frac{\sum_{tc \in S} P(tc \text{ fails})}{|S|} \in [0, 1] \tag{3}$$

The intention of this objective is to reveal faults. We abbreviate this objective as $FP(S)$.

- **Execution time:** Due to limited time budgets the resulting test suite should have a low execution time:

$$D(S) = \sum_{tc \in S} d(tc) \in \mathbb{R}_{\geq 0} \tag{4}$$

where $d(\cdot)$ is the approximated duration of a test case. We use the mean value of previous executions as an estimate. We call this objective $D(S)$.

It is worth mentioning that the estimations of the failure probability and execution time can be retrieved rather easily with modern testing tools. If for example test-frameworks such as *pytest* or *google test* are used, the testing results can be stored in a *junit xml* which contains the duration and outcome of each executed test.

Throughout this work we encode a test suite as a binary vector of length $m$ where $m$ is the number of available test cases. If an entry is set to 1 then the test is used and if it is set to 0 then the test is not used. For example if there are 5 tests available and only the third bit is set to one, then only the third test is used and the other four not.

With the formal definitions of the objectives and solution encoding we are able to introduce the optimization problem at hand:

$$\max \quad Cov(S), \quad \max \quad FP(S), \quad \min \quad D(S)$$
$$D(S) \leq B \tag{5}$$
$$S \in \{0, 1\}^m$$

where $B$ is a predefined execution time boundary. It indicates how much time is available for a test suite to run.

Some of the objectives are opposed to each other. For example if more tests are added to enlarge the coverage then the execution time is also increased.

Furthermore, if the coverage and average fault probability should be enlarged simultaneously, an optimization algorithm has to add tests with a high likelihood to fail that also verify previously uncovered requirements. If only tests are added that nearly never fail but cover a lot of requirements, then the average fault probability will decline. Further, if a test suite solely focuses on tests with a high likelihood to fail but only tests a few features then this will lead to a rather small coverage.

Solutions of a multi-objective optimization problem can be compared using *Pareto-domination* [20]. A solution $\mathbf{x}$ is said to Pareto-dominate another solution $\mathbf{y}$ if and only if it is superior in at least one objective but not worse in the others. Within this work this is the case if one of the following conditions is fulfilled:

1. $Cov(\mathbf{x}) > Cov(\mathbf{y}) \wedge FP(\mathbf{x}) \geq FP(\mathbf{y}) \wedge D(\mathbf{x}) \leq D(\mathbf{y})$
2. $Cov(\mathbf{x}) \geq Cov(\mathbf{y}) \wedge FP(\mathbf{x}) > FP(\mathbf{y}) \wedge D(\mathbf{x}) \leq D(\mathbf{y})$
3. $Cov(\mathbf{x}) \geq Cov(\mathbf{y}) \wedge FP(\mathbf{x}) \geq FP(\mathbf{y}) \wedge D(\mathbf{x}) < D(\mathbf{y})$

Condition 1 is valid if $\mathbf{x}$ has a higher requirement coverage whilst not having a greater execution time or worse average fault probability. A satisfied condition 2 implies that the tests of $\mathbf{x}$ have a higher chance of failing whilst not covering less requirements and having a higher runtime. The last condition is fulfilled if $\mathbf{x}$ has no worse coverage and average failure probability whilst having a lower runtime. We abbreviate the relation as $\mathbf{x} >_p \mathbf{y}$.

# 4    Germinal Center Artificial Immune System

GCAIS is a population-based, randomised search heuristic that is based on the immune system of vertebrates. The heuristic has been influenced by recent insights about germinal centre reaction [12]. *Germinal centres* (GC) are regions where the invading *antigen* (Ag) is presented to immune cells. If an invasion occurs, the cells produce *antibodies* (Ab) that try to bind the pathogen and eradicate it. The GCs start to grow and try to find the best Abs. GCs communicate with each other in order to exchange their Abs. The latter can be improved by proliferation, mutation, and selection of immune cells. In this analogy the Ag is the optimization problem, the B cell is a solution population and the Abs are the solutions that are developed by the heuristic. For a more detailed description the reader is referred to Joshi et al. [11].

GCAIS employs a standard bitwise mutation. Hence the algorithm flips individual bits with a probability of $\frac{1}{m}$. For example, if there are 10 test cases and a solution uses the fifth test then the mutation operator deletes the test from the solution with a probability of 0.1.

During each iteration GCAIS mutates its entire population. The mutated population is merged with the original one. Afterwards those solutions of the newly created population are deleted that are Pareto-dominated by another population member. This is repeated until a stopping criterion such as a maximum number of search iterations is met. We describe the procedure in Algorithm 1.

---

**Algorithm 1:** Germinal centre artificial immune system (GCAIS).

---

   **input** : problem with binary encoded solutions, problem dimension $m$

1  P = initialize_population()

2  **while** *stopping criterion is not met* **do**

3     |   P' = {}

4     |   **for** $x$ *in* $P$ **do**

5     |     |   $\mathbf{y}$ = mutate $\mathbf{x}$

6     |     |   insert $\mathbf{y}$ to P'

7     |   **end**

8     |   // merge populations

9     |   P = P ∪ P'

10    |   // sort out dominated solutions

11    |   P = $\{\mathbf{x} \in P |\ \nexists \mathbf{y} \in P : \mathbf{y} >_p \mathbf{x}\}$

12 **end**

---

---

**Algorithm 2:** Multi-axis initialization.

---

   **input** : test cases, $m$, fault probabilities, time budgets $t_k$, max retries $r$

   **output**: a population of solutions

1  P = {}

2  // create random solutions

3  **for** *each time budget $t_k$* **do**

4     |   S = **0**

5     |   count = 0

6     |   **while** *count $\leq r$ and $D(S) \leq t_k$* **do**

7     |     |   draw test case i

8     |     |   **if** *time left to add i* **then**

9     |     |     |   add i to S

10    |     |     |   count = 0

11    |     |   **else**

12    |     |     |   count += 1

13    |     |   **end**

14    |   **end**

15    |   insert S to P

16 **end**

17 S = **0**

18 // create greedy solutions

19 **while** $\exists i \in \{0, 1, ..., m - 1\} : S[i] = 0 \wedge P(tc\ i\ fails) > 0$ **do**

20    |   index = $i \in \{0, 1, ..., m - 1\} : S[i] = 0 \wedge P(\text{tc i fails}) > 0$

21    |   S[index] = 1

22    |   insert a copy of S to P

23 **end**

24 **return** P

---

It is worth mentioning that GCAIS also allows unfeasible solutions (in our case solutions with an execution time higher than the available execution time).

The function initialize_population creates a starting population. Joshi et al. [12] generate a population consisting solely of the zero vector. However, we took another route. We create random solutions for a series of fixed time budgets $t_k$ (within this use case for 5, 10,...,100% of the total execution time). If we try to create a random solution for a fixed time budget then we draw a test uniformly at random. If the remaining budget is higher than the test case's execution time we add the test case. If not then we draw another test case. If we cannot add a new test after a maximum number of retries $r$ (in our later experiments we used $r = 5$) then we stop adding new tests. We also add the test cases collected by a greedy algorithm that takes tests according to their failure probability (see Algorithm 2). Thus we create a diverse starting population for the metaheuristic. We coin this method *multi-axis initialization* as it creates solutions alongside a random and a greedy axis.

It is worth mentioning that Joshi et al. [13] developed a mechanism based on *epsilon dominance* to keep the population size in bounds whilst keeping a level of diversity. In a follow-up study Rosenbauer et al. [25] showed that this can be detrimental for covering problems and proposed another way. They limited the number of solutions that have the same values for the objectives. If the number is exceeded, random solutions of this class of solutions are deleted. For example if the boundary is set to 100 then only hundred solutions of an execution time of one hour, a requirement coverage of 50% and an average failure probability of 80% are allowed. We use the method of Rosenbauer et al. [25] since one of our optimization objectives (requirements coverage) corresponds to a set covering problem.

## 5   Evaluation

For our experiments we acquired datasets from BSH Hausgeräte GmbH which is an international producer of home appliances. The company produces various devices ranging across dishwashers, ovens to washing machines. Each dataset contains the results of several test sessions (a *test session* is a run of a test suite). Tests are executed iteratively since software is developed iteratively within BSH which is a common practice. Hence a time series of test executions is available. We use the first 5 sessions to compute the initial estimations for the fault probabilities as described in Eq. 2 and update them as we progress through the test sessions.

An overview of the datasets is found in Table 1. They are from two different oven projects and a dishwasher project. In total we can rely on the results of about 2,000 test cases and more than 200,000 verdicts. The datasets include a project with a higher percentage of failed tests (Oven 1) and a rather low number of failed test cases (Dishwasher and Oven 2). Furthermore the number of test cases between the two product groups is also vastly different. Hence the considered datasets show some degree of diversity.

In order to evaluate our GCAIS variant, we compare the method against the following algorithms:

- **Random selection:** In software validation it is a common approach to select a test suite entirely at random [5]. Random tests are greedily added until a given predefined execution time budget for the test suite is exhausted. We added the method in order to compare GCAIS to a widespread selection mechanism and to see if it is better than pure randomness.
- **NSGA-II:** Lachmann et al. [15] successfully used a NSGA-II for black box testing. We employ the same variant which uses a bitflip mutation, binary tournament selection and a HUX-crossover (probability of 0.9).
- **SEMO:** We widen our comparison by using the *simple evolutionary multiobjective optimizer* (SEMO) [20] as well. Thus we are able to explore if reasonable results may already be achieved with this rather simple metaheuristic since the NSGA-II based approaches were only compared against a pure random selection [1,15].

For the GCAIS we use a population boundary of 200. We chose the hyperparameter in a preliminary evaluation which we leave out here (due to space restrictions). We selected the hyperparameter by evaluating the possible boundary values $\{100, 200, ..., 500\}$ on the dishwasher dataset's first test session considered. There our GCAIS variant performed the best with the aforementioned boundary.

Within our evaluation we use the available data to simulate the use case. Before each test session we run the considered algorithms on the corpus of available tests. Then we only consider found solutions that are within a given execution time budget $B$. We choose the solution $\mathbf{x}$ that maximizes $FP(\mathbf{x}) + Cov(\mathbf{x})$. Thus we focus on a high mixture of coverage and failure probability. We consider time budgets of 5, 10, ..., 100% of the execution time of all tests. We repeat all experiments a hundred times and focus on averaged results.

**Table 1.** Examined datasets.

|          | Oven 1  | Oven 2 | Dishwasher |
|----------|---------|--------|------------|
| sessions | 39      | 36     | 45         |
| test cases | 486   | 477    | 1499       |
| verdicts | 22,350  | 17,349 | 186,195    |
| failed   | 10.94%  | 3.41%  | 3.44%      |

We give each algorithm a search time of five minutes. Furthermore if the encountered Pareto-Frontier does not change over 100 iterations then we regard this as convergence and let the algorithm terminate.

For our experiments we used a Dell OptiPlex XE3 with 32 GB RAM and an Intel i7 8700 processor with exclusive use for the experiment. We also published our source code and datasets in order to ease the reproducibility of results[1].

---

[1] Available here: https://github.com/LagLukas/moa_testing.

## 5.1   Failure Revealing Capabilities

The goal of testing is finding errors and not proving their absence. Hence we examined how well the found test suites perform in this task. We compared the solutions of SEMO, NSGA-II and the random selection on every considered dataset with GCAIS' solutions. In order to compare the found test suites we rely on statistical tests. We employ a series of one-sided Wilcoxon tests[2] and use a significance level of 0.05. We use the statistical tests to evaluate the null hypothesis of the form "the solutions of algorithm $x$ find more errors than GCAIS's solutions on dataset $y$". We display the corresponding $p$-values in Table 2. All of them are below the significance level and hence we reject every null hypothesis and infer that the computed test suites of GCAIS are better in finding errors.

**Table 2.** $p$-values for one-sided Wilcoxon tests. The columns represent a dataset and the rows an algorithm to compare with. The entry of row x and column y is the $p$-value of the null hypothesis "the solutions of algorithm $x$ find more errors than GCAIS's solutions on dataset $y$".

|         | Oven 1     | Oven 2      | Dishwasher |
|---------|------------|-------------|------------|
| NSGA-II | 0.0        | 6.26e−274   | 1.70e−10   |
| Random  | 1.19e−110  | 0.0         | 8.98e−25   |
| SEMO    | 1.18e−81   | 1.41e−262   | 0.04       |

We switch our focus to analysing how high the time budget needs to be until the first error is detected. We display the minimum, maximum, quartiles and average execution time that were necessary to find a first fault in Table 3. There we can see that, in three experimental settings, GCAIS produces solutions that have on average the lowest time budget.

We can also see a large range for the time budgets needed to detect the first error. This variety of values may disturb the mean value. Thus we decided to examine the median for each algorithm and dataset. There we can observe that GCAIS has the lowest median. SEMO and the pure random selection lead to similar results on the Oven 1 dataset (regarding the median). We also added the first and third quartiles into Table 3 in order to describe the time budget distribution more accurately. Once more we can make the observation of rather low values for GCAIS. Furthermore the necessary time budget seems to be rather stable as in many cases the first quartile, the median, and the third quartile are the same. For the other examined methods this not the case and their distributions are more diverse. We consider these close quartiles as an indicator for the robustness of the solutions produced by GCAIS.

---

[2] It is worth mentioning that for this statistical test no preconditions must be checked. Furthermore, the one-sided variant checks if the median of a random variable $X$ is higher than the median of a random variable $Y$.

**Table 3.** Overview of the time budget needed to reveal the first error. The table contains the mean values ± σ, medians, maxima and minima. The best values are marked bold.

|  |  | Oven 1 | Oven 2 | Dishwasher |
|---|---|---|---|---|
| GCAIS | mean | **0.145 ± 0.28** | **0.296 ± 0.39** | **0.024 ± 0.06** |
|  | third quartile | **0.05** | **0.36** | **0.01** |
|  | median | **0.05** | **0.05** | **0.01** |
|  | first quartile | 0.05 | **0.04** | **0.01** |
|  | max | 0.65 | 0.64 | 0.3 |
|  | min | 0.02 | 0.02 | **0.01** |
| NSGA-II | mean | 0.316 ± 0.36 | 0.415 ± 0.45 | 0.239 ± 0.25 |
|  | third quartile | 0.48 | 1.0 | 0.55 |
|  | median | 0.1 | 0.1 | 0.05 |
|  | first quartile | 0.1 | 0.05 | 0.05 |
|  | max | 0.69 | **0.41** | 0.75 |
|  | min | 0.02 | **0.01** | **0.01** |
| RANDOM | mean | 0.182 ± 0.27 | 0.313 ± 0.38 | 0.056 ± 0.02 |
|  | third quartile | 0.19 | 0.41 | 0.05 |
|  | median | **0.05** | 0.1 | 0.05 |
|  | first quartile | 0.05 | 0.05 | 0.05 |
|  | max | **0.63** | 0.74 | **0.15** |
|  | min | 0.04 | **0.01** | 0.05 |
| SEMO | mean | 0.222 ± 0.32 | 0.35 ± 0.37 | 0.062 ± 0.04 |
|  | third quartile | 0.35 | 0.48 | 0.05 |
|  | median | **0.05** | 0.185 | 0.05 |
|  | first quartile | **0.04** | 0.08 | 0.05 |
|  | max | 0.8 | 0.59 | 0.25 |
|  | min | **0.01** | **0.01** | **0.01** |

**Table 4.** $P$-values for one-sided Wilcoxon tests. The columns represent a dataset and the rows an algorithm to compare with. The entry of row x and column y is the $p$-value of the null hypothesis "algorithm $x$'s solutions find the first error earlier than GCAIS's solutions on dataset $y$".

|  | Oven 1 | Oven 2 | Dishwasher |
|---|---|---|---|
| NSGA-II | $6.52e{-}46$ | $5.95e{-}13$ | $6.46e{-}06$ |
| Random | $7.56e{-}11$ | $9.89e{-}06$ | $2.28e{-}17$ |
| SEMO | $1.03e{-}60$ | $5.80e{-}55$ | $2.14e{-}50$ |

We deem a discussion solely based on central tendencies such as mean values or medians as insufficient and once more decided to use statistical tests to take a deeper look at our results. We reuse a significance level of 0.05 and rely on one-sided Wilcoxon tests. We examine the nullhpyothesis "algorithm $x$'s solutions need a lower time budget to find a first error than GCAIS's solutions on dataset $y$" for each considered algorithm and dataset. We show the corresponding $p$-values in Table 4. The $p$-values are below our significance level and we reject all null hypotheses and accept the alternative hypothesis. Thus we infer that solutions found by GCAIS are capable of finding the first error earlier than NSGA-II, a pure random selection, and SEMO. It is worth mentioning that the comparably good results of GCAIS are not only due to the initialization. An additional comparison between our immune system and the initialization showed that GCAIS also performs better in this scenario. The same is valid on these datasets if we compare our variant with the vanilla version of Joshi et al. [12].

We additionally examined the objective failure probability $F(\cdot)$ and there we could see that in all but one combination the GCAIS approach is significantly better than the other approaches. This might be seen as an indicator for why GCAIS has been found to be superior in detecting errors on the considered datasets.

## 5.2   Detection of Broken Features

In the previous subsection we solely focused on the detection of failures. It lacks the link to the requirements the tests cover. For example if two features are broken and one test suite leads to a lot of failed test exclusively for one feature and another one has a failed test for each broken feature then the first test suite would be better even though it would have failed to identify all broken features. Thus within this subsection we focus on the evaluation of how well the considered methods recognize broken features.

We examine hypotheses of the form "the solutions of algorithm $x$ detect more broken features than GCAIS's solutions on dataset $y$" using a series of one-sided Wilcoxon tests and a significance level of 0.05. The $p$-values are displayed in Table 5. We can reject the null hypothesis in 8 out of 9 cases and accept the alternative hypothesis (GCAIS detects more broken features). However, on our dishwasher dataset the $p$-value for NSGA-II is rather low (about 0.1751) but still not significant which means that we cannot reject the null hypothesis. Thus we decided to perform an additional two-sided Wilcoxon test to check if on average both methods perform equally well on this dataset and this statistical test indicated that this is the case.

We decided to investigate the difference between GCAIS and NSGA-II on the dishwasher dataset more deeply. We plotted the average difference of the percentage of detected broken features in Fig. 1. The x-axis displays the test session index, the y-axis the relative time budget and the z-axis the difference in detected broken features. We can see a clear superiority of GCAIS for high budgets and very low budgets (less than 10%). NSGA-II is ahead for time budgets of about 20% for very early test sessions. For succeeding test sessions this

**Table 5.** $p$-values for one-sided Wilcoxon tests. The columns represent a dataset and the rows an algorithm to compare with. The entry of row $x$ and column $y$ is the $p$-value of the null hypothesis "algorithm $x$'s solutions detect more broken features than GCAIS's solutions on dataset $y$".

|         | Oven 1      | Oven 2      | Dishwasher  |
|---------|-------------|-------------|-------------|
| NSGA-II | $1.38e{-}309$ | $3.38e{-}12$ | $0.1751$    |
| Random  | $0.0$       | $9.50e{-}200$ | $1.09e{-}109$ |
| SEMO    | $5.31e{-}245$ | $5.02e{-}85$ | $1.399e{-}43$ |

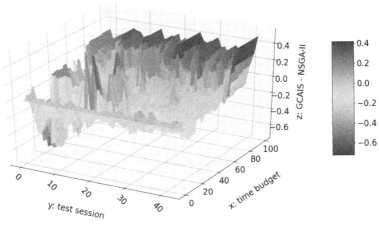

Difference of GCAIS and NSGA-II in detecting broken Features

**Fig. 1.** Difference in the percentage of found broken features between GCAIS and NSGA-II on the dishwasher dataset. A positive value indicates that GCAIS found more broken features (a negative one indicates the opposite).

gap is tightening. After session index 5 GCAIS becomes significantly superior. Hence if more test outcomes are available then GCAIS's performance increases on this dataset.

Our previous evaluation exclusively focused on the question if GCAIS performs better than the other methods. We also intend to give a total overview instead of only this relative consideration. Thus we also visualized the raw numbers in Fig. 2. Figure 2 a) shows the performance on the dishwasher dataset and there performance is generally slowly rising with an increasing time budget. There a few outliers became apparent where GCAIS finds even more broken functionalities, especially at the start of testing. In this phase there are the most faults in our dataset which explains this observation. Figures 2 b) and c) display the performance on the two oven datasets. There generally more errors occurred overall and also in a higher number during later stages of testing. This explains why there are several test sessions where we detected high numbers of broken requirements. The succeeding test sessions often show less issues found since

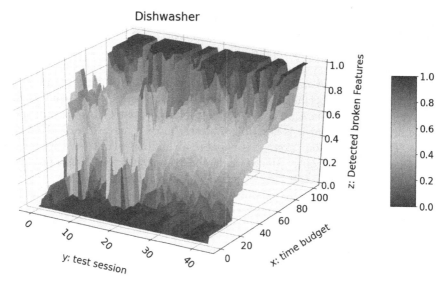

(a) Percentage of found broken features using GCAIS on the dishwasher dataset.

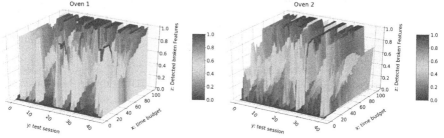

(b) Percentage of found broken features using GCAIS on the oven 1 dataset

(c) Percentage of found broken features using GCAIS on the oven 2 dataset.

**Fig. 2.** Percentages of found broken features across the considered datasets.

between the two sessions the found oven software bugs have been fixed. It is worth mentioning that the two oven projects share some generic components hence they show some similarity in their error behaviour.

## 6   Conclusion and Future Work

Black box testing is a scenario where only limited knowledge of the underlying system is available. This is for example often the case in the system test level or if software from third parties is used. For such environments there are situations where it is not feasible to run all tests. Hence sets of important tests must be known in order to get a quick insight into the system when a new software version is examined.

We adapted a recent metaheuristic called *germinal center artificial immune system* (GCAIS) to search for critical tests. The system is aware of previous failures, test execution times, and requirement coverage. Based on this knowledge the artificial immune system is able to compile crucial test suites for a variety of time budgets. We compared our approach with other algorithms including a NSGA-II variant specialized for black box test case selection using several industrial datasets. There we could not only observe that GCAIS is capable of finding errors earlier but it is also able to detect more errors within a given time budget. Furthermore, our system is capable of finding failing tests for several requirements simultaneously. Hence we demonstrated that GCAIS is a promising candidate for black box test case selection.

In the future we intend to examine white box testing objectives such as code coverage metrics as well. We think that GCAIS performs equally well in such scenarios.

# References

1. Arrieta, A., Wang, S., Arruabarrena, A., Markiegi, U., Sagardui, G., Etxeberria, L.: Multi-objective black-box test case selection for cost-effectively testing simulation models. In: Proceedings of the Genetic and Evolutionary Computation Conference, New York, NY, USA, pp. 1411–1418. GECCO 2018. Association for Computing Machinery (2018)

2. Azuaje, F.: Review of "Artificial Immune Systems: A New Computational Intelligence Approach" by L.N. de Castro and J. Timmis (Eds) Springer, London, 2002. Neural Net. **16**(8), 1229 (2003)

3. Bath, G., McKay, J.: The Software Test Engineer's Handbook: A Study Guide for the ISTQB Test Analyst and Technical Test Analyst Advanced Level Certificates 2012. Rocky Nook Computing, Rocky Nook (2014)

4. Dinur, I., Steurer, D.: Analytical approach to parallel repetition. In: Proceedings of the Forty-sixth Annual ACM Symposium on Theory of Computing, New York, NY, USA, pp. 624–633. STOC 2014. ACM (2014)

5. Duran, J.W., Ntafos, S.C.: an evaluation of random testing. IEEE Trans. Softw. Eng.SE **10**(4), 438–444 (1984)

6. Dustin, E., Rashka, J., Paul, J.: Automated Software Testing: Introduction, Management, and Performance. Addison-Wesley Longman Publishing Co., Inc., New York (1999)

7. Gotlieb, A., Marijan, D.: FLOWER: optimal test suite reduction as a network maximum flow. In: Proceedings of the 2014 International Symposium on Software Testing and Analysis, New York, NY, USA, pp. 171–180. ISSTA 2014. ACM (2014)

8. Haga, H., Suehiro, A.: Automatic test case generation based on genetic algorithm and mutation analysis. In: 2012 IEEE International Conference on Control System, Computing and Engineering, pp. 119–123 (2012)

9. Jeffrey, D., Gupta, N.: Improving fault detection capability by selectively retaining test cases during test suite reduction. IEEE Trans. Softw. Eng. **33**(2), 108–123 (2007)

10. Jia, Y., Harman, M.: Constructing subtle faults using higher order mutation testing. In: 2008 Eighth IEEE International Working Conference on Source Code Analysis and Manipulation, pp. 249–258 (2008)

11. Joshi, A.: The Germinal Centre Artificial Immune System. Ph.D. thesis, University of Birmingham (2017)
12. Joshi, A., Rowe, J.E., Zarges, C.: An immune-inspired algorithm for the set cover problem. In: Bartz-Beielstein, T., Branke, J., Filipič, B., Smith, J. (eds.) PPSN 2014. LNCS, vol. 8672, pp. 243–251. Springer, Cham (2014). https://doi.org/10.1007/978-3-319-10762-2_24
13. Joshi, A., Rowe, J.E., Zarges, C.: Improving the performance of the germinal center artificial immune system using epsilon-dominance: a multi-objective knapsack problem case study. In: Ochoa, G., Chicano, F. (eds.) EvoCOP 2015. LNCS, vol. 9026, pp. 114–125. Springer, Cham (2015). https://doi.org/10.1007/978-3-319-16468-7_10
14. Joshi, A., Rowe, J., Zarges, C.: On the effects of incorporating memory in GC-AIS for the set cover problem. In: MIC 2015: The XI Metaheuristics International Conference (2015)
15. Lachmann, R., Felderer, M., Nieke, M., Schulze, S., Seidl, C., Schaefer, I.: Multi-objective black-box test case selection for system testing. In: Proceedings of the Genetic and Evolutionary Computation Conference, New York, NY, USA, pp. 1311–1318. GECCO 2017. Association for Computing Machinery (2017)
16. Luo, W., Liu, R., Jiang, H., Zhao, D., Wu, L.: Three branches of negative representation of information: a survey. IEEE Trans. Emerg. Top. Comput. Intell. 2(6), 411–425 (2018)
17. Marijan, D., Gotlieb, A., Sen, S.: Test case prioritization for continuous regression testing: an industrial case study. In: 2013 IEEE International Conference on Software Maintenance, pp. 540–543 (2013)
18. McMaster, S., Memon, A.M.: Call stack coverage for test suite reduction. In: 21st IEEE International Conference on Software Maintenance (ICSM 2005), pp. 539–548 (2005)
19. Mondal, D., Hemmati, H., Durocher, S.: Exploring test suite diversification and code coverage in multi-objective test case selection. In: 2015 IEEE 8th International Conference on Software Testing, Verification and Validation (ICST), pp. 1–10 (2015)
20. Neumann, F., Witt, C.: Bioinspired Computation in Combinatorial Optimization: Algorithms and Their Computational Complexity. Natural Computing Series. Springer-Verlag, Heidelberg (2010). https://doi.org/10.1007/978-3-642-16544-3. ISBN 978-3-642-16543-6
21. Nguyen, A., Le, B., Nguyen, V.: Prioritizing automated user interface tests using reinforcement learning. In: Proceedings of the Fifteenth International Conference on Predictive Models and Data Analytics in Software Engineering, New York, NY, USA, pp. 56–65. PROMISE 2019. Association for Computing Machinery (2019)
22. Note Narciso, E., Delamaro, M., Nunes, F.: Test case selection: a systematic literature review. Int. J. Softw. Eng. Knowl. Eng. 24, 653–676 (2014)
23. Rodrigues, D.S., Delamaro, M.E., Corrêa, C.G., Nunes, F.L.S.: Using genetic algorithms in test data generation: a critical systematic mapping. ACM Comput. Surv. 51(2) (2018)
24. Rosenbauer, L., Stein, A., Hähner, J.: An artificial immune system for adaptive test selection. In: 2020 IEEE Symposium Series on Computational Intelligence (SSCI) (2020)
25. Rosenbauer, L., Stein, A., Hähner, J.: A Germinal Centre Artificial Immune System for Software Test Suite Reduction. Artificial Life (2020)

26. de Souza, L.S., de Miranda, P.B.C., Prudencio, R.B.C., Barros, F.D.A.: A multi-objective particle swarm optimization for test case selection based on functional requirements coverage and execution effort. In: 2011 IEEE 23rd International Conference on Tools with Artificial Intelligence, pp. 245–252 (2011)
27. Spieker, H., Gotlieb, A., Marijan, D., Mossige, M.: Reinforcement Learning for Automatic Test Case Prioritization and Selection in Continuous Integration. CoRR abs/1811.04122 (2018)
28. Vaysburg, B., Tahat, L.H., Korel, B.: Dependence Analysis in Reduction of Requirement Based Test Suites. In: Proceedings of the 2002 ACM SIGSOFT International Symposium on Software Testing and Analysis, pp. 107–111. ISSTA 2002, Association for Computing Machinery, New York, NY, USA (2002)
29. Whalen, M.W., Rajan, A., Heimdahl, M.P., Miller, S.P.: Coverage metrics for requirements-based testing. In: Proceedings of the 2006 International Symposium on Software Testing and Analysis, New York, NY, USA, pp. 25–36. ISSTA 2006. Association for Computing Machinery (2006)
30. Xiao, L., Miao, H., Shi, T., Hong, Y.: LSTM-based deep learning for spatial-temporal software testing. Distrib. Parallel Databases **38**(3), 687–712 (2020)
31. Yoo, S., Harman, M.: Regression testing minimization, selection and prioritization: a survey. Softw. Test. Verif. Reliab. **22**(2), 67–120 (2012)
32. Yu, Y., Yao, X., Zhou, Z.H.: On the approximation ability of evolutionary optimization with application to minimum set cover. Artif. Intell. **S180–181** (2010)

# Symmetry Breaking for Voting Mechanisms

Preethi Sankineni[✉] and Andrew M. Sutton

University of Minnesota Duluth, Duluth, MN, USA
sanki002@d.umn.edu

**Abstract.** Recently, Rowe and Aishwaryaprajna [FOGA 2019] introduced a simple majority vote technique that efficiently solves JUMP with large gaps, ONEMAX with large noise, and any monotone function with a polynomial-size image. In this paper, we identify a pathological condition for this algorithm: the presence of spin-flip symmetry. Spin-flip symmetry is the invariance of a pseudo-Boolean function to complementation. Many important combinatorial optimization problems admit objective functions that exhibit this pathology, such as graph problems, Ising models, and variants of propositional satisfiability. We prove that no population size exists that allows the majority vote technique to solve spin-flip symmetric functions with reasonable probability. To remedy this, we introduce a symmetry-breaking technique that allows the majority vote algorithm to overcome this issue for many landscapes. We prove a sufficient condition for a spin-flip symmetric function to possess in order for the symmetry-breaking voting algorithm to succeed, and prove its efficiency on generalized TwoMax and families of constructed 3-NAE-SAT and 2-XOR-SAT formulas. We also prove that it fails on the one-dimensional Ising model, and suggest different techniques for overcoming this. Finally, we present empirical results that explore the tightness of the runtime bounds and the performance of the technique on randomized satisfiability variants.

## 1 Introduction

Voting crossover is a recombination operator requiring multiple parents in which each position of the resulting offspring is decided by a majority vote. This technique has existed as an evolutionary operator for decades [4], but only recently has received attention from the theory community [7,17]. Notably, a recent paper by Rowe and Aishwaryaprajna [17] introduced a simple evolutionary algorithm in which a population of $\mu$ individuals is constructed by performing $\mu$ rounds of tournament selection on randomly selected parents, and then produces a final string by performing a single majority vote over the entire population. Despite the relative simplicity of this approach, the authors proved it solves ONEMAX and JUMP in $O(n \log n)$ time, even when the JUMP function has a gap of size $(1 - \epsilon)n$. It also attains a $O((n + \sigma^2) \log n)$ upper bound on ONEMAX perturbed by any additive noise with variance $\sigma^2$, which is so far the best known running time for any randomized search heuristic on noisy ONEMAX.

© Springer Nature Switzerland AG 2021
C. Zarges and S. Verel (Eds.): EvoCOP 2021, LNCS 12692, pp. 185–201, 2021.
https://doi.org/10.1007/978-3-030-72904-2_12

Rowe and Aishwaryaprajna also pointed out that the voting algorithm always samples strings with close to $n/2$ ones, which is a limitation on many functions such as LEADINGONES, or even on linear functions with largely unbalanced weights like BINVAL. Nevertheless, they proved that any monotone function $f$ can be optimized using only $O(|\text{IM} f|^2 \log n)$ tournaments to create the voting population, and introduced a modification that focuses on a single bit at a time and successfully solves LEADINGONES in $O(n \log n)$ time.

In this paper, we want to extend the analysis of this surprisingly simple yet effective approach to understand how it performs on other nonlinear, nonmonotone function classes. In particular, we investigate the particular pathology known as *spin-flip* symmetry, which is invariance of a function under complementation. We show that the voting algorithm fails on every spin-flip symmetric function with probability exponentially close to one, irrespective of the population size. This issue arises for the Voting Algorithm because it constructs its population by performing tournaments over a uniform distribution of the search space. Thus, even for relatively simple functions, the event the tournament winner points in the direction of a global optimum $x^\star$ is obscured by the equally likely event that a complementary string pair points in the direction of $\bar{x}^\star$. To address this, we create a technique that effectively breaks the spin-flip symmetry in the space so that the distribution is no longer uniform over the space, but will prefer certain strings over their complements. One important feature that such a technique must exhibit is that it has no prior knowledge of the location of the global optima.

The concept of *symmetry breaking* for search algorithms arose first from the constraint programming community [2]. In this framework, the set of symmetries of a propositional formula corresponds to the set of permutations on variables that belong to the automorphism group of the formula. In other words, the symmetries are exactly the variable permutations to which the truth of the formula is invariant. Symmetry breaking in this context is a method of adding extra constraints that eliminate such symmetries, and thus reduce the decision tree complexity, which can be used as a technique in constraint solvers [16, 18]. The effect of symmetry breaking constraints on local search algorithms was investigated by Prestwich and Roli [15] who empirically demonstrated that the extra constraints have a detrimental effect on local search by increasing the relative size of basins of attraction for suboptimal solutions, while simultaneously decreasing the relative size of basins of attraction for optimal solutions.

In the related context of evolutionary optimization, Naudts and Naudts [12] examined the presence of symmetry in fitness functions for genetic algorithms. They investigated a classical model of magnetism for which discrete spin-value orientations are assigned to each site of a lattice. The energy function of this model (called the Ising model [10]) exhibits *spin-flip* symmetry, which is an invariance to the inversion of spin-values. They found empirically that the simple genetic algorithm (SGA) is incapable of solving the spin-flip symmetric one-dimensional Ising model. Symmetry breaking techniques, involving systematically fixing spin orientations [13], can solve the one-dimensional problem in quadratic time.

One challenge for traditional crossover on spin-flip symmetric functions comes from the presence of *synchronization problems* [9]. Loosely speaking, synchronization problems arise from attempting to recombine two relatively fit individuals with complementary building blocks. In this situation, a search heuristic must rely on something other than recombination, such as niching, fitness sharing, or specialized mutation operators. Naudts and Naudts [12] also identified difficult configurations that a pure hill-climber would have trouble navigating.

The interest in the effect of spin-flip symmetry on hardness was also picked up by the theory community. Fischer [5] proved that despite the perceived hardness of the one-dimensional Ising model, the two-dimensional Ising model (where the underlying structure is a toroidal lattice) can be solved by the Metropolis algorithm in polynomial time. Briest et al. [1] also studied the two-dimensional Ising model, as well as a number of other structures including cliques connected by bridge edges and the hypercubic lattice. Fischer and Wegener [6] studied the one-dimensional Ising model more closely and proved that pure hill-climbing or mutation strategies such as RLS and the $(1 + 1)$ EA can solve it in expected cubic time. They also considered a number of other techniques such as the Gene Invariant Genetic Algorithm [3] and fitness sharing. Sudholt [20] investigated the efficacy of crossover on the Ising model with an underlying tree topology. He proved that a mutation-only evolutionary algorithm with constant population size needs exponential time to solve this problem due to the presence of synchronization problems. On the other hand, a $(2 + 2)$ GA employing two-point crossover and fitness sharing solves the problem in expected time $O(n^3)$. Sutton [21] studied a set of spin-flip symmetric functions that are computationally easy while problematic for stochastic hill-climbing algorithms like the $(1 + 1)$ EA.

The remainder of this paper is structured as follows. In the next section we describe the Voting Algorithm and provide some background and discuss intractability of symmetric functions. In Sect. 3 we present a symmetry-breaking strategy and prove efficient bounds for a number of problems. In Sect. 4 we report experiments that study the tightness of the bounds and the dependence of success on other problem parameters. We conclude the paper in Sect. 5.

## 2    Search by Majority Vote

Majority vote crossover was introduced in 1994 as *occurrence-based scanning* by Eiben, Raué and Ruttkay [4]. In this technique, a set of $\mu > 2$ parents $\{x^{(1)}, x^{(2)}, \ldots, x^{(\mu)}\} \subseteq \{0,1\}^n$ produce an offspring $y$ where the $i$-th position $y_i$ is chosen deterministically by the majority function $y_i = \mathrm{Maj}(x_i^{(1)}, x_i^{(2)}, \ldots, x_i^{(\mu)})$, where $\mathrm{Maj}(b_1, \ldots, b_m)$ chooses the element of $\{0, 1\}$ that occurs most frequently among its arguments[1]. The idea behind the design of this operator is that good values are more prevalent in a set of parents selected by fitness.

This crossover was analyzed as a component in traditional evolutionary algorithms by Friedrich et al. [7] on the JUMP benchmark function, and instances

---

[1] The tie-breaking rule, when $m$ is even, depends on the setting.

of the vertex cover problem. Whitley et al. [22] showed that JUMP with gap $O(\log n)$ could be solved in linear time by producing three parents for majority vote crossover using next ascent hill-climbing.

Motivated by this work, Rowe and Aishwaryaprajna [17] introduced the Voting Algorithm, listed in Algorithm 1, that works by first generating a population of size $\mu$ by conducting $\mu$ repeated rounds of tournament selection applied to two parents generated uniformly at random. At the end of the $\mu$ rounds, a single output string is generated by majority vote. Note that this algorithm is not a traditional evolutionary algorithm employing majority crossover, but a time-limited sampling procedure that applies majority crossover to all of its samples.

---

**Algorithm 1:** Voting Algorithm

---

1  Let $p \leftarrow (0, \ldots, 0)$;

2  **repeat** $\mu$ *times*

3      Let $x, y \in \{0, 1\}^n$ be chosen uniformly at random;

4      **if** $f(x) > f(y)$ **then** $p \leftarrow p + x$;

5      **else** $p \leftarrow p + y$;

6  **for** $1 \leq i \leq n$ **do**

7      **if** $p_i = \mu/2$ **then** choose $z_i$ uniformly at random from $\{0, 1\}$;

8      **else** $z_i \leftarrow [p_i > \mu/2]$;

9  **return** $z$;

---

The strong performance of the Voting Algorithm on ONEMAX and ONEMAX with additive noise comes from the fact that tournament selection can reveal a clear signal in the sampling process that ensures the majority vote is in the right direction. This is also the case with JUMP. In fact, for gaps of $(1 - \epsilon)n$, where $1/2 < \epsilon < 1$ is a constant, the algorithm is unable to even detect a difference between ONEMAX and JUMP as it is extremely unlikely to even generate a string within the gap in polynomial time.

### 2.1   Functions with Spin-Flip Symmetry

Spin-flip symmetric functions pose a particular challenge to Algorithm 1, because, in contrast to simple hill-climbing algorithms that follow a local fitness signal, the Voting Algorithm relies on a majority of the population to agree on which part of the space the target for optimization lies. If the function has spin-flip symmetry, then any string in $\{0, 1\}^n$ has the same probability of winning the tournament in lines 4 and 5 (and hence participating in the vote) as its complement. The intuition is that each member of the population has an equal chance to vote for a global optimum or its binary complement, which obscures the signal toward either. We formalize this in the following theorem. Unfortunately, our proof only works for sublinear population sizes. However, we conjecture that the claim would actually hold for any population size.

**Theorem 1.** *Let $f$ be a spin-flip symmetric function with a polynomial number of global optima and let $0 < \epsilon < 1$. Then for any population size $\mu = O(n^{1-\epsilon})$, with probability $1 - o(1)$, the Voting Algorithm fails to find an optimal solution.*

*Proof.* Without loss of generality, suppose $x^\star = 1^n$ is a global optimum of $f$. We define the sequence of random variables $X_1, X_2, \ldots, X_\mu$, where $X_t$ is the Hamming weight of the winner of the tournament in lines 4 and 5 of Algorithm 1.

The Hamming weight is the number of ones in the bit string. By Chernoff bounds, the probability that the Hamming weight of a binary string drawn uniformly at random from $\{0,1\}^n$ exceeds $n/2 + \sqrt{n}\log n$ is at most $e^{-\frac{2\log^2 n}{3}}$. Let $\mathcal{E}$ be the event that no string in the population bears a Hamming weight of larger than $n/2 + \sqrt{n}\log n$. Taking a union bound over all $2\mu$ random strings generated during the execution of the algorithm, $\Pr(\mathcal{E}) \geq 1 - 2\mu e^{-\frac{2\log^2 n}{3}} = 1 - e^{-\Omega(\log^2 n)}$.

By spin-flip symmetry, for any string $x \in \{0,1\}^n$, the probability that $x$ is the winner of the tournament is equal to the probability that the binary complement $\bar{x}$ is the winner of the tournament. Therefore, for all $k \in \{0, 1, \ldots n/2\}$, define $\Pr(X_t = n/2 - k) = \Pr(X_t = n/2 + k) =: P_k$. Thus we have $E[X_t] = \sum_{k=0}^{n/2} ((n/2 - k) + (n/2 + k)) P_k = n/2$, and setting $X := \sum_{t=1}^{\mu} X_t$, $E[X] = n\mu/2$. Conditioning on $\mathcal{E}$, we have

$$\Pr(X \geq n\mu/2 + n \mid \mathcal{E}) \leq \exp\left(\frac{-2n}{\mu \log^2 n}\right)$$

by Hoeffding's inequality.

But $X \geq n\mu/2 + n$ is a necessary condition for $z_i = 1$ for all $i \in \{1, 2, \ldots, n\}$, since for this to occur, we must have every $p_i > \mu/2$, $X = \sum_{i=1}^{n} p_i$ and $\mu = O(n^{1-\epsilon})$. Therefore, the probability that the global optimum $1^n$ is generated is at most $e^{-\Omega\left(\frac{n}{\mu \log^2 n}\right)} + 1 - \Pr(\mathcal{E})$. Taking a union bound over all of the global optima of $f$ completes the proof. □

## 3  A Symmetry-Breaking Strategy

Theorem 1 demonstrates that the voting mechanism can fail to optimize functions with spin-flip symmetry when the sample size is too low. Moreover, we conjecture that this result can be extended to higher sample sizes. The failure seems to arise because the fitness-distance coupling is perfectly mirrored everywhere in the search space. The consequence is that information about distance to a global optimum is canceled out. In order to overcome this effect, it is necessary to implement a symmetry-breaking strategy to introduce a bias toward a global optimum or its complement. Moreover, the strategy must not rely on information about the true location of a global optimum. We take advantage of the following symmetry property.

*Property 1.* Let $f: \{0,1\}^n \to \mathbb{R}$ be a function with spin-flip symmetry. Then there is a global optimum $x^\star \in \{0,1\}^n$ with $x_1^\star = 1$.

This property follows from the invariance of $f$ under complementation. In the case that the first bit of a global optimum $x_1^\star = 0$, by spin-flip symmetry $f(x^\star) = f(\overline{x}^\star)$, so $\overline{x}^\star$ is also globally optimal, and $\overline{x}_1^\star = 1$. The modified voting algorithm is listed in Algorithm 2. The only modification occurs after the parent generation step by complementing a generated string if it contains a 0 in the first element.

---

**Algorithm 2:** Voting Algorithm with Symmetry Breaking

---

1  Let $p \leftarrow (0, \dots, 0)$;
2  **repeat** $\mu$ *times*
3  |   Let $x, y \in \{0, 1\}^n$ be chosen uniformly at random;
4  |   **if** $x_1 = 0$ **then** $x \leftarrow \overline{x}$;
5  |   **if** $y_1 = 0$ **then** $y \leftarrow \overline{y}$;
6  |   **if** $f(x) > f(y)$ **then** $p \leftarrow p + x$;
7  |   **else** $p \leftarrow p + y$;

8  **for** $1 \leq i \leq n$ **do**
9  |   **if** $p_i = \mu/2$ **then** choose $z_i$ uniformly at random from $\{0, 1\}$;
10 |   **else** $z_i \leftarrow [p_i > \mu/2]$;

11 **return** $z$;

---

We analyze what properties a spin-flip symmetric function $f$ must have so that Algorithm 2 could optimize $f$. Let $x^\star$ be a global maximum of $f$ such that $x_1^\star = 1$. We define two indicator functions

$$1_{x,y}^> := \begin{cases} 1 & \text{if } f(x) > f(y), \\ 0 & \text{otherwise.} \end{cases} \qquad 1_{x,y}^= := \begin{cases} 1 & \text{if } f(x) = f(y), \\ 0 & \text{otherwise.} \end{cases}$$

Define the set $Q = \{(x, y) \in \{0, 1\}^n \times \{0, 1\}^n : x_1 = y_1 = 1\}$. Algorithm 2 generates a pair of candidates by drawing a pair uniformly from $Q$. For $k > 1$, define the set

$$S_k^\triangle := \{(x, y) \in Q : x_k = x_k^\star, y_k \neq x_k^\star\}$$

to be the pairs of strings that are not equal in the $k$-th component, but the left element of the pair is correct with respect to $x^\star$.  □

**Lemma 1.** *For any spin-flip symmetric function $f$ with global maximum $x^\star$, suppose there is a bound $\xi_k(n) > 0$ such that*

$$\frac{1}{2^{2n-4}} \sum_{(x,y) \in S_k^\triangle} \left( 1_{x,y}^> - 1_{x,y\oplus e_k}^> + \frac{1_{x,y}^= - 1_{x,y\oplus e_k}^=}{2} \right) \geq \xi_k(n),$$

*where $e_k$ is the $k$-th standard basis vector of $\{0, 1\}^n$ and $k$ is an arbitrary index in $\{2, \dots, n\}$. The $\oplus$ denotes the component-wise exclusive-or operation. Then the string $z$ returned by Algorithm 2 is correct (with respect to $x^\star$) in position $k$ with probability $\Pr(z_k = x_k^\star) \geq 1 - \exp\left(-\mu \xi_k(n)^2/2\right)$.*

*Proof.* In a tournament decided by $f$, the element $x$ of the pair $(x, y)$ wins with probability one if $f(x) > f(y)$ and with probability $1/2$ if $f(x) = f(y)$. Thus,

$$\Pr(x \text{ wins} \mid x_k = x_k^\star, y_k \neq x_k) = \frac{1}{2^{2n-4}} \sum_{(x,y) \in S_k^\triangle} \left( 1_{x,y}^{>} + \frac{1_{x,y}^{=}}{2} \right),$$

since $\left| S_k^\triangle \right| = (2^{(n-2)})^2$. Similarly,

$$\Pr(x \text{ wins} \mid x_k = y_k = x_k^\star) = \frac{1}{2^{2n-4}} \sum_{(x,y) \in S_k^\triangle} \left( 1_{x,y \oplus e_k}^{>} + \frac{1_{x,y \oplus e_k}^{=}}{2} \right).$$

By this, and the assumption in the claim,

$$\Pr(x \text{ wins} \mid x_k = x_k^\star, y_k \neq x_k) - \Pr(x \text{ wins} \mid x_k = y_k = x_k^\star) \geq \xi_k(n).$$

Since $f$ is spin-flip symmetric, $\Pr(x \text{ wins} \mid x_k = y_k = x_k^\star) = 1/2$, therefore

$$\Pr(x \text{ wins} \mid x_k = x_k^\star, y_k \neq x_k) \geq \frac{1}{2} + \xi_k(n).$$

The remainder of the proof is similar to the proof that the original Voting Algorithm succeeds on ONEMAX [17]. In particular, we appeal to Bayes' Theorem and the fact that $\Pr(x_k = x_k^\star) = \Pr(x \text{ wins}) = 1/2$.

$$\Pr(x_k = x_k^\star \mid x \text{ wins}) = \frac{\Pr(x \text{ wins} \mid x_k = x_k^\star) \Pr(x_k = x_k^\star)}{\Pr(x \text{ wins})} = \Pr(x \text{ wins} \mid x_k = x_k^\star)$$

$$= \frac{1}{2} \Pr(x \text{ wins} \mid x_k = x_k^\star, y_k \neq x_k) + \frac{1}{2} \Pr(x \text{ wins} \mid x_k = y_k = x_k^\star)$$

$$\geq \frac{1}{2} + \xi_k(n)/2. \tag{1}$$

At the end of the main loop at line 7 of Algorithm 2, $\mathrm{E}[p_k] \geq \frac{\mu}{2} + \mu \xi_k(n)/2$. The probability that $z_k$ is correct in the final string $z$ can be bounded by Hoeffding's inequality. $\Pr(z_k \neq x_k^\star) = \Pr(p_k \leq \mu/2) \leq \exp\left(-\mu \xi_k(n)^2/2\right)$.  □

## 3.1  Generalized TwoMax

Perhaps the simplest function exhibiting spin-flip symmetry is the TWOMAX function [8,14], the generalized form of which can be defined as follows. Given an arbitrary string $z \in \{0,1\}^n$, $\text{TWOMAX}_z(x) = \max\{d(x, z), n - d(x, z)\}$.

Here $d(x, z)$ denotes the Hamming distance between strings $x$ and $y$. There are exactly two global optima for the TWOMAX function: $z$ and its complement $\bar{z}$.

**Theorem 2.** *For any $\mu \geq 8\pi^2(c + 1)n^2 \ln n$, Algorithm 2 correctly finds an optimum of generalized* TWOMAX$_z$ *with probability at least $1 - n^{-c}$.*

*Proof.* Let $k > 1$ be an arbitrary index. Without loss of generality, we assume the $\text{TwoMax}_z$ target is $z = x^\star = 1^n$. By this assumption, for any $(x, y) \in S_k^\triangle$, $(y \oplus \mathbf{e}_k)_k - y_k = 1$. Moreover, for any $x \in \{0, 1\}^n$, designating $\text{TwoMax}_z$ as $f$, if $f(x) \neq (n + 1)/2$, then $|f(x) - f(x \oplus \mathbf{e}_k)| = 1$ (otherwise it is zero).

For any pair of strings $(x, y) \in S_k^\triangle$, we define the quantity

$$\psi_k(x, y) := 1^>_{x,y} - 1^>_{x,y \oplus \mathbf{e}_k} + \frac{1^=_{x,y} - 1^=_{x,y \oplus \mathbf{e}_k}}{2} \in \{-1/2, 0, 1/2\}. \qquad (2)$$

Note that $\psi_k(x, y) \in \{-1/2, 0, 1/2\}$ since $\psi_k(x, y) = 1/2$ if and only if (a) $f(x) = f(y)$ and $f(x) < f(y \oplus \mathbf{e}_k)$, or (b) $f(x) > f(y)$ and $f(x) = f(y \oplus \mathbf{e}_k)$. Case (a) occurs only when $f(x) = |y|_1$, and case (b) only when $f(x) = |y|_1 + 1$. Similarly, $\psi_k(x, y) = -1/2$ if and only if (c) $f(x) = f(y)$ and $f(x) > f(y \oplus \mathbf{e}_k)$, or (d) $f(x) < f(y)$ and $f(x) = f(y \oplus \mathbf{e}_k)$. Case (c) occurs only when $f(x) = n - |y|_1$, and case (d) when $f(x) = n - |y|_1 - 1$.

To bound $\xi_k(n)$, it is helpful to partition the search space into sets $L_i$ according to fitness value. Denote by $L_i = |\{x \in \{0, 1\}^n : x_1 = x_k = 1, f(x) = i\}|$ for $i \in \{\lceil n/2 \rceil, \ldots, n\}$. Then[2]

$$L_i = \binom{n-2}{i-2} + \binom{n-2}{i}, \text{ for } i > n/2, \text{ and } L_{n/2} = \binom{n-2}{n/2}, \text{ when } n \text{ is even.}$$

This follows from the fact that, fixing $x_1 = x_k = 1$, we can either choose the remaining $i - 2$ ones or choose $i$ zeros among the $n - 2$ free positions. If $i = n/2$ for even $n$, there are exactly $n/2$ ways to choose the remaining positions.

We explicitly calculate the sum of $\psi_k(x, y)$ over elements $(x, y) \in S_k^\triangle$ where $f(x) = i$ by multiplying the count $L_i$ of elements with fitness value $i$ by the number of different ways to construct cases (a)–(d). In particular, since $y_1 = 1$ and $y_k = 0$ are fixed, there are exactly $\binom{n-2}{j-1}$ ways to construct a $y$ so that $|y|_1 = j$. For any $f(x) = i \geq \lceil n/2 \rceil$, $\sum_{(x,y) \in S_k^\triangle \,:\, f(x)=i} \psi_k(x, y)$ is equal to

$$\frac{L_i}{2} \left( \underbrace{\binom{n-2}{i-1}}_{\text{case (a)}} + \underbrace{\binom{n-2}{i-2}}_{\text{case (b)}} - \underbrace{\binom{n-2}{(n-i)-1}}_{\text{case (c)}} - \underbrace{\binom{n-2}{(n-i-1)-1}}_{\text{case (d)}} \right)$$

$$= \frac{L_i}{2} \left( \binom{n-2}{i-2} - \binom{n-2}{i} \right) = \frac{1}{2} \binom{n-2}{i-2}^2 - \frac{1}{2} \binom{n-2}{i}^2. \qquad (3)$$

Note that when $n$ is even, $\binom{n-2}{n/2-2} = \binom{n-2}{n/2}$. Furthermore, cases (b) and (c) cannot occur since we require $f(y) \geq n/2$ and $f(y \oplus \mathbf{e}_k) \geq n/2$. Thus, when $n$ is even and $f(x) = n/2$,

---

[2] We implicitly use the convention that $\binom{a}{b} = 0$ for $a < b$.

$$\sum_{(x,y)\in S_k^{\triangle} \,:\, f(x)=\frac{n}{2}} \psi_k(x,y) = \frac{1}{4} L_{n/2} \left( \underbrace{\binom{n-2}{n/2-1}}_{\text{case (a)}} - \underbrace{\binom{n-2}{n/2}}_{\text{case (d)}} \right) = \frac{L_{n/2} C_{n/2-1}}{4},$$

(4)

where $C_n$ is the $n$-th Catalan number.

Rewriting the sum over indicator functions by summing over all fitness values for $f(x)$, we obtain

$$\sum_{(x,y)\in S_k^{\triangle}} \psi_k(x,y) = \frac{1}{2} \sum_{i=\lceil n/2 \rceil}^{n} \left( \binom{n-2}{i-2}^2 - \binom{n-2}{i}^2 \right) + \frac{[n \equiv 0 \bmod 2] L_{n/2} C_{n/2-1}}{4},$$

by (3) and (4), where we have used $[\cdot]$ to denote the Iverson bracket. The sum telescopes, hence

$$\geq \frac{1}{2} \left( \binom{n-2}{\lfloor n/2 \rfloor}^2 + \binom{n-2}{\lfloor n/2 \rfloor - 1}^2 \right) \geq \frac{1}{2} \left( \frac{4^{n/2-1}}{\sqrt{\pi n}} \right)^2 = \frac{4^{n-2}}{2\pi n} =: 2^{2n-4} \xi_k(n),$$

where the final inequality follows from Stirling's formula. Applying Lemma 1, the probability that $z_k$ is correct in the final string $z$ is bounded as $\Pr(z_k = 1) \geq 1 - \exp\left(-\frac{2\mu}{16\pi^2 n^2}\right) \geq 1 - n^{-(c+1)}$. Taking a union bound over all $n-1$ positions $k > 1$ completes the proof. $\qquad\square$

### 3.2 Satisfiability Problems

We now consider special cases of the propositional satisfiability problem (SAT) that exhibit spin-flip symmetry. In the classical propositional satisfiability problem, we have a propositional formula $F$ constructed as a conjunction of disjunctive clauses over a set of Boolean variables. Another family of SAT instances is the conjunction of XOR clauses, which are clauses containing an exclusive disjunction of literals. XOR clauses are useful for describing cryptographic primitives for applying SAT solvers to problems in cryptography [19]. When every clause in the formula is an exclusive disjunction of $k$ literals, the problem is called $k$-XOR-SAT. Deciding the satisfiability of a $k$-XOR-SAT formula is not NP-complete, as it can be expressed as solving a system of linear equations over a finite field, and resolved with Gaussian elimination. If $k$ is even, then the objective function counting the number of satisfied clauses is spin-flip symmetric because the truth of each XOR clause containing an even number of literals is invariant to complementation.

We define an infinite family of satisfiable 2-XOR-SAT formulas $\{F_i^{\text{XOR}}\}_{i \in \mathbb{N} \setminus \{1\}}$ as $F_n^{\text{XOR}} := \bigwedge_{i \in [n]} \bigwedge_{j \in [i-1]} ((x_i \oplus \neg x_j) \wedge (\neg x_i \oplus x_j))$. The formula is constructed in such a way that it is always satisfiable by both $1^n$ and $0^n$.

The fitness of an individual $f \colon \{0,1\}^n \to \mathbb{N}$ to $F_n^{\text{XOR}}$ is the number $f(x)$ of clauses in that are satisfied by the string $x$. Note that $f$ exhibits spin-flip symmetry.

Not-all-equal 3-satisfiability (3-NAE-SAT) is a variant of Boolean 3-satisfiability (3-SAT) in which each clause is a disjunction of three literals. 3-NAE-SAT requires the three truth values in each clause are not all equal to each other: at least one is true, and at least one is false. A clause in which all three literals are true would be considered unsatisfied in 3-NAE-SAT (unlike in 3-SAT, where such a clause would be satisfied). NAE-SAT is particularly useful in problem reductions, and is used for NP-completeness proofs for classes of problems with similar symmetry [11, Section 5.3.1].

Similar to 2-XOR-SAT, we define an infinite family of satisfiable 3-NAE-SAT formulas as $\{F_3^{\mathrm{NAE}}, F_4^{\mathrm{NAE}}, \ldots\}$ as follows. We define the set of all 3-CNF clauses on the variables $u, v, w$ with only one negated literal as $C_{u,v,w}^{(1)} := (\neg u \vee v \vee w) \wedge (u \vee \neg v \vee w) \wedge (u \vee v \vee \neg w)$.

Similarly, we define the set of all 3-CNF clauses on the variables $u, v, w$ with only two negated literals as $C_{u,v,w}^{(2)} := (\neg u \vee \neg v \vee w) \wedge (\neg u \vee v \vee \neg w) \wedge (u \vee \neg v \vee \neg w)$. A 3-NAE-SAT formula is constructed as $F_n^{\mathrm{NAE}} := \bigwedge_{i<j<k} \left( C_{x_i,x_j,x_k}^{(1)} \wedge C_{x_i,x_j,x_k}^{(2)} \right)$. Note that again, $F_n^{\mathrm{NAE}}$ is constructed in such a way that both $1^n$ and $0^n$ are satisfying assignments. Furthermore the fitness function counting the number of satisfied clauses exhibits spin-flip symmetry since the NAE satisfiability of any $k$ clause is invariant to assignment complementation.

**Lemma 2.** *For any $k \in \{2, \ldots, n\}$, the fitness functions for both $F_n^{\mathrm{XOR}}$ and $F_n^{\mathrm{NAE}}$ yield a bound on the tournament probability in the claim of Lemma 1, which is $\xi_k(n) = \Omega(1/n)$.*

*Proof.* For any $x \in \{0,1\}^n$, $x$ satisfies exactly $2\binom{n}{2} - 2|x|_1 (n - |x|_1) = n^2 + 2|x|_1^2 - n(2|x|_1 + 1)$ clauses in $F_n^{\mathrm{XOR}}$ since there are $2\binom{n}{2}$ clauses in $F_n^{\mathrm{XOR}}$, but any clause $(x_i \oplus \neg x_j)$ is not satisfied if $x_i = 1$ and $x_j = 0$ (similarly, any clause $(\neg x_i \oplus x_j)$ is not satisfied if $x_i = 0$ and $x_j = 1$). In the case of the 3-NAE-SAT formulas, an arbitrary $x \in \{0,1\}^n$ satisfies exactly $6\binom{n}{3} - 2|x|_1\binom{n-|x|_1}{2} - 2(n - |x|_1)\binom{|x|_1}{2} = (n-2)\left(n^2 + |x|_1^2 - n(|x|_1 + 1)\right)$ clauses in $F_n^{\mathrm{NAE}}$, since there are $6\binom{n}{3}$ clauses in the formula, but a clause containing only true (respectively, false) literals under $x$ must be subtracted from the count.

Define the set $\mathcal{L}_i := \{x : |x|_1 = i\}$. Then it follows that (i) the number of clauses in $F_n^{\mathrm{XOR}}$ satisfied by any $x \in \mathcal{L}_i$ is $n^2 + 2i^2 - n(2i+1)$, and (ii) the number of clauses in $F_n^{\mathrm{NAE}}$ satisfied by any $x \in \mathcal{L}_i$ is $(n-2)(n^2 + i^2 - n(i+1))$. Since $2i^2 - n(2i+1) = 2(n-i)^2 - n(2(n-i)+1)$ and $i^2 - n(i+1) = (n-i)^2 - n(n-i+1)$, $\mathcal{L}_i \cup \mathcal{L}_{n-i}$ partitions $\{0,1\}^n$ into fitness levels.

There are exactly $\binom{n-2}{i-2} + \binom{n-2}{i}$ strings $x \in \{0,1\}^n$ with $x_1 = x_k = 1$ on fitness level $\mathcal{L}_i \cup \mathcal{L}_{n-i}$. For any string $y \in \mathcal{L}_j$ with $y_1 = 1$ and $y_k = 0$, $y \oplus \mathbf{e}_k$ is in $\mathcal{L}_{j+1}$. Thus it either jumps up one fitness level, or falls down one fitness level. We derive an appropriate bound for claim in Lemma 1 as

$$\frac{1}{2^{2n-4}} \sum_{(x,y)\in S_k^{\triangle}} \left( 1_{x,y}^{>} - 1_{x,y\oplus e_k}^{>} + \frac{1_{x,y}^{=} - 1_{x,y\oplus e_k}^{=}}{2} \right)$$

$$= \frac{1}{2^{2n-4}} \sum_{i=\lceil n/2 \rceil}^{n} |(\mathcal{L}_i \cup \mathcal{L}_{n-i}) \cap \{x : x_k = 1\}|$$

$$\times \left( \binom{n-2}{i-2} - \binom{n-2}{i-1} - \binom{n-2}{i} + \binom{n-2}{i-2} \right)$$

$$\geq \frac{1}{2^{2n-3}} \left[ \binom{n-2}{\lceil n/2 \rceil}^2 + \binom{n-2}{\lceil n/2 \rceil - 1}^2 \right] = \Omega(1/n),$$

where the final inequality is the same as in the proof of Theorem 2.    $\square$

Setting $\mu$ sufficiently large and applying the result of Lemma 2 via a union bound over all variables proves the following theorem.

**Theorem 3.** *Algorithm 2 with $\mu = \Omega(n^2 \log n)$ finds a satisfying assignment to $F_n^{\mathrm{XOR}}$ or $F_n^{\mathrm{NAE}}$ with probability $1 - o(1)$.*

Ideally, we would like to see how the success of Algorithm 2 depends on the constraint density in random models of $F_n^{\mathrm{XOR}}$ and $F_n^{\mathrm{NAE}}$, but so far have not been able to do so. Instead, we investigate this dependency empirically in Sect. 4

### 3.3   Failure Case: 1D Ising Model

The Ising model was originally developed in statistical physics [10], but more recently developed into a model for studying the behavior of evolutionary algorithms, especially in the presence of spin-flip symmetry. Van Hoyweghen, Goldberg and Naudts [8] point out that in the class of spin-flip symmetric functions, TwoMax and the Ising model are at two extremes. The former can be optimized by a simple hill-climber, whereas the latter is confounded by a highly neutral landscape.

The Ising model is defined by a set of variables that correspond to nuclear magnetic moments, or spins, of a set of sites arranged on a graph structure. Each variable takes one of two states $\{-1, +1\}$, and neighboring spins with the same value have lower energy than neighbors with opposite values.

Given an undirected graph $G = (V, E)$, the energy of a configuration $s \in \{-1, +1\}^{|V|}$ is given by the Hamiltonian function

$$H(s) = -\sum_{(i,j)\in E} J_{ij} s_i s_j - \sum_{j=1}^{|V|} h_j s_j,$$

where $J_{ij}$ is the interaction between neighboring sites $i$ and $j$, and $h_j$ is an external magnetic field that interacts with site $j$. The *ground state* configuration is the sequence that minimizes $H$.

In the context of evolutionary algorithms, one usually ignores the linear term (e.g., by setting $h_j = 0$), and takes the interactions as $J_{ij} = -1/2$, and performs an affine transform from $\{-1, +1\}^n \rightarrow \{0, 1\}$ to obtain the quadratic pseudo-Boolean function

$$f_{Ising} : \{0, 1\}^n := x \mapsto \sum_{(i,j)\in E} (x_i x_j + (1 - x_i)(1 - x_j)).$$

Note when $h_j = 0$, the maxima of $f_{Ising}$ correspond to the ground states of $H$. We will also follow this convention, and consider the maximization of $f_{Ising}$ rather than the minimization of $H$.

We consider the *one-dimensional Ising model* (see Fig. 1), that is, the function $f_{Ising}$ defined above on an undirected cycle graph $G = (V, E)$ of $n$ nodes, i.e., $V = \{1, 2, \ldots, n\}$ and $E = \{(i, j) : i < j\} \cup \{(1, n)\}$. We prove that the bias toward the ground state solution $1^n$ introduced by Algorithm 2 is not sufficient to provide information to the majority vote.

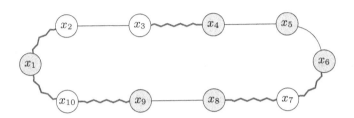

**Fig. 1.** One-dimensional Ising model on 10 sites for a given configuration $x = (1001110110)$. Shaded sites have spin 1 and unshaded sites have spin 0. Each edge incident on nodes with opposite spin contributes one violation, so the pictured configurations has 6 violations yielding a fitness value $f_{Ising}(x) = 4$.

**Lemma 3.** *For the 1D Ising model on $n$ vertices, if $n$ is even, then for any odd index $k > 1$ it holds that $\Pr(z_k = x_k^* = 1) = 1/2$.*

*Proof.* We prove the case of $k = 3$, and note that the proof strategy is similar for all odd $k$. We say an edge $(i, j)$ is *violated* for a length-$n$ binary string $x$ when $x_i \neq x_j$. Thus $f_{Ising}(x) = n - |\{(i, j) \in E : (i, j) \text{ is violated for } x\}|$. We refer to a violated edge as a *violation*.

We show that for a pair $(x, y)$ drawn uniformly at random from the set $Q$, $\Pr(x \text{ wins} \mid x_k = 1) = 1/2$. As with the proof of Lemma 1, we have $\Pr(x \text{ wins} \mid x_k = 1, y_k = 0) - \Pr(x \text{ wins} \mid x_k = y_k = 1)$ is equal to

$$\frac{1}{2^{2n-4}} \sum_{(x,y) \in S_k^{\triangle}} \left( 1_{x,y}^{>} - 1_{x,y \oplus e_k}^{>} + \frac{1_{x,y}^{=} - 1_{x,y \oplus e_k}^{=}}{2} \right).$$

To determine this sum, we count the pairs $(x, y) \in S_k^{\triangle}$ for which the inequality (or equality) of $f_{Ising}(x)$ and $f_{Ising}(y)$ changes when flipping bit $k$. Note that $f_{Ising}(y) - f_{Ising}(y \oplus e_k) \in \{-2, 0, 2\}$, since changing a single bit in $y$ can only introduce two violations (if $y_{k-1} = y_k = y_{k+1}$), remove two violations (if $y_{k-1} = y_{k+1}$, but $y_{k-1} \neq y_k$), or leave the count of violations unchanged. As $k = 3$, violations are only produced by $y \oplus e_k$ in strings with prefix 1000, and violations are only removed by $y \oplus e_k$ in strings with prefix 1101.

Fix $x = (1 \ldots)$. We claim that the sum $\sum\limits_{y \,:\, y_1=1, y_k=0} \mathbf{1}^>_{x,y} - \mathbf{1}^>_{x,y \oplus e_k}$ is equal to

$$|\{y = (1101 \ldots) : f(y) = f(x) - 2\}| - |\{y = (1000 \ldots) : f(y) = f(x)\}|$$

$$= \binom{n-3}{n-f(x)} - \binom{n-3}{n-f(x)-1} \tag{5}$$

The positive terms in the LHS of the sum in Eq. (5) correspond to strings $y$ with $f(x) > f(y)$ but $f(x) \leq f(y \oplus e_k)$. Such strings have more violations than $x$, but must have no more violations than $x$ once the $k$-th bit is flipped. Since the only strings whose violations are reduced when flipping the $k$-th bit from 0 to 1 must contain the prefix $(1101 \ldots)$, this set of strings corresponds to $\{y = (1101 \ldots) : f(y) = f(x) - 2\}$, i.e., have exactly $n - f(x) + 2$ violations. There are two violations already coming from the length-4 prefix of these strings. Hence we enumerate the length-$(n - 4)$ suffixes that contain exactly $n - f(x)$ violations. There are $\binom{n-3}{n-f(x)}$ such suffixes.

The negative terms of the sum in Eq. (5) correspond to strings $y$ with $f(x) \leq f(y)$ but $f(x) > f(y \oplus e_k)$. Since the fitness of $y$ can only be decreased by introducing violations with changing the $k$-th bit from 0 to 1, this set can be described as $\{y = (1000 \ldots) : f(y) = f(x)\}$. Such strings have exactly one violation in the length-4 prefix, thus we count strings of length $n - 4$ with $n - f(x) - 1$ violations.

Similarly, we claim the sum $\sum\limits_{y \,:\, y_1=1, y_k=0} \mathbf{1}^=_{x,y} - \mathbf{1}^=_{x,y \oplus e_k}$ is equal to

$$(|\{y = (1101 \ldots) : f(y) = f(x)\}| + |\{y = (1000 \ldots) : f(y) = f(x)\}|)$$

$$- (|\{y = (1101 \ldots) : f(y) = f(x) - 2\}| + |\{y = (1000 \ldots) : f(y) = f(x) + 2\}|)$$

$$= \binom{n-3}{n-f(x)-2} + \binom{n-3}{n-f(x)-1} - \binom{n-3}{n-f(x)} - \binom{n-3}{n-f(x)-3}$$

$$= \binom{n-3}{f(x)-1} + \binom{n-3}{n-f(x)-1} - \binom{n-3}{n-f(x)} - \binom{n-3}{f(x)}. \tag{6}$$

The positive terms of the sum in Eq. (6) correspond to strings $y$ with $f(x) = f(y)$ but $f(x) \neq f(y \oplus e_k)$. Such strings must have $n - f(x)$ violations. If $y = (1101 \ldots)$, two violations already occur in the prefix, so we enumerate the $\binom{n-3}{n-f(x)-2}$ length-$(n-4)$ suffixes with $n - f(x) - 2$ violations. If $y = (1000 \ldots)$, one violation already occurs in the prefix, so we count the $\binom{n-3}{n-f(x)-1}$ length-$(n-4)$ suffixes with $n - f(x) - 1$ violations. The negative terms are accounted for similarly.

Let $L_{2i} = |\{x : x_1 = x_k = 1 \text{ and } f(x) = 2i\}|$ denote the count of strings $x = (1x_2 1 \ldots)$ on fitness level $2i$. For these strings, either $x = (111 \ldots)$ or $x = (101 \ldots)$. In the first case, the number of strings on fitness level $2i$ is equal to the number of strings of length $n - 3$ starting at $x_{k+1}$ with $n - 2i$ violations. In the second case, we must account for the two violations already in the length-3 prefix. It follows that

$$L_{2i} = \binom{n-2}{n-2i} + \binom{n-2}{n-2(i+1)} = \binom{n-2}{n-2i} + \binom{n-2}{2i}.$$

Combining (5) and (6), $\sum_{(x,y) \in S_k^\triangle} \left( 1_{x,y}^{>} - 1_{x,y \oplus e_k}^{>} + \frac{1_{x,y}^{=} - 1_{x,y \oplus e_k}^{=}}{2} \right)$ is exactly

$$\frac{1}{2} \sum_{i=0}^{n/2} L_{2i} \left( \binom{n-3}{n-2i} - \binom{n-3}{n-2i-1} + \binom{n-3}{2i-1} - \binom{n-3}{2i} \right) = 0,$$

since the $i$-th term of the sum is the negative of the $(n/2 - i)$-th term. We conclude that $\Pr(x \text{ wins} \mid x_k = 1, y_k = 0) = \Pr(x \text{ wins} \mid x_k = y_k = 1)$. By spin-flip symmetry, $\Pr(x \text{ wins} \mid x_k = y_k = 1) = 1/2$, and by the law of total probability, $\Pr(x \text{ wins} \mid x_k = 1) = 1/2$. Applying Bayes' Theorem, $\Pr(x_k = 1 \mid x \text{ wins}) = 1/2$. The value of $p_k$ at the end of the loop at line 7 of Algorithm 2, is binomially distributed with median $\mu/2$, which completes the proof. □

Thus for the 1D Ising model, there is no bias toward correct values, and the substring matches an optimal solution only with exponentially low probability. This is captured as follows.

**Theorem 4.** *Let $n$ be an even positive integer. For any $\mu$, Algorithm 2 fails to generate an optimal solution for the 1D Ising model on $n$ vertices with probability at least $1 - 2^{-n/2}$.*

*Proof.* The two optimal solutions for the 1D Ising model are $1^n$ and $0^n$. Let $z$ be the string generated by Algorithm 2. Since Algorithm 2 only generates strings with $x_1 = 1$, $\Pr(z = 0^n) = 0$. By Lemma 3, for any odd $k > 1$, $\Pr(z_k = 1) = 1/2$. Thus $\Pr(z = x^\star) \leq \Pr\left( \bigcap_{k \in \{3,5,\dots,n-1\}} z_k = 1 \right) = 2^{-n/2}$. □

The pathology of the Ising model search space is that the Hamming weight of the candidate solutions do not give strong hints to the location of global optima. In addition to spin-flip symmetry, there are vast classes of equal fitness that correspond to strings with the same count of zero blocks. One technique to overcome this would be to add a so-called *external field* to the Ising model energy function. In this case, the energy of the system would have an additional linear term to break the symmetry, i.e., $f_{Ising}(x) + \sum_i x_i$. With the additional term, the spin-flip symmetry is also broken, and Algorithm 1 will succeed for sufficiently large $\mu$, since the fitness is now biased by Hamming weight. However, this leverages the fact that $1^n$ is now the unique optimum, and is not as interesting from the perspective of randomized optimization heuristics.

## 4   Experiments

To investigate the tightness of the runtime bounds and study the effect of constraint density on the performance of Algorithm 2 on satisfiability problems, we report the results of a number of simulations. For generalized TwoMax, we measure the empirical cumulative distribution of solved cases as a function of population size $\mu$. For each value of $\mu$, we performed 100 trials of uniformly generating a random hidden string, and counting the number of trials in which

Algorithm 2 was able to locate the string or its complement. Figure 2a reports the result of this experiment. Empirically, we see that already all of the trials are solved well below the population size required by Theorem 2. For example, to obtain a success probability of at least $1/2$ when $n = 100$, Theorem 2 requires $\mu \approx 4.18 \times 10^6$ (i.e., by setting $c = \ln 2 / \ln 100 \approx 0.15$).

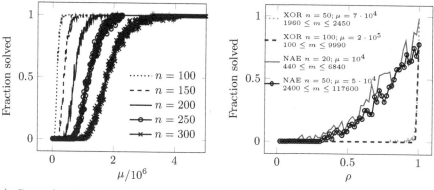

(a) Generalized TwoMax vs pop. size $\mu$.    (b) SAT formulas vs. constraint density.

**Fig. 2.** Runtime distribution for the Voting Algorithm with Symmetry Breaking (Algorithm 2) on various problem instances.

For the SAT problems, we generate random XOR- or NAE-SAT formula on $n$ variables and $m$ clauses conditioned on satisfiability by a randomly chosen assignment. This is identical to the so-called random planted distribution over formulas. To investigate the effect of constraint density as captured by the relative number of clauses to variables, we fix the population size $\mu$ and vary the clause parameter $m$ for the random planted distributions. For each value of $m$, we generate 100 random formulas and record whether Algorithm 2 finds any optimal solution. The proportion solved by the algorithm as a function of constraint density $\rho$ for each class is plotted in Fig. 2b, and the details of the generated formula classes are listed in the plot legend. Note that we calculate total constraint density as a ratio of selected clauses to total clauses in $F_n^{\mathrm{XOR}}$ or $F_n^{\mathrm{NAE}}$. Thus for XOR-2-SAT, $\rho = \frac{m}{n(n-1)}$ and for NAE-3-SAT, $\rho = \frac{m}{n(n-1)(n-2)}$. We mention here that due to the large number of clauses for high-density formulas (especially for NAE-SAT), we are restricted to smaller values of $n$ in our experiments. There is a sharp transition from unsolved to solved for fixed $\mu$ on XOR-2-SAT formulas. On NAE-3-SAT formulas, the transition is smoother. This phenomenon illustrates the rate at which the constraint density of formulas provides enough signal for Algorithm 2 to be successful.

## 5    Conclusion

We studied the effect of spin-flip symmetry on an algorithm that employs majority vote crossover on a population constructed by binary tournament selection from $\{0,1\}^n$. We rigorously proved that the algorithm fails to solve spin-flip symmetric functions with at most $2^{o(n)}$ global optima. The presence of symmetry convolutes the selective advantage of tournament winners by the existence of their equally fit complements, and this is problematic for majority vote crossover.

To ameliorate this problem, we introduced a symmetry breaking technique that does not require *a priori* knowledge of the global optimum. It operates by sampling strings uniformly from the set with a single position fixed. We proved that this small modification results in a $O(n^2 \log n)$ performance guarantee on easy spin-flip symmetric functions such as generalized TwoMax and variants of propositional satisfiability. We also proved that the technique cannot handle the highly neutral landscape of the one-dimensional Ising model.

## References

1. Briest, P., et al.: The Ising model: simple evolutionary algorithms as adaptation schemes. In: Yao, X., et al. (eds.) PPSN 2004. LNCS, vol. 3242, pp. 31–40. Springer, Heidelberg (2004). https://doi.org/10.1007/978-3-540-30217-9_4
2. Crawford, J.M., Ginsberg, M.L., Luks, E.M., Roy, A.: Symmetry-breaking predicates for search problems. Proc. KR **96**(1996), 148–159 (1996)
3. Culberson, J.: Genetic invariance: a new paradigm for genetic algorithm design. Technical Report TR92-02, University of Alberta, June 1992
4. Eiben, A.E., Raué, P.-E., Ruttkay, Z.: Genetic algorithms with multi-parent recombination. In: Davidor, Y., Schwefel, H.-P., Männer, R. (eds.) PPSN 1994. LNCS, vol. 866, pp. 78–87. Springer, Heidelberg (1994). https://doi.org/10.1007/3-540-58484-6_252
5. Fischer, S.: A polynomial upper bound for a mutation-based algorithm on the two-dimensional ising model. In: Deb, K. (ed.) GECCO 2004. LNCS, vol. 3102, pp. 1100–1112. Springer, Heidelberg (2004). https://doi.org/10.1007/978-3-540-24854-5_108
6. Fischer, S., Wegener, I.: The one-dimensional Ising model: mutation versus recombination. Theor. Comput. Sci. **344**(2–3), 208–225 (2005)
7. Friedrich, T., Kötzing, T., Krejca, M.S., Nallaperuma, S., Neumann, F., Schirneck, M.: Fast building block assembly by majority vote crossover. In: Proceeding of GECCO 2016 (2016)
8. Hoyweghen, C.V., Goldberg, D.E., Naudts, B.: From TwoMax to the Ising model: easy and hard symmetrical problems. In: Proceeding of GECCO (2002)
9. Hoyweghen, C.V., Naudts, B., Goldberg, D.E.: Spin-flip symmetry and synchronization. Evol. Comput. **10**(4), 317–344 (2002)
10. Ising, E.: Beitrag zur Theorie des Ferromagnetismus. Zeitschrift für Physik **31**(1), 253–258 (1925)
11. Moore, C., Mertens, S.: The Nature of Computation. Oxford University Press, Oxford (2011)

12. Naudts, B., Naudts, J.: The effect of spin-flip symmetry on the performance of the simple GA. In: Eiben, A.E., Bäck, T., Schoenauer, M., Schwefel, H.-P. (eds.) PPSN 1998. LNCS, vol. 1498, pp. 67–76. Springer, Heidelberg (1998). https://doi.org/10.1007/BFb0056850

13. Naudts, B., Verschoren, A.: SGA search dynamics on second order functions. In: Hao, J.-K., Lutton, E., Ronald, E., Schoenauer, M., Snyers, D. (eds.) AE 1997. LNCS, vol. 1363, pp. 207–221. Springer, Heidelberg (1998). https://doi.org/10.1007/BFb0026602

14. Pelikan, M., Goldberg, D.E.: Genetic algorithms, clustering, and the breaking of symmetry. In: Schoenauer, M., et al. (eds.) PPSN 2000. LNCS, vol. 1917, pp. 385–394. Springer, Heidelberg (2000). https://doi.org/10.1007/3-540-45356-3_38

15. Prestwich, S., Roli, A.: Symmetry breaking and local search spaces. In: Barták, R., Milano, M. (eds.) CPAIOR 2005. LNCS, vol. 3524, pp. 273–287. Springer, Heidelberg (2005). https://doi.org/10.1007/11493853_21

16. Puget, J.: Symmetry breaking revisited. Constraints 10(1), 23–46 (2005)

17. Rowe, J.E.: Aishwaryaprajna: the benefits and limitations of voting mechanisms in evolutionary optimisation. In: Proceeding of FOGA (2019)

18. Shlyakhter, I.: Generating effective symmetry-breaking predicates for search problems. Discrete Appl. Math. 155(12), 1539–1548 (2007)

19. Soos, M., Nohl, K., Castelluccia, C.: Extending SAT solvers to cryptographic problems. In: Kullmann, O. (ed.) SAT 2009. LNCS, vol. 5584, pp. 244–257. Springer, Heidelberg (2009). https://doi.org/10.1007/978-3-642-02777-2_24

20. Sudholt, D.: Crossover is provably essential for the Ising model on trees. In: Proceeding of GECCO (2005)

21. Sutton, A.M.: Superpolynomial lower bounds for the (1+1) EA on some easy combinatorial problems. Algorithmica 75(3), 507–528 (2016)

22. Whitley, D., Varadarajan, S., Hirsch, R., Mukhopadhyay, A.: Exploration and exploitation without mutation: solving the *Jump* function in $\Theta(n)$ time. In: Auger, A., Fonseca, C.M., Lourenço, N., Machado, P., Paquete, L., Whitley, D. (eds.) PPSN 2018. LNCS, vol. 11102, pp. 55–66. Springer, Cham (2018). https://doi.org/10.1007/978-3-319-99259-4_5

# A Heuristic Algorithm for School Bus Routing with Bus Stop Selection

Monique Sciortino$^{(\boxtimes)}$, Rhyd Lewis, and Jonathan Thompson

School of Mathematics, Cardiff University, Cardiff CF24 4AG, Wales
{sciortinom,lewisr9,thompsonjm1}@cardiff.ac.uk

**Abstract.** In this paper a heuristic algorithm is proposed for a school bus routing problem which is formulated as a capacitated and time-constrained open vehicle routing problem with a homogeneous fleet and single loads. The algorithm determines the selection of bus stops from a set of potential stops, the assignment of students to the selected bus stops, and the routes along the selected bus stops. Its goals are to minimize the number of buses used, the total route journey time and the student walking distances. It also aims at balancing route journey times between buses. The performance of the algorithm is evaluated on a set of twenty real-world problem instances and compared against solutions achieved by a mixed integer programming model. Reported results indicate that the heuristic algorithm finds high-quality solutions in very short amounts of computational time.

**Keywords:** School bus routing · Bus stop selection · Local search · Heuristics · Set covering · Mixed integer programming

## 1 Introduction

The school bus routing problem (SBRP) is a combinatorial optimization problem which was first investigated over 40 years ago [18]. In various countries, school bus transportation forms part of the government's administrative mechanism and is funded through local taxes. Students who live at least a certain distance from the school they attend are entitled to free or subsidized transport to and from school. In Malta, for example, school transport is provided free of charge to all state school students residing at least 1km from their school. Additional restrictions are also typically imposed on the distance that students are expected to walk between their homes and their designated bus stops.

For the academic year 2019–2020, the Maltese government announced that €27 million was to be allocated to provide free transport for over 26,000 students. Given the large amount of funds being invested, it is crucial that governments make efforts to minimize the total cost required to provide these services. One of the highest priorities is to limit the number of buses used, since each bus has an acquisition cost and a driver employment cost. Moreover, it is critical to minimize operational costs by ensuring that route journey times are kept as

C. Zarges and S. Verel (Eds.): EvoCOP 2021, LNCS 12692, pp. 202–218, 2021.
https://doi.org/10.1007/978-3-030-72904-2_13

short as possible. This also promotes positive well-being of students, particularly younger ones. In Wales, for example, a maximum 45 min and 60 min journey time is recommended for primary and secondary school pupils, respectively.

The SBRP falls into a larger class of problems called vehicle routing problems (VRPs). These involve designing optimal delivery or collection routes from one or more depots to a set of geographically scattered customers, subject to a variety of side constraints [15]. Typical constraints in VRPs include maximum capacity restrictions on vehicles (capacitated VRP (CVRP)) and maximum time/distance restrictions on routes. The CVRP first appeared over six decades ago in the seminal paper by Dantzig and Ramser [9]. Sariklis and Powell [21] proposed the open VRP (OVRP) in which routes do not start and end at a depot (as in the classical VRP), but rather either start or end at a depot. The SBRP can be modelled as a capacitated and time/distance-constrained OVRP (e.g. [4]). A taxonomic review of the VRP and its variants is presented by Braekers et al. [6].

Desrosiers et al. [10] decompose the SBRP into five subproblems. In the first subproblem, *data preparation*, a network containing the schools, student residences, potential bus stop locations and bus depots is generated. Information on the number of students at each residence, the school destination of each student, the number of buses available and their capacities is also specified. The second subproblem, *bus stop selection*, seeks to select a subset of bus stops from a set of potential bus stops and assign students to these stops. *Route generation* deals with designing routes that optimize operational efficiency without sacrificing bus safety and service quality. These objectives are often conflicting in nature since an improvement in the level of service quality can increase the cost of provision. The last two subproblems, *school bell time adjustment* and *route scheduling*, adjust the schools' opening/closing times to allow buses to service multiple schools and establish chains of routes that can be executed by the same vehicle.

In this paper, we focus on the single-school SBRP in which a series of routes is constructed for each school. This is because mixed loads (students from different schools travelling on the same bus simultaneously) are not permitted in the locations considered. The majority of the publications on school bus routing also deal with the single-school SBRP (e.g. [20,22,24]). Here, we cover the first three subproblems stated above. Park and Kim [19] and Ellegood et al. [13] note that bus stop selection is often omitted in the literature; however, they classify the solution approaches developed for the SBRP with bus stop selection. Here, we employ the location-allocation-routing (LAR) strategy, in which bus stops are first selected, students are assigned to stops, and then route generation follows. Park and Kim [19] and Ellegood et al. [13] also observe that most studies assume a homogeneous (same capacity) fleet. We take the same assumption here and dedicate our research to the morning problem, whereby students are picked up from stops and dropped off at school. A solution to the afternoon problem, whereby students are picked up from school and dropped off at stops, can be found by reversing the routes.

The remainder of the paper is organized as follows. Section 2 defines our SBRP whereas Sect. 3 describes our heuristic algorithm developed for this problem. Section 4 presents the set of real-world problem instances considered as well as the computational results. Finally, Sect. 5 provides the concluding remarks.

## 2 Problem Definition

In our SBRP, we define a parameter $m_w$ which indicates the maximum walking distance that a student is expected to walk to get to a bus stop. We also define parameter $m_e$ which specifies the minimum walking distance that students should live from the school to be eligible for school transportation. As in [16], our problem can be represented via two sets of vertices, $V_1$ and $V_2$, and two sets of edges, $E_1$ and $E_2$. The vertex set $V_1$ consists of one school $v_0$ and $n$ potential bus stops $v_1, v_2, \ldots, v_n$ and the edge set $E_1$ contains all $n(n+1)$ directed edges $(u, v)$ with $u, v \in V_1$ and $u \neq v$. Each edge $(u, v)$ in the complete directed graph $(V_1, E_1)$ is weighted by the shortest driving time $t(u, v)$ from $u$ to $v$. Meanwhile, the vertex set $V_2$ consists of eligible student addresses, with each address $w \in V_2$ being weighted by the number $s(w)$ of students living at address $w$ who require school transportation. $E_2$ is the set $\{\{w, v\} : w \in V_2 \wedge v \in V_1 \setminus \{v_0\} \wedge d(w, v) \leq m_w\}$, where $d(w, v)$ gives the shortest walking distance from address $w$ to bus stop $v$.

For this problem we can assume that the undirected bipartite graph $(V_2, V_1 \setminus \{v_0\}, E_2)$ has no isolated vertices. Otherwise, either an address has no bus stop within walking distance $m_w$ (and therefore a new bus stop must be added to $V_1$), or a bus stop has no address within walking distance $m_w$ and can thus be removed from $V_1$. Moreover, a bus stop $v \in V_1 \setminus \{v_0\}$ for which there exists an address $w \in V_2$ with just one incident edge $\{w, v\}$ shall be referred to as a *compulsory stop*. This is because such a stop $v$ is the only stop within walking distance $m_w$ to students living in address $w$ and must therefore be present in a solution.

A feasible solution to our SBRP is given by a set of routes $\mathcal{R} = \{R_1, R_2, \ldots\}$, as illustrated in Fig. 1. Each route $R \in \mathcal{R}$ uses one bus of capacity $C$ which visits a subset of bus stops and terminates at the school $v_0$. The subset of bus stops traversed by all routes is denoted by $V_1' \subseteq V_1 \setminus \{v_0\}$. This set should *cover* each address $w \in V_2$ at least once, meaning that students in each address $w$ will have at least one bus stop in $V_1'$ within walking distance $m_w$. Such a covering shall be referred to as a *complete covering* of $V_2$ whereas a covering which does not satisfy this property shall be referred to as an *incomplete covering* of $V_2$. In addition, the total number $s(R)$ of students boarding the bus on route $R$ should not exceed the capacity $C$, and the journey time $t(R)$ of route $R$ should not exceed the maximum journey time $m_t$. These constraints can be expressed as follows:

$$\bigcup_{R \in \mathcal{R}} R = V_1' \tag{1}$$

$$\forall w \in V_2, \ \exists v \in V_1' \mid \{w, v\} \in E_2 \tag{2}$$

$$s(R) \leq C \qquad\qquad \forall\, R \in \mathcal{R} \qquad (3)$$
$$t(R) \leq m_{\mathrm{t}} \qquad\qquad \forall\, R \in \mathcal{R}. \qquad (4)$$

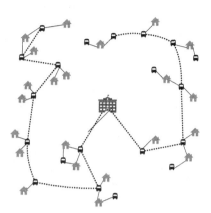

⌾ Potential Bus Stop   ⌂ Student Address   ▦ School   — Student Walk   ··· Bus Route

**Fig. 1.** A feasible solution with $|\mathcal{R}| = 2$, $|V_1| = 21$, $|V_1'| = 14$ and $|V_2| = 22$.

It is important to note that bus stops in $V_1'$ are not restricted to feature in exactly one route in $\mathcal{R}$. For example, there may not be enough spare capacity in a bus to serve all students waiting at a bus stop $v \in V_1'$. In that case, bus stop $v$ must be visited by more than one bus and we call such a stop a *multistop*. In VRP literature, this characteristic is referred to as the allowance of split deliveries [11,12]. Each student boarding at a multistop is only permitted to board one specific route serving that stop since, otherwise, a bus stopping at that multistop may possibly be too full to serve a subsequent stop in its path.

As in [16], the calculation of the journey time $t(R)$ of route $R \in \mathcal{R}$ is composed of two components; the total *bus travel time* and the total *bus dwell time*. Each dwell time within a route captures the time spent servicing a designated bus stop, i.e. the time spent to halt the bus, open the doors, board the students and merge back into traffic. In our case, we estimate the dwell time at stop $v$ in route $R$ by $d(v, R) = d_1 + d_2 s(v, R)$, where $s(v, R)$ represents the number of boarding students at stop $v$ onto route $R$, $d_2$ represents the boarding time per student, and $d_1$ is a parameter which accounts for the remaining servicing time. Here, $d_1$ and $d_2$ are taken to be 15 and 5 s, respectively. Therefore, given a route $R = (v_1, v_2, \ldots, v_l, v_0)$, the route journey time $t(R)$ is given by

$$t(R) = \left( \sum_{i=1}^{l-1} t(v_i, v_{i+1}) + t(v_l, v_0) \right) + \left( \sum_{i=1}^{l} \left( d_1 + d_2 s(v_i, R) \right) \right), \qquad (5)$$

where the first component gives the total bus travel time and the second component gives the total bus dwell time.

As previously mentioned, the primary objective of our SBRP is to identify an appropriate subset of bus stops $V_1'$ in order to minimize the number $|\mathcal{R}|$ of routes (buses) included in a solution. In our case, this is achieved by attempting to produce feasible solutions that use the lower bound of $\lceil \sum_{w \in V_2} s(w)/C \rceil$ routes needed to serve all students. A solution satisfying constraints (1)–(3) and meeting this lower bound of $|\mathcal{R}|$ is always guaranteed since multistops are allowed; however, any one of the routes could potentially violate the maximum riding time constraint (4). Thus, there may be cases where additional routes are needed. In the generation of feasible solutions, two secondary objectives and a tertiary objective are also considered. The first secondary objective deals with the effectiveness of the school transportation service, whereby we seek to minimize the total student walking distance. Through the other secondary objective, we aim to target efficiency of the service by minimizing the total route journey time in a solution. The tertiary objective is employed whenever multiple feasible solutions have the same minimum total route journey time. Here, we encourage equity of service by minimizing the discrepancy between the longest and shortest routes.

## 3    Algorithm Description

Our heuristic algorithm uses the following overall strategy. Initially, a subset of bus stops is selected and a nearest neighbour heuristic is employed to construct an initial solution using a fixed number $|\mathcal{R}|$ of routes. As mentioned, $|\mathcal{R}|$ is initially taken to be the lower bound stated in the previous section. Note also that the initial assignment of stops to routes allows the violation of (4). A local search routine involving six improvement heuristics is then invoked on this solution to try and shorten the routes using the current subset of bus stops. After this routine has completed, a procedure is performed whereby the current subset of selected bus stops is altered, the current solution is repaired, and the local search routine is re-applied. This is repeated until a time limit is reached, leading to an approach similar to iterated local search. If no solution satisfying (1)–(4) is achieved at this limit, the number $|\mathcal{R}|$ of routes is increased by one and the algorithm is restarted.

### 3.1    Construction of Initial Solution

In our approach, the initial subset of bus stops $V_1'$ is selected as follows. First, all compulsory stops are included in $V_1'$. The non-compulsory stops are arranged in non-increasing order according to the number of currently uncovered addresses they serve. The stop with the largest such value is then added to $V_1'$, breaking ties randomly. This ordering and selection procedure is repeated until a complete covering of $V_2$ is obtained. Each address in $V_2$ is then assigned to the closest bus stop in $V_1'$. The assignment of addresses to stops determines the number $s_{V_1'}(v)$ of boarding students at each stop $v \in V_1'$. It may also be the case that some stops have no boarding students, in which case they are removed from $V_1'$. Next, each bus stop in $V_1'$ is assigned to one of the $|\mathcal{R}|$ routes such that each bus is not overloaded. This assignment follows a parallel backward implementation of the nearest neighbour constructive heuristic. To start, $|\mathcal{R}|$ empty routes are defined and the remaining capacity $c_i$ of each route $R_i \in \mathcal{R}$ is set to $C$. The $|\mathcal{R}|$ closest stops to the school are then added at the front of the routes, one in each route. Closeness to school is measured by the dwell time at the stop plus the shortest driving time from the stop to the school. In order to calculate the dwell time at stop $v \in V_1'$ in route $R_i$, the minimum of $c_i$ and $s_{V_1'}(v)$ is considered as there may be more than $c_i$ students boarding at stop $v$. In this case, a multistop is created since the remaining $s_{V_1'}(v) - c_i$ students boarding at stop $v$ will need to be assigned to a different route $R_j$. The remaining capacities $c_i$ are then updated accordingly. This iterative procedure of determining the closest stop to the most recently added stop in route $R_i$, adding it to the front of the route and updating the remaining capacity $c_i$ is repeated until all stops in $V_1'$ are assigned to a route. On completion, an initial solution $\mathcal{R}$ will have been generated and can be evaluated according to the cost function described presently.

## 3.2   Cost Function

A solution $\mathcal{R} = \{R_1, R_2, \ldots\}$ is evaluated according to the cost function

$$f(\mathcal{R}) = \sum_{R \in \mathcal{R}} t'(R), \tag{6}$$

where

$$t'(R) = \begin{cases} t(R) & \text{if } t(R) \leq m_t, \\ m_t + m_t(1 + t(R) - m_t) & \text{otherwise.} \end{cases} \tag{7}$$

This means that if the journey time of a route $R \in \mathcal{R}$ exceeds $m_t$, then this journey time is scaled up heavily via a penalty. Otherwise, it is unaltered. The addition of the value 1 in the second case of (7) guarantees that two routes both with journey time at most $m_t$ are always preferred over one route with journey time exceeding $m_t$.

## 3.3   Local Search Routine

As mentioned, the intention of our local search routine is to shorten the journey times of routes in $\mathcal{R}$ while maintaining the satisfaction of (1)–(3). Our routine uses a combination of three intra-route and three inter-route operators, with the former being applied to a single route $R_1 \in \mathcal{R}$ and the latter being applied to a pair of routes $R_1, R_2 \in \mathcal{R}$. Without loss of generality, assume that $R_1 = (v_1, v_2, \ldots, v_{l_1}, v_0)$ and $R_2 = (u_1, u_2, \ldots, u_{l_2}, v_0)$. Note that this local search acts on a solution using a fixed subset of bus stops $V_1'$. The six operators considered are the following:

- *Exchange*: Choose two stops $v_i, v_j$ in $R_1$, where $1 \leq i < j \leq l_1$, and swap their position.

- *Two-Opt*: Choose two stops $v_i, v_j$ in $R_1$, where $1 \leq i < i + 3 \leq j \leq l_1$, and invert sub-route $v_i, \ldots, v_j$.
- *Generalized Or-Opt*: Choose stops $v_i, v_j, v_k$ in $R_1$, where $1 \leq i \leq j \leq l_1$ and $(1 \leq k < i$ or $j + 1 < k \leq l_1 + 1)$. Remove sub-route $v_i, \ldots, v_j$ and transfer it before stop $v_k$, possibly also inverting the sub-route if this yields a better cost. If $k = l_1 + 1$, then the sub-route is transferred before school $v_0$.
- *Or-Exchange*: Choose stops $v_i, v_j$ in $R_1$, where $1 \leq i \leq j \leq l_1$, and stop $u_k$ in $R_2$, where $1 \leq k \leq l_2 + 1$. Remove sub-route $v_i, \ldots, v_j$ from $R_1$ and transfer it before stop $u_k$ in $R_2$, possibly also inverting the sub-route if this yields a better cost. If $k = l_2 + 1$, then the sub-route is transferred before school $v_0$.
- *Cross-Exchange*: Choose stops $v_{i_1}, v_{j_1}$ in $R_1$, where $1 \leq i_1 \leq j_1 \leq l_1$, and stops $u_{i_2}, u_{j_2}$ in $R_2$, where $1 \leq i_2 \leq j_2 \leq l_2$. Swap sub-routes $v_{i_1}, \ldots, v_{j_1}$ and $u_{i_2}, \ldots, u_{j_2}$, possibly inverting either sub-route if this yields a better cost.
- *Creating Multistops*: If routes $R_1, R_2$ satisfy $t(R_1) > m_t$ and $s(R_2) < C$, then choose stop $v_i$ in $R_1$, where $1 \leq i \leq l_1$, for which $s(v_i, R_1) \geq 2$. If $v_i$ is not already in $R_2$, then insert a copy of $v_i$ in $R_2$ before the stop $u_k$, where $1 \leq k \leq l_2$, (or school $v_0$) which causes the smallest increase in $t(R_2)$. Next transfer $z = \min\{s(v_i, R_1) - 1, C - s(R_2)\}$ students from the occurrence of stop $v_i$ in $R_1$ to the occurrence of stop $v_i$ in $R_2$. Here, the value $z$ gives the maximum number of students who can be transferred (hence, decreasing $t(R_1)$ as much as possible) such that both occurrences of $v_i$ have at least one boarding student and both routes $R_1$ and $R_2$ satisfy (3).

The neighbourhood sizes corresponding to the above operators are $\mathcal{O}(|V_1'|^2)$, $\mathcal{O}(|V_1'|^2)$, $\mathcal{O}(|V_1'|^3)$, $\mathcal{O}(|V_1'|^3)$, $\mathcal{O}(|V_1'|^4)$, and $\mathcal{O}(|V_1'|^3)$, respectively. These operators are the same as those used in [16]. The exchange, two-opt and cross-exchange operators are also used in a similar context in [8], while the generalized Or-opt and Or-exchange are extensions (case $i \neq j$) of operators used in [8] and [22]. Note that some Or-exchange and cross-exchange moves can lead to a violation of (3). Such moves are therefore not evaluated. Moreover, these two operators can result in duplicate stops in the same route, which are removed as follows. Without loss of generality, assume that sub-route $v_i, \ldots, v_j$ is being transferred from route $R_1$ to route $R_2$ and that one stop $v_h, i \leq h \leq j$, is already present in $R_2$. Then stop $v_h$ is removed from the sub-route and the students boarding this occurrence of $v_h$ are all transferred to the occurrence of $v_h$ in $R_2$.

Our local search routine follows the direction of steepest descent. In each iteration, all moves in the union of the six neighbourhoods are evaluated and the move which gives the largest reduction in cost is performed. If multiple moves give the largest reduction in cost, the one which yields the smallest discrepancy between the longest and shortest routes in the solution is performed. Such a breakage of ties aims at balancing the journey times between buses. The local search routine terminates when a solution whose neighbourhoods contain no improving moves is reached.

## 3.4  Generation of Alternative Solutions

Recall that the subset of bus stops $V_1'$ is fixed during our local search routine. For this reason, our algorithm also contains an operator that generates a new subset of bus stops $V_1''$, assigns students to these bus stops, and then creates a set of routes that use these stops. We designed four variants of the algorithm, which differ in the way they generate $V_1''$. These are:

(I) Generating $V_1''$ from scratch;

(II) Generating $V_1''$ from the subset $V_1'$ used in the previous iteration;

(III) Generating $V_1''$ from the most recent subset $V_1'$ that yields a feasible solution with the lowest cost found so far;

(IV) Generating $V_1''$ via a trade-off between Variants II and III, whereby $V_1''$ has 50% chance of being generated according to Variant II and 50% chance of being generated according to Variant III.

Note that in Variant III, the subset of stops generated in the previous iteration is used if no subset has yielded a feasible solution so far.

In Variant I, the generation of $V_1''$ follows the same selection strategy as that discussed in Sect. 3.1 and new routes are again produced via nearest neighbour construction. In Variants II to IV, the non-compulsory stops in $V_1'$ are identified and a random selection of these is removed. Assuming a total number $\alpha$ of non-compulsory stops, in our case the number of removals is selected according to a Binomial distribution with parameters $\alpha$ and $3/\alpha$ so that three stops are removed on average. Upon removal, if we have an incomplete covering of $V_2$, then additional stops must be added to $V_1'$. If all addresses not covered by the stops in $V_1'$ are covered by stops which were not originally in $V_1'$, then, at each stage, a stop from the latter set of stops which serves the largest number of uncovered addresses is added, breaking ties randomly. If, on the other hand, some address is also uncovered by the stops which were not originally in $V_1'$, then at least one of the removed stops must be added back. The same selection strategy is applied in this case and the whole procedure is repeated until a new complete covering $V_1''$ of $V_2$ is achieved. Each address is then reassigned to the closest stop in $V_1''$ and stops with no addresses assigned to them are removed from $V_1''$.

Having determined a new subset of bus stops, repairs are then made to $\mathcal{R}$ so that only bus stops in $V_1''$ feature in the solution. To do this, all occurrences of stops in $V_1' \setminus V_1''$ are first removed from $\mathcal{R}$. For stops $v \in V_1'' \cap V_1'$ for which $s_{V_1''}(v) < s_{V_1'}(v)$, $s_{V_1'}(v) - s_{V_1''}(v)$ students are removed from occurrences of $v$ in $\mathcal{R}$. If this results in an occurrence of $v$ with no boarding students, then this occurrence is removed from $\mathcal{R}$. For stops $v \in V_1'' \cap V_1'$ for which $s_{V_1''}(v) > s_{V_1'}(v)$, an attempt is made to add students to occurrences of $v$ in $\mathcal{R}$. If not all $s_{V_1''}(v) - s_{V_1'}(v)$ students can be added, then a new occurrence of $v$ must be added to $\mathcal{R}$. Stops $v \in V_1'' \setminus V_1'$ must also be added to the solution. A new stop is inserted in a route having the lowest load, at the position which causes the least increase in the route journey time. If this insertion does not cater for all students boarding that stop, then the procedure is repeated.

Having repaired solution $\mathcal{R}$ (or generated completely new routes in the case of Variant I), the local search routine is then re-invoked. This repair-and-improve process is repeated until the time limit is reached.

## 4    Computational Experiments

A total of twenty real-world problem instances are considered here, summarized in Table 1. The problem instances pertaining to the UK and Australia originate from [16] and can be downloaded at [1]. The remainder were generated by us using the Bing Maps API and can be downloaded at [2]. Each problem instance was generated as follows. The location of the school was first identified and a number of random student addresses were selected on/outside a circle of radius $m_e$ from the school. The number of students living at each address was generated randomly according to the following distribution: 1, 2, 3 and 4 with probabilities 0.45, 0.4, 0.14 and 0.01, respectively. As mentioned in [16], this distribution

**Table 1.** Summary statistics for the twenty real-world problem instances, listed in increasing order of $|V_1|$. The number $S$ represents the total number of students, calculated as $\sum_{w \in V_2} s(w)$. Distances $m_e$ and $m_w$ are given in km.

| Location | Country/State | $|V_1|$ | $|V_2|$ | $S$ | $m_e$ | $m_w$ | $C$ |
|---|---|---|---|---|---|---|---|
| Mġarr | Malta | 60 | 110 | 190 | 1.0 | 1.0 | 40 |
| Mellieħa | Malta | 83 | 98 | 171 | 1.0 | 1.0 | 40 |
| Porthcawl | Wales | 153 | 42 | 66 | 3.2 | 1.6 | 70 |
| Qrendi | Malta | 161 | 150 | 255 | 1.0 | 1.0 | 40 |
| Suffolk | England | 174 | 123 | 209 | 4.8 | 1.6 | 70 |
| Senglea | Malta | 186 | 158 | 266 | 1.0 | 1.0 | 40 |
| Victoria | Gozo | 292 | 99 | 171 | 1.0 | 1.0 | 40 |
| Handaq | Malta | 298 | 170 | 285 | 1.0 | 1.0 | 40 |
| Pembroke | Malta | 329 | 200 | 335 | 1.0 | 1.0 | 40 |
| Canberra | ACT | 331 | 296 | 499 | 4.8 | 1.0 | 70 |
| Valletta | Malta | 419 | 159 | 268 | 1.0 | 1.0 | 40 |
| Birkirkara | Malta | 422 | 181 | 306 | 1.0 | 1.0 | 40 |
| Ħamrun | Malta | 519 | 321 | 192 | 1.0 | 1.0 | 40 |
| Cardiff | Wales | 552 | 90 | 156 | 4.8 | 1.6 | 70 |
| Milton Keynes | England | 579 | 149 | 274 | 4.8 | 1.6 | 70 |
| Bridgend | Wales | 633 | 221 | 381 | 4.82 | 1.6 | 70 |
| Edinburgh-2 | Scotland | 917 | 190 | 320 | 1.6 | 1.6 | 70 |
| Edinburgh-1 | Scotland | 959 | 409 | 680 | 1.6 | 1.6 | 70 |
| Adelaide | South Australia | 1188 | 342 | 565 | 1.6 | 1.6 | 70 |
| Brisbane | Queensland | 1817 | 438 | 757 | 3.2 | 1.6 | 70 |

approximates the relevant statistics in the locations considered. Potential bus stops were then identified through public records such that each stop has at least one address within walking distance $m_w$ and each address has at least one stop within walking distance $m_w$. Shortest driving times between each bus stop pair and shortest walking distances between each bus stop and address pair were then determined. Here, we use $C \in \{40, 70\}$, depending on the location of the problem instance under study, and $m_t = 2700$ s (45 min).

Our heuristic algorithm was coded in C++ and run on a 3.6 GhZ 8-Core Intel Core i9 processor with 8GB RAM. Variants I to IV were each run 25 times on each instance. The time limit for each run was taken to be five minutes. Overall, we found that feasible solutions using the lower bound of $\lceil \sum_{w \in V_2} s(w)/C \rceil$ routes were achieved in nineteen of the twenty instances in all runs. The only instance which required one additional route was the rural-based Suffolk instance. This was also observed in [16].

Statistics on the results achieved by our algorithm are summarized in Table 2. Columns 3 to 6 display the average number of iterations performed in each

**Table 2.** Number of iterations performed by our algorithm. All figures are averaged across the 25 runs, rounded to the nearest integer, plus/minus the standard deviation.

| Location | $|\mathcal{R}|$ | Variant I | Variant II | Variant III | Variant IV |
|---|---|---|---|---|---|
| Mġarr | 5 | 12777 ± 112 | 32131 ± 367 | 25137 ± 360 | 27482 ± 248 |
| Mellieħa | 5 | 1323 ± 9 | 6951 ± 53 | 5359 ± 107 | 5848 ± 74 |
| Porthcawl | 1 | 91916 ± 1091 | 96843 ± 1210 | 85351 ± 2465 | 92171 ± 3915 |
| Qrendi | 7 | 9602 ± 68 | 28176 ± 230 | 22880 ± 656 | 23901 ± 621 |
| Suffolk | 4 | 3384 ± 37 | 8076 ± 95 | 6436 ± 193 | 6660 ± 191 |
| Senglea | 7 | 16026 ± 105 | 45776 ± 317 | 42425 ± 1177 | 43792 ± 896 |
| Victoria | 5 | 2124 ± 31 | 6956 ± 90 | 5135 ± 206 | 5367 ± 254 |
| Handaq | 8 | 2439 ± 30 | 8438 ± 120 | 7433 ± 495 | 7731 ± 362 |
| Pembroke | 9 | 4208 ± 29 | 16109 ± 160 | 14170 ± 621 | 14702 ± 468 |
| Canberra | 8 | 2431 ± 42 | 10073 ± 167 | 8038 ± 290 | 8456 ± 247 |
| Valletta | 7 | 6639 ± 6 | 21976 ± 78 | 20915 ± 577 | 21405 ± 403 |
| Birkirkara | 8 | 4691 ± 70 | 17519 ± 276 | 16039 ± 635 | 16466 ± 491 |
| Hamrun | 9 | 3233 ± 54 | 13681 ± 242 | 13210 ± 258 | 13404 ± 259 |
| Cardiff | 3 | 16241 ± 102 | 25386 ± 181 | 25330 ± 185 | 25345 ± 131 |
| Milton Keynes | 4 | 8601 ± 7 | 20039 ± 40 | 18699 ± 637 | 19196 ± 272 |
| Bridgend | 6 | 3932 ± 57 | 11066 ± 106 | 9944 ± 119 | 10277 ± 85 |
| Edinburgh-2 | 5 | 5681 ± 66 | 5692 ± 55 | 5709 ± 52 | 5680 ± 60 |
| Edinburgh-1 | 10 | 1792 ± 20 | 6489 ± 83 | 6365 ± 141 | 6417 ± 128 |
| Adelaide | 9 | 1745 ± 24 | 5943 ± 72 | 5530 ± 173 | 5668 ± 134 |
| Brisbane | 11 | 742 ± 18 | 3304 ± 71 | 3241 ± 70 | 3262 ± 70 |

algorithm variant. For each variant, this number was compared with the average number of iterations that resulted in solutions satisfying (1)–(4). For Bridgend, approximately 74%, 61%, 26% and 32% of the iterations performed in Variants I to IV, respectively, produced infeasible solutions, on average. For Brisbane, the four average proportions of infeasible solutions were all less than 0.2%. Some infeasible solutions were also produced in Variants II and IV for Porthcawl and Variants I, II and IV for Suffolk. The average proportions were both less than 0.04% for Porthcawl and all less than 0.7% for Suffolk. All remaining instances saw (1)–(4) satisfied in all runs.

According to Table 2, Variant II performed the highest average number of iterations for all instances except Edinburgh-2. On the contrary, Variant I performed the lowest average number of iterations for all instances except Porthcawl and Edinburgh-2. This was expected given that this variant does not use information from previous iterations when altering the current subset of bus stops. Consequently, applications of local search take longer in each iteration. It is also evident that Variant IV performed a higher average number of iterations than Variant III for all instances except Edinburgh-2.

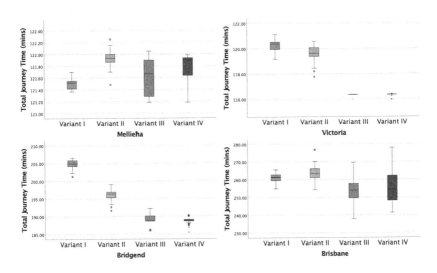

**Fig. 2.** Boxplots displaying the performance of the variants on the Mellieħa, Victoria, Bridgend and Brisbane instances.

For each instance, a Kruskal-Wallis test on the 100 total journey times (in minutes) reached by our heuristic algorithm revealed statistically significant differences between the algorithm variants ($p < 0.001$). Post-hoc Bonferroni-adjusted pairwise comparison tests indicated that, for 16 instances, the total journey times of Variants I and II are significantly different at the 0.05 level than those of Variants III and IV. Half of these instances also saw a significant difference between Variants I and II. Two instances (Porthcawl and Edinburgh-2) saw significant differences between Variant III and all other variants, whereas the Mġarr instance saw significant differences between Variant I and all other variants. For the Mellieħa instance, significant differences were found between Variants I and II, I and IV, and II and III. Boxplots displaying the performance of the variants on four selected instances are displayed in Fig. 2.

Moving to Table 3, Columns 4 to 7 display the best total journey times for the different algorithm variants. Each instance's best reported result across all variants is displayed in bold and the number of runs giving that result is shown in brackets. According to Table 3, Variants I to IV produced best reported results in 2, 3, 16 and 14 instances, respectively. Moreover, Variants I to IV produced best total journey times that are at most 14.37%, 8.76%, 4.28% and 2.14% (respectively) worse than the best reported results. It is also evident that our best reported result for eight instances was achieved by only one run. For the other twelve instances, multiple runs reached the best reported result. Some or all multiple runs for all these instances except Porthcawl and Edinburgh-2 have different corresponding subsets of bus stops. The total number of alternative subsets of bus stops is given in Column 3 of Table 3.

The best reported results from our heuristic algorithm are also compared with those of Lewis and Smith-Miles [16], in Column 2. The algorithm used in [16] is similar to Variant II of our algorithm, but makes use of a first-fit-decreasing bin-packing heuristic for the assignment of bus stops to routes rather than a nearest neighbour heuristic. Our best reported results for two instances (Porthcawl and Cardiff) match those in [16]. Additionally, our best reported results for six of the remaining eight instances are better than those in [16].

An attempt was also made to improve each instance's best reported result from our heuristic algorithm. For this purpose, a mixed integer programming (MIP) model was formulated as shown in the appendix. This model was executed using Gurobi 9.0 with a run time limit of one hour per bus stop subset. Each run was also seeded with the best feasible solution found by our heuristic algorithm for that bus stop subset. The MIP results are presented in Columns 8 to 10 of Table 3. Column 8 gives the total route journey time of the best incumbent solution reached. Column 9 displays the percentage improvement between the best reported result from our heuristic algorithm and the best incumbent result. Finally, Column 10 gives the relative MIP optimality gap between the best incumbent result and the best known lower bound on the optimal total route journey time. Note that Gurobi was able to find a solution having a better total route journey time for four instances and the percentage improvements of these range between 0.19% and 1.58%. On the other hand, Gurobi was not able to improve the heuristic algorithm solution of ten instances within the time limit. For the remaining six instances, the solver did not provide any results before the time limit was reached. The optimality gaps for the achieved MIP results range between 8.35% and 69.17%.

**Table 3.** Best (across the 25 runs) total route journey times (in minutes) achieved by our heuristic algorithm. Column 2 gives the best reported results in [16] for the instances pertaining to the UK and Australia. Column 8 gives the total route journey time of the best incumbent solution achieved by Gurobi. Column 9 displays the percentage improvement between the best reported result from our heuristic algorithm and the result in Column 8. The last column gives the relative gap between the result in Column 8 and the best known lower bound on the optimal total route journey time.

| Location | Best [16] | Heuristic Algorithm | | | | | MIP (Gurobi) | | |
|---|---|---|---|---|---|---|---|---|---|
| | | Subsets | Variant I | Variant II | Variant III | Variant IV | Incumbent | Improvement | MIP Gap |
| Mgarr | - | 3 | 90.70 | 87.77 (2) | 87.77 (2) | 87.77 (3) | 87.42 | 0.40% | 8.35% |
| Mellieha | - | 2 | 121.37 | 121.25 | 121.18 (2) | 121.18 (1) | 120.95 | 0.19% | 11.65% |
| Porthcawl | 26.87 | 1 | 28.02 | 27.35 | 28.02 | 26.87 (2) | 26.87 | - | 44.35% |
| Qrendi | - | 7 | 97.67 | 96.63 | 95.80 (6) | 95.80 (5) | 95.80 | - | 40.52% |
| Suffolk | 113.43 | 12 | 118.77 | 116.08 | 112.75 (20) | 112.75 (21) | 112.75 | - | 55.54% |
| Senglea | - | 4 | 84.85 | 84.57 | 83.18 (7) | 83.18 (7) | 81.87 | 1.58% | 63.36% |
| Victoria | - | 4 | 119.18 | 117.77 | 115.97 (2) | 115.97 (4) | 115.97 | - | 38.66% |
| Handaq | - | 12 | 142.47 | 140.28 | 138.18 (11) | 138.18 (7) | - | - | - |
| Pembroke | - | 1 | 132.82 | 133.37 | 128.80 (1) | 129.48 | 128.80 | - | 52.89% |
| Canberra | - | 1 | 198.27 | 193.13 | 188.25 (1) | 188.37 | 188.25 | - | 57.00% |
| Valletta | - | 1 | 125.28 | 122.50 | 117.82 | 117.28 (1) | 117.28 | - | 66.12% |
| Birkirkara | - | 4 | 116.27 | 113.87 | 108.25 (1) | 108.25 (3) | 108.25 | - | 55.40% |
| Hamrun | - | 2 | 118.68 | 108.75 | 104.27 (1) | 104.27 (1) | 104.27 | - | 62.12% |
| Cardiff | 57.52 | 3 | 57.52 (13) | 57.52 (16) | 57.52 (25) | 57.52 (25) | 57.52 | - | 69.17% |
| Milton Keynes | 63.48 | 1 | 68.57 | 65.45 | 61.63 (1) | 62.95 | 61.10 | 0.87% | 54.58% |
| Bridgend | 185.52 | 1 | 201.28 | 191.67 | 186.00 | 185.48 (1) | - | - | - |
| Edinburgh-2 | 56.08 | 1 | 55.87 (25) | 55.87 (25) | 55.90 | 55.87 (25) | - | - | - |
| Edinburgh-1 | 151.40 | 1 | 174.50 | 165.95 | 152.58 (1) | 153.60 | - | - | - |
| Adelaide | 136.80 | 1 | 147.83 | 143.23 | 135.47 (1) | 136.22 | - | - | - |
| Brisbane | 235.15 | 1 | 254.97 | 254.30 | 237.63 (1) | 241.38 | - | - | - |

## 5     Conclusions and Future Developments

In this paper a real-world SBRP has been studied, which incorporates several features found in the literature such as bus capacities, student eligibility, maximum student riding time, maximum student walking distance, multistops (multiple buses visiting a single bus stop) and bus dwell times. A heuristic algorithm has been developed which encompasses the first three subproblems of the SBRP, as defined in [10].

Experiments conducted on twenty problem instances from Malta, the UK and Australia demonstrate the success of the heuristic algorithm on a variety of real-sized instances. For all instances, our algorithm was able to find high-quality solutions in a very short computational time. It also copes with large-scale instances of more than 1800 potential bus stops and 750 students. Through different variants, our algorithm has provided multiple subsets of bus stops yielding the best reported total journey time for ten instances. This extension to what has been done in [16] is beneficial since government administrators can liaise with bus operators to identify the most appropriate subset of bus stops based on factors such as bus depot locations and bus stop accessibility.

The performance of our heuristic algorithm was also compared against solutions achieved through a branch-and-cut method. Only four of the twenty best reported results from the heuristic algorithm turned out to be slightly worse (by less than two minutes) than the results achieved by MIP. The MIP optimality gaps turned out to be higher than expected. These may be improved by allowing a longer time limit.

The proposed heuristic algorithm as well as the MIP model formulation presented in this paper can be extended to other SBRP variants. One such variant is the heterogeneous fleet, in which buses are characterized by different capacities, as studied in [17,20,24] amongst others. Another potential future development is the consideration of multi-tripping where several routes, possibly pertaining to different schools, are merged so that buses are able to perform multiple routes successively. Recent work in this area is discussed in [5,23]. It is also suggested that future work should address uncertainty in the bus travel times as this will make the heuristic algorithm more applicable in real-world settings (e.g. [3,7,25]).

**Acknowledgement.** The research work disclosed in this publication is supported by the Tertiary Education Scholarships Scheme (TESS, Malta).

## Appendix

The MIP model presented here produces solutions consisting of cycles that start and end at the school. The arc from the school to the first bus stop in each route is then excluded. This is possible by assuming that the driving time from the school to any stop is zero.

The decision variables of our model are as follows. Binary variable $x_{uvR}$ indicates whether route $R \in \mathcal{R}$ travels from $u \in V_1$ to $v \in V_1 \setminus \{u\}$. Binary variable $y_{vR}$ indicates whether route $R \in \mathcal{R}$ visits $v \in V_1$. Also, binary variable $z_{wv}$ indicates whether students in address $w \in V_2$ walk to stop $v \in V_1 \setminus \{v_0\}$. Variable $s_{vR} \in \{0, 1, \ldots, C\}$ gives the number of students boarding route $R \in \mathcal{R}$ from stop $v \in V_1 \setminus \{v_0\}$. Moreover, variable $l_{vR} \in \{0, 1, \ldots, C\}$ gives the total load of route $R \in \mathcal{R}$ just after visiting stop $v \in V_1 \setminus \{v_0\}$. Finally, variable $t_R \in [0, m_t]$ specifies the total journey time of route $R \in \mathcal{R}$. The MIP formulation is as follows:

$$\min \sum_{R \in \mathcal{R}} t_R \tag{8}$$

$$\text{s.t.} \quad \sum_{u \in V_1} x_{uvR} = y_{vR} \quad \forall v \in V_1, R \in \mathcal{R} \tag{9}$$

$$\sum_{u \in V_1} x_{vuR} = y_{vR} \quad \forall v \in V_1, R \in \mathcal{R} \tag{10}$$

$$y_{v_0 R} \geq y_{vR} \quad \forall v \in V_1 \setminus \{v_0\}, R \in \mathcal{R} \tag{11}$$

$$\sum_{\substack{v \in V_1 \setminus \{v_0\} \\ d(w,v) \leq m_w}} z_{wv} = 1 \quad \forall w \in V_2 \tag{12}$$

$$\sum_{R \in \mathcal{R}} y_{vR} \geq z_{wv} \quad \forall v \in V_1 \setminus \{v_0\}, w \in V_2 \tag{13}$$

$$\sum_{w \in V_2} s(w) z_{wv} - \sum_{R \in \mathcal{R}} s_{vR} = 0 \quad \forall v \in V_1 \setminus \{v_0\} \tag{14}$$

$$y_{vR} \leq s_{vR} \quad \forall v \in V_1 \setminus \{v_0\}, R \in \mathcal{R} \tag{15}$$

$$C y_{vR} \geq s_{vR} \quad \forall v \in V_1 \setminus \{v_0\}, R \in \mathcal{R} \tag{16}$$

$$l_{uR} + s_{vR} - C(1 - x_{uvR}) \leq l_{vR} \quad \forall u, v \in V_1 \,|\, v \neq v_0, R \in \mathcal{R} \tag{17}$$

$$l_{uR} + s_{vR} + C(1 - x_{uvR}) \geq l_{vR} \quad \forall u, v \in V_1 \,|\, v \neq v_0, R \in \mathcal{R} \tag{18}$$

$$\sum_{u,v \in V_1} t(u,v) x_{uvR} + \sum_{v \in V_1 \setminus \{v_0\}} (d_1 y_{vR} + d_2 s_{vR}) = t_R \quad \forall R \in \mathcal{R}. \tag{19}$$

Objective function (8) minimizes the total journey time of all routes. Constraints (9)–(11) relate to stop and school visits. Constraints (9)–(10) guarantee that if route $R \in \mathcal{R}$ visits $v \in V_1$, then route $R$ should enter and leave $v$ exactly once. Next, Constraints (11) force each route $R \in \mathcal{R}$ to visit school $v_0$ whenever it visits at least one stop $v \in V_1 \setminus \{v_0\}$. Constraints (12)–(14) relate to student walks and pickups. Constraints (12) ensure that students living in each address $w \in V_2$ walk to exactly one stop within walking distance $m_w$. Constraints (13) assure that no student walks to an unvisited stop, while Constraints (14) guarantee that the total number of students boarding from stop $v \in V_1 \setminus \{v_0\}$ is equal to the total number of students walking to that stop. Constraints (15)–(16) relate to student boardings. These constraints force the number of students boarding route $R \in \mathcal{R}$ from stop $v \in V_1 \setminus \{v_0\}$ to be 0 if route $R$ does not visit stop $v$. If route $R$ visits stop $v$, then (15) also updates the lower bound on the number of boarding students to 1. In addition, Constraints (17)–(18) relate to route loads and also serve as subtour elimination constraints as proposed in [14]. Note that $l_{v_0 R} = 0 \; \forall R \in \mathcal{R}$. These constraints guarantee that no route contains a subtour disconnected from school $v_0$ and that each route load increases in accordance to the number of students boarding the bus on that route. In fact, if route $R \in \mathcal{R}$ goes from $u \in V_1$ to stop $v \in V_1 \setminus \{u, v_0\}$, then the load of route $R$ just after visiting stop $v$ is set equal to the sum of the load of route $R$ just after visiting $u$ and the number of students boarding route $R$ from stop $v$. Finally, Constraints (19) calculate the total journey time of each route $R \in \mathcal{R}$.

# References

1. http://rhydlewis.eu/resources/busprobs.zip
2. https://github.com/MoniqueSciortino/sbrpMaltaInstances
3. Babaei, M., Rajabi-Bahaabadi, M.: School bus routing and scheduling with stochastic time-dependent travel times considering on-time arrival reliability. Comput. Ind. Eng. **138**, 106125 (2019)
4. Bektaş, T., Elmastaş, S.: Solving school bus routing problems through integer programming. J. Oper. Res. Soc. **58**(12), 1599–1604 (2007)
5. Bertsimas, D., Delarue, A., Martin, S.: Optimizing schools' start time and bus routes. Proc. Nat. Acad. Sci. **116**(13), 5943–5948 (2019)
6. Braekers, K., Ramaekers, K., Nieuwenhuyse, I.V.: The vehicle routing problem: state of the art classification and review. Comput. Ind. Eng. **99**, 300–313 (2016)
7. Caceres, H., Batta, R., He, Q.: School bus routing with stochastic demand and duration constraints. Transp. Sci. **51**(4), 1349–1364 (2017)
8. Chen, X., Kong, Y., Dang, L., Hou, Y., Ye, X.: Exact and metaheuristic approaches for a bi-objective school bus scheduling problem. PLoS One **11**(4), e0153614 (2015)
9. Dantzig, G.B., Ramser, J.H.: The truck dispatching problem. Manag. Sci. **6**(1), 80–91 (1959)
10. Desrosiers, J., Ferland, J.A., Rousseau, J.M., Lapalme, G., Chapleau, L.: An overview of a school busing system. Sci. Manag. Transp. Syst. 235–243 (1981)
11. Dror, M., Trudeau, P.: Savings by split delivery routing. Transp. Sci. **23**(2), 141–145 (1989)
12. Dror, M., Trudeau, P.: Split delivery routing. Naval Res. Logistics **37**(3), 383–402 (1990)
13. Ellegood, W.A., Solomon, S., North, J., Campbell, J.F.: School bus routing problem: contemporary trends and research directions. Omega **95**, 1–18 (2020)
14. Kek, A.G.H., Cheu, R.L., Meng, Q.: Distance-constrained capacitated vehicle routing problems with flexible assignment of start and end depots. Math. Comput. Model. **47**(1–2), 140–152 (2008)
15. Laporte, G., Nobert, Y., Taillefer, S.: Solving a family of multi-depot vehicle routing and location-routing problems. Transp. Sci. **22**(3), 161–172 (1988)
16. Lewis, R., Smith-Miles, K.: A heuristic algorithm for finding cost-effective solutions to real-world school bus routing problems. J. Discrete Algorithms **52–53**, 2–17 (2018)
17. Lima, F.M., Pereira, D.S., Conceição, S.V., Ramos Nunes, N.T.: A mixed load capacitated rural school bus routing problem with heterogeneous fleet: algorithms for the Brazilian context. Expert Syst. Appl. **56**, 320–334 (2016)
18. Newton, R.M., Thomas, W.H.: Design of school bus routes by computer. Socio Econ. Plann. Sci. 75–85 (1969)
19. Park, J., Kim, B.I.: The school bus routing problem: a review. Eur. J. Oper. Res. **202**, 311–319 (2010)

20. Sales, L.D., Melo, C.S., Bonates, T.D., Prata, B.D.: Memetic algorithm for the heterogeneous fleet school bus routing problem. J. Urban Plann. Dev. **144**(2), 04018018 (2018)
21. Sariklis, D., Powell, S.: A heuristic method for the open vehicle routing problem. J. Oper. Res. Soc. **51**(5), 564–573 (2000)
22. Schittekat, P., Kinable, J., Sörensen, K., Sevaux, M., Spieksma, F., Springael, J.: A metaheuristic for the school bus routing problem with bus stop selection. Eur. J. Oper. Res. **229**(2), 518–528 (2013)
23. Shafahi, A., Wang, Z., Haghani, A.: Solving the school bus routing problem by maximizing trip compatibility. Transp. Res. Rec. J. Transp. Res. Board **2667**(1), 17–27 (2017)
24. Siqueira, V.S., Silva, E.N., Silva, R.V., Rocha, M.L.: Implementation of the meta-heuristic GRASP applied to the school bus routing problem. Int. J. E-Educ. E-Bus. E-Manag. E-Learn. **6**(2), 137–145 (2016)
25. Sun, S., Duan, Z., Xu, Q.: School bus routing problem in the stochastic and timede-pendent transportation network. PLoS ONE **13**(8), e0202618 (2018)

# Hybrid Heuristic and Metaheuristic for Solving Electric Vehicle Charging Scheduling Problem

Imene Zaidi[1]([⊠])(iD), Ammar Oulamara[2](iD), Lhassane Idoumghar[1](iD), and Michel Basset[1](iD)

[1] Université de Haute -Alsace, IRIMAS UR 7499, 68100 Mulhouse, France
{imene.zaidi,lhassane.idoumghar,michel.basset}@uha.fr
[2] Université de Lorraine, LORIA Laboratory UMR7503,
54506 Vandoeuvre-lès-Nancy, France
ammar.oulamara@loria.fr

**Abstract.** The electric vehicle (EV) charging scheduling problem has become a research focus to mitigate the impact of large-scale deployment of EV in the near future. One of the main assumptions in literature is that there are enough charging points (CP) in the charging station to meet all charging demands. However, with the deployment of EVs, this assumption is no longer valid. In this paper, we address the electric vehicle charging problem in a charging station with a limited number of heterogeneous CPs and a limited overall power capacity. Before arriving at the station, the EV drivers submit charging demands. Then, the scheduler reserves a suitable CP for each EV and allocates the power efficiently so that the final state-of-charge at the departure time is as close as possible to the requested state-of-charge. We present two variants of the problem: a constant output power model and a variable power model. To solve these problems, heuristic and simulated annealing (SA) combined with linear programming are proposed. Simulation results indicate that the proposed approaches are effective in terms of maximizing the state-of-charge by the departure time for each EV.

**Keywords:** Electric vehicle · Charging scheduling · Optimization · Heuristic · Simulated annealing

## 1 Introduction

The adoption of EVs has been growing rapidly over the past decade, mainly as a result of ambitious government policies to reduce environmental pollution and advances in the EV industry. In 2019, worldwide EV sales reached 2.1 million, bringing the global EV fleet to 7.2 million, an increase of 40% compared to 2018 [6]. However, the future large-scale adoption of EV raises concerns about charging service quality since charging an EV is time-consuming and requires a considerable amount of electrical power. Nowadays, EV drivers tend to choose

© Springer Nature Switzerland AG 2021
C. Zarges and S. Verel (Eds.): EvoCOP 2021, LNCS 12692, pp. 219–235, 2021.
https://doi.org/10.1007/978-3-030-72904-2_14

the nearest available CPs to plug in their EVs and start charging immediately. This can easily overload charging infrastructures in a large-scale scenario resulting in poor service quality and EV drivers' dissatisfaction.

Recently, there has been growing interest in the development of EV charging scheduling strategies that focus on economical objectives such as minimizing the electricity costs or on power grid reliability by minimizing the power losses [3] and voltage deviations [7]. However, many studies assume that there is a sufficient number of CPs thus focus on the power allocation and neglect the assignment of EV to a suitable CP. Besides, they assume an identical output power of CPs. Yet, in real life, CPs with different charging output power are installed in the same charging station to meet the various type of charging demands and improve the quality of service [12].

In this work, we consider a charging station that regroups several CPs with different charging power levels. Moreover, this charging station has a maximum power capacity that the distribution-level transformer poses. Each EV driver can submit a charging reservation before arriving to avoid queuing. We developed a heuristic based on interval scheduling [9] and a Simulated Annealing (SA) to assign each EV to a suitable CP and allocate the electrical power over the plugin time. The objective of the scheduling is to ensure a final state-of-charge at the departure that is as close as possible to the requested state-of-charge.

The remainder of this paper is organized as follows. In Sect. 2, we present a brief review of the main works on electric vehicle charging scheduling problems. In Sect. 3, we describe in detail the investigated problem and formulate it as a mixed-integer linear programming (MILP). We then propose our optimization methods in Sect. 4 and evaluate their performance in Sect. 5. Finally, we conclude the paper in Sect. 6.

## 2   Related Work

Several studies have been conducted on the EV charging scheduling problem (EVCSP). Here, we discuss some of the relevant literature that addresses the EVCSP in charging stations and parking lots. From the perspective of these charging service providers, the main objectives are to reduce costs [21–23,25], improve service quality charging by maximizing the energy delivered [14,16,23], or maximizing charging incomes [14] while maintaining the physical constraints of the charging infrastructure. One largely used constraint is the charging infrastructure capacity that defines the overall power limit to avoid potential transformer overload and feeder congestion [14,16,21,25]. For the charging power of CPs, some papers consider variable power in which the charging rate varies over time [14,16,25] others consider fixed constant power [4].

EV charging demands are usually defined by arrival and departure times and the requested charging energy. [4,22,23] consider uncertainly in the arrival time. [21] consider that EVs may arrive with or without an appointment. Departure times can be provided by the EVs drivers [4,21,24,25] or it can be estimated based on historical behavior [23]. In [16], an EV is allowed to leave before its initially provided departure time. Regarding the requested charging energy, [23] assumes that the EVs drivers will provide their desired state-of-charge when they plug-in their EVs. [20,21] assume that the EV owners will directly request the energy demand expressed in kWh. Other papers consider charging EVs to the rated battery capacity [14,16,24,25]. The requested charging energy can either be a hard constraint where the desired energy must be reached [4,20,21] or a soft constraint where the scheduler tries to achieve satisfactory energy by the departure time [25].

Different optimization methods were used for tackling these problems. [24] developed a charging scheme based on binary programming for demand respond application and a convex relaxation method is proposed to solve the charging scheduling problem in real-time. Authors in [25] propose a two-stage approximate dynamic programming strategy for charging a large number of EVs. [20] provide a model predictive control based algorithm. We can also find different metaheuristics, for example, particle swarm optimization [16,21], a GRASP-like algorithm [4], memetic algorithm [4].

Although the above-mentioned studies have examined various aspects of the EV charging scheduling, they have essentially assumed there are a sufficient number of chargers for all EVs and thus the scheduler doesn't decide to which CP each EV is assigned. In this paper, hybrid heuristics and meta-heuristic are proposed to jointly assign the EVs to CP and schedule the EV charging. We consider different charging levels.

## 3   Problem Description and Formulation

We consider a charging station with $m$ charging points (CPs). The switching on and off of each CP can be controlled. An electric vehicle (EV) can be connected to the CP but does not necessarily have to charge immediately. Each charging point $i$, $i = 1, .., m$ has a constant output power $p_i$ (kW). We also consider the variable power model where the output power of each CP $i$ can vary over time from 0 to $p_i$. The charging station has a maximum power supply of $P^{max}$ (kW) which is insufficient to sustain simultaneous activation of all CPs. Thus, the sum of the output power of the CPs cannot exceed $P^{max}$ (kW) at any time. The scheduling time horizon of one day and is divided into $H$ time slots of length $\tau$ (minutes). A set of $n$ EVs that need charging in the whole day. Each EV $j$, $j = 1, .., n$, submits a charging demand by providing the following information: the desired arrival time to the station $r_j$, the estimated initial state-of-charge at the arrival $(e_j^0)$, the desired state-of-charge at the departure $(e_j^d)$, the battery capacity $B_j$ (kWh), and the departure time $d_j$. The scheduler collects all charging demands and determines a day ahead optimized charging schedule by assigning EV to each CP. Since there are a limited number of CP, the actual starting time for each EV can exceed the desired arrival time. An EV will occupy a CP from the assigned stating time until its departure time and cannot be plugged out during this period. The preemption of charging operation is allowed. Ideally, each EV should be charged to its desired state-of-charge by the departure time, however, this may not be possible due to the limited number of CP and the charging station capacity limit. So, the objective of the scheduler is to minimize the difference between the desired state-of-charge and the final state-of-charge at the departure time.

In the following, a MILP model is proposed to jointly optimize the assigned of the EV to CP and the power allocation at each time slot.

In the case of a constant power model, the decision variables are:

$$x_{ij}^t = \begin{cases} 1 \text{ if EV } j \text{ is charging by the CP } i \text{ at time slot } t \\ 0 \text{ otherwise.} \end{cases}$$

$e_j^f$ : final state-of-charge of EV $j$ at it departure time

- Objective: minimize the difference between the state-of-charge at the departure $(e_j^f)$ and the desired state-of-charge $e_j^d$.

$$\min \quad \sum_{j=1}^{n}(e_j^d - e_j^f)$$

- Constraints:

$$\sum_{i=1}^{m} x_{ij}^t \leq 1 \quad \forall j, t \tag{1}$$

Constraints (1) ensure that each EV $j$ is assigned to one CP at each time slot $t$.

$$\sum_{j=1}^{n} x_{ij}^t \leq 1 \quad \forall i, t \tag{2}$$

Constraints (2) ensure that each CP $i$ charges one EV at each time slot $t$.

$$x_{i'j}^{t'} \leq 1 - x_{ij}^t \quad \forall i, j, t, i' \neq i \quad and \quad t' < t \tag{3}$$

Constraints (3) ensure that each EV $j$ is charged by one CP $i$ i.e., The EV assigned to a CP cannot be moved to another CP.

$$e_j^0 \leq e_j^f \leq e_j^d \quad \forall j \tag{4}$$

$$e_j^f = e_j^0 + \frac{\tau \sum_{t=r_j}^{d_j} p_i \times x_{ij}^t}{B_j} \quad \forall j \tag{5}$$

Constraints (4) and (5) calculate the final state-of-charge of each EV $j$.

$$x_{ij}^t = 0 \quad \forall i, j \quad \forall t, t < r_j \quad and \quad t \geq d_j \tag{6}$$

Constraints (6) ensure each EV $j$ can only be charged between its desired arrival time $r_j$ and departure time $d_j$.

$$x_{ij}^t + x_{ij'}^{t'} \leq 1 \quad \forall i, j, t, j' : j' \neq j, t' \in [t, d_j] \tag{7}$$

Constraints (7) ensure that the CP $i$ is reserved to the EV $j$ from the time it begins to charge to its departure time $d_j$.

$$\sum_{i=1}^{m}\sum_{j=1}^{n} p_i \times x_{ij}^t \leq P^{max} \qquad \forall t \tag{8}$$

Constraints (8) ensure that the total output power doesn't exceed the charging station limit at each time slot.

When considering variable power model, we add the decision variables $p_{ij}^t$ that represent the power delivered by the CP $i$ to EV $j$ at time $t$. The constraints (5) and (8) will be replaced by the following constraints:

$$e_j^f = e_j^0 + \frac{\tau \sum_{t=r_j}^{d_j} p_{ij}^t}{B_j} \qquad \forall j \tag{9}$$

$$\sum_{i=1}^{m}\sum_{j=1}^{n} p_{ij}^t \leq P^{max} \qquad \forall t \tag{10}$$

We also add the following constraints to ensure that the delivered power to an EV $j$ by CP $i$ doesn't exceed its maximum output power $p_i$ :

$$p_i \times x_{ij}^t \geq p_{ij}^t \qquad \forall i, j, t \tag{11}$$

## 4   Proposed Methods

Solving the optimal charging scheduling problem with an exact method cannot be done in polynomial time [18]. Thus, heuristics and the Simulated Annealing (SA) metaheuristic combined with an exact method were developed.

### 4.1   Solution Representation

Solving the scheduling problem consists of determining the assignment of each Ev to CP and then, in case of a constant charging model, choose the appropriate time slots of charging. In the case of a variable power model, choose the appropriate charging rate at each time slot. Therefore, a feasible solution consists of the assignment of EVs to the CPs and the power allocation. The assignment of EVs to CPs is represented as a vector $(\pi_1, .., \pi_n)$ where $\pi_i$ is the sequence of EVs assigned to CP $i$. Once we have the EV-CP assignment, we solve the power allocation by determining the amount of power delivered by each CP to each EV at each time slot. To this end, we define the solution variables $a_{ij}^t$ of the EV-CP assignment, which is equal to 1 if the EV $j$ is plugged to the CP $i$ at time $t$. The $a_{ij}^t$ values will be used as inputs for solving the power allocation problem. To get $a_{ij}^t$ from $(\pi_1, .., \pi_n)$, we simply schedule all EVs sequentially without idles times while respecting their arrival times. For each EV $j$ in the sequence $\pi_i$, we select the earliest possible starting time $st_j = \max(r_j, d_{j'})$ where $j'$ is the EV scheduled before the EV $j$ in $\pi_i$. In this case, the variable $a^t_{ij}$ will be equal to 1 for all $t \in [st_j, d_j]$. If $st_j > d_j$, the charging demand of EV $j$ will be rejected but penalized in the objective function.

The proposed heuristics and neighborhood search in the SA deals with the EV-CP assignment problem. Then, for each solution, the power allocation problem is formulated as a linear programming model.

## 4.2   Heuristics for Solving the Assignment of EVs to CPs

In the following, we define greedy rules for the assignment of EVs to CPs.

**First Come Fist Served (FCFS) Heuristic.** First-come-fist-served (FCFS) rule is a popular approach to schedule the EVs charging. It assigns the EV with the smallest arrival time to the first available CP.

**Interval Graph Coloring Based Heuristic (IGCH).** Consider the interval graph $G = (V, E)$ where each vertex $v \in V$ represents a charging demand of an EV $j$, $j \in n$ defined by its arrival time $r_j$ and its departure time $d_j$. There is an edge $e \in E$ between two vertices if and only if their associated intervals have a nonempty intersection i.e. $(j, j') \in E$ if $[r_j, d_j] \cap [r_{j'}, d_{j'}] \neq \emptyset$. Assigning a set of EVs to a given CP is equivalent to the k-coloring problem of the graph $G$. The k-coloring problem is to assign a color $c \in \{1, .., k\}$ to each vertex of $G$ so that no adjacent vertices have the same color. The set of vertices colored with the same color corresponds to the set of EVs assigned to the same CP and it is called a color class. Since an interval graph is a chordal graph, the greedy coloring algorithm delivers an optimal coloring on a chordal graph following the perfect elimination orderings [5]. A perfect elimination ordering in a graph is an ordering of the vertices of the graph such that, for each vertex $v$ and the neighbors of $v$ that occur after $v$ in the order form a clique. We use the lexicographic breadth-first (LexBFS) search proposed by [17] to find the perfect elimination ordering in linear time. We add randomness to the algorithm to generate difference perfect elimination orders by adding a random weight $w$ to each vertex. Therefore, when two vertices have the same label, we choose the vertex with maximum weight $w$. Algorithm 1 shows the pseudocode of LexBFS ordering.

In the case where we have the chromatic number $k$, i.e. the number of color classes is less than or equal to the number of CP, we assign the EVs with the same color class to the same CP. We start with the color classes that have the greater cardinally (the highest number of vertices with the same color) and assign them to the CP with the greater charging output power. Otherwise, when $k > m$, each remaining non assigned EV $j$ will be assigned to the CP that has the largest available time from $r_j$ to $d_j$. The overall procedure of the IGCH algorithm is depicted in Algorithm 2.

---

**Algorithm 1: LexBFS**

---

   **input** : Interval graph $G = (V, E)$
   **output**: A perfect elimination orderings $\sigma = (v_1, .., v_n)$
1  **for** $v \in V$ **do**
2     |  label $(v) \leftarrow []$
3     |  $w(v) \leftarrow$ random()
4  **end**
5  **for** $i = |V|$ down to 1 **do**
6     |  choose a vertex $v \in V$ with lexicographically maximal label with ties being
       |  broken by $w(v)$
7     |  $\sigma(i) \leftarrow v$
8     |  **for** $u \in Neighborhood(v)$ **do**
9     |    |  label $(u) \leftarrow$ label $(u)$.concatenate($u$)
10    |  **end**
11 **end**
12 **return** $\sigma$

---

### 4.3  Exact Methods for Solving the Power Allocation Problem

After determining the assignment of the EVs to CPs, the objective here is to decide the amount of electric power delivered by each CP at each time slot. We formulate both constant and variable power models as an integer linear programming (ILP) model.

**A Linear Programming for Constant Power Model**: We define the binary decision variables $y_{ij}^t$.

$$y_{ij}^t = \begin{cases} 1 \text{ if EV j is charging by the CP i at time t} \\ 0 \text{ otherwise.} \end{cases}$$

Let $H_j = \{t | t \in H, a_{ij}^t = 1\}$. We set $y_{ij}^t = 0 \qquad \forall t \notin H_j$

– Objective :

$$\min \quad \sum_{j=1}^{m}(e_j^d - e_j^f)$$

– Constraints:

$$e_j^0 \le e_j^f \le e_j^d \qquad \forall j \tag{12}$$

$$e_j^f = e_j^0 + \frac{\sum_{t=r_j}^{d_j} y_{ij}^t \times \tau \times p_i}{B_j} \qquad \forall j \tag{13}$$

Constraints (12) and (13) calculate the final state-of-charge of each EV $j$.

$$\sum_{i=1}^{n}\sum_{j=1}^{m} y_{ij}^t \times p_i \le P^{max} \qquad \forall t \tag{14}$$

Constraints (14) ensure that the total output power doesn't exceed the charging station limit at each time slot.

---

**Algorithm 2:** Heuristic using interval graph coloring

---

    **input** : Scenario of $n$ EVs and $m$ CPs

    **output**: assignment of EVs to CPs

1  Construct the interval graph $G$ of the scenario

2  Get a perfect elimination ordering $\sigma$ of $G$ using Algorithm 1

3  $k \leftarrow 1$

4  $Color(\sigma(0)) \leftarrow k$

5  **for** $i = n$ down to 1 **do**

6      $v \leftarrow \sigma(i)$

7      **for** $c = 0$ to $k$ **do**

8          **if** color $c \notin Color(Neighborhood(v))$ **then**

9             $Color(v) \leftarrow c$

10        **end**

11      **end**

12      **if** all colors $c \in Color(Neighborhood(v))$ **then**

13         $k \leftarrow k + 1$

14         $Color(v) \leftarrow c$

15      **end**

16  **end**

17  Sort CPs in non-decreasing order of their output power

18  Sort the color classes in non-decreasing order their cardinality

19  Assign the EVs of the k first classes to the k first CPs

20  **if** $k > m$ **then**

21      **for** each EV $j$ in the remaining non assigned classes **do**

22         Choose the CP $i$ that have the largest sub-interval of $[r_j, d_j]$ where the CP is free (no EV is plugged in) and assign the EV CP $i$

23      **end**

24  **end**

---

**A Linear Programming for the Variable Power Model**: We define the continuous decision variables $p_{ij}^t$ which represents the power delivered by the CP $i$ to EV $j$ at time $t$.

$$\min \quad \sum_{j=1}^{m}(e_j^d - e_j^f)$$

Constraints :

$$e_j^0 \le e_j^f \le e_j^d \qquad \forall j \tag{15}$$

$$e_j^f = e_j^0 + \frac{\sum_{t=r_j}^{d_j} p_{ij}^t \times \tau}{B_j} \qquad \forall j \tag{16}$$

Constraints (15) and (16) calculate the final state-of-charge of each EV $j$.

$$a_{ij}^t \times p_i \ge p_{ij}^t \qquad \forall i, j, t \in H_j \tag{17}$$

Constraints (17) ensure that the delivered power by each CP $i$ doesn't exceed its maximum rated power $p_i$ at each time $t$.

$$\sum_{i=1}^{n}\sum_{j=1}^{m} p_{ij}^{t} \leq P^{max} \qquad \forall t \tag{18}$$

Constraints (18) ensure that the total output power doesn't exceed the charging station limit.

## 4.4   Simulated Annealing

Simulated annealing algorithm, initially proposed by [8], has been successfully adapted to address several optimal resource scheduling problems. The pseudo-code of the SA algorithm is shown in Algorithm 3.

A candidate solution for SA represents the EV-CP assignment part described in Sect. (4.1). We obtain its objective value by solving the LP problem representing the power allocation as described in Sect. (4.3). In preliminary experiments, the population-based meta-heuristics turned out to be worse than SA and time-consuming due to solving an LP for each newly generated solution in the population. Thus, we pursue with SA only.

First, an initial solution $S_0$ is generated using the FCFS rule. Starting from the initial solution $S_0$, SA generates a neighborhood solution $S'$. The difference in the objective value between the new solution $S'$ and the current solution $S$ is calculated as $\Delta f = f(S') - f(S)$. The neighborhood solution $S'$ replaces the current solution based on the Metropolis criteria; It will replace the current solution if there is an improvement i.e. $\Delta f < 0$. If it also improves the best solution found so far, it will become the new global best solution $S_{best}$. Otherwise, a random number $r$ is generated following the uniform distribution $U[0,1]$ and the neighborhood solution $S'$ will become the current solution if $r \leq e^{-\Delta f/T}$ where $T$ is a parameter called temperature, which regulates the probability of accepting worsening solutions. The temperature is initially set to a value $T_0$ proportional to the objective function value of the initial solution $T_0 = \mu f(S_0)$ where $\mu$ is a fixed parameter.

At each iteration $l$, the temperature is gradually decreased by a cooling scheme. The authors in the original SA paper [8] propose the following decreasing geometric cooling scheme:

$$T_{l+1} = \alpha T_l$$

Where $0 < \alpha < 1$. Typically, when $\alpha$ is set to high implying a slow decrease of the temperature. We also consider the Lundy-Mees cooling scheme proposed by [11]:

$$T_{l+1} = \frac{T_l}{a + bT_l}$$

Connolly in [1] develops a variant of the Lundy-Mees scheme that set the parameter $a$ to 1 and $b$ in dependence of the initial temperature $T_0$, the final temperature $T_f$ and the size of the neighborhood $M$:

$$b = \frac{T_0 - T_f}{MT_0T_f}$$

The algorithm will stop if it reaches the maximum number of iteration or after a fixed number of moves that did not result in accepted solutions. When the stopping criterion is met, the algorithm terminates returning the best solution $S_{best}$ found so far.

**Neighborhood Generation.** A neighborhood solution is obtained through a perturbation on the assignment of Evs to CPs in the current solution by one of the following moves:

- Swap in the same CP: a neighborhood solution is generated by randomly choosing a CP and interchange the position of two EVs scheduled in this CP. The positions of EVs are randomly selected.
- Swap from two different CP: a neighborhood solution is generated by randomly swapping two EVs between two CPs. The CPs and EVs are randomly selected.
- Insert: a neighborhood solution is generated by randomly choosing an EV in a CP and move it to another position in another CP. The CPs and the position of the EV are randomly selected.
- Shift left: a neighborhood solution is generated by moving an EV at position $p_1$ from a CP to position $p_2$ in the same CP where $p_2 < p1$. The positions $p_1$ and $p_2$ are randomly selected.
- Shift right: a neighborhood solution is generated by moving an EV at position $p_1$ from a CP to position $p_2$ in the same CP where $p_2 > p_1$. The positions $p_1$ and $p_2$ are randomly selected.

After each move, the power allocation is solved using the ILP models described in Sect. 4.3.

## 5    Experimental Analysis

To evaluate the performance of the proposed methods, several simulations are conducted, and the relevant results are discussed in this section. Note that all algorithms are implemented in C++ programming language. We use CPLEX 12.7 as a solver for the LP models in our heuristics and SA. Note that the CPLEX was not used to solve the MILP presented in Sect. (3) since we couldn't get results for small instances vehicles even after turning for several days due to out of memory errors.

### 5.1    Parameters Tuning for SA

The tuning of the particular optimization problem is essential to obtain good results. We use the IRACE (Iterated Racing for Automatic Algorithm Configuration) package [10] which allows an automatic parameter configuration using the Iterated F-race method. Two scenarios of each case were selected to be the training instances for IRACE. Table 1 presents the parameters of the SA along with the range of values that tested for each parameter. The final choices of the parameter values are also presented and they are used in all experiments in the following sections.

---

**Algorithm 3:** Simulated annealing

**input** : Initial solution $S_0$, maxNeighbours, maxAccepted, final Temperature $T_f$, maxTrials, $\mu$

**output**: best solution found $S_{best}$

1   $S_{best} \leftarrow S_0$ , $S \leftarrow S_0$ , $T \leftarrow \mu f(S_0)$,   $M \leftarrow \frac{maxTrails}{maxNeighbours}$,   $trail \leftarrow 0$

2 **do**

3     $accepted \leftarrow 0$,   $generated \leftarrow 0$

4     **while** $generated \leq maxNeighbours$ **and** $accepted \leq maxAccepted$ **do**

5        $S' \leftarrow Neighbour(S)$

6        $\Delta f \leftarrow f(S') - f(S)$

7        $generated \leftarrow generated + 1$

8        $trail \leftarrow trail + 1$

9        **if** $f(S') < f(S)$ **or** $U(0,1) \leq e^{-\Delta f/T}$ **then**

10          $S \leftarrow S'$,   $accepted \leftarrow accepted + 1$

11          **if** $f(S) < f(S_{best})$ **then**

12            $S_{best} \leftarrow S$

13          **end**

14        **end**

15     **end**

16     update T according to the cooling scheme

17 **while** $trail \leq maxTrails$ **and** $accepted > 0$;

18 **return** $S_{best}$

---

**Table 1.** Values tested for the SA parameters tuning.

| Parameter | value range | Best |
|---|---|---|
| $\mu$ | [0.01, 0.9] | 0.12 |
| Max neighbours | 20,50,100 | 50 |
| Cooling technique | LandyMees, Geometric | LandyMees |
| $\alpha$ | [0.01,0.9] | - |
| final temperature | [0.001,0.1] | 0.01 |
| max accepted | 0.1,0.2,0.3,0.4 | 0.1 |

## 5.2   Scenarios Generation

Regarding the charging station, we consider three cases with different number of CPs $m = \{10, 20, 40\}$. For each case, 30% of CPs deliver 3.7 kW, 30% deliver 11 kW, 30% deliver 22 kW and 10% deliver 43 kW. This charging rates are chosen from the standard IEC 61851 [19] that defines the classification of the different charging modes. The charging station maximum capacity $P^{max}$ is set to 70% of $\sum_{i=1}^{n} p_i$. The scheduling time horizon is one day and the time slot $\tau$ is set to 6 min.

To test the model implemented, we need to model the stochastic EVs charging demands. The EV arrivals are randomly occurring and independent events.

**Table 2.** Comparison of results with $m = 10$.

| scenario | $n$ | k | FCFS | | IGCH | | | | SA | | | |
|---|---|---|---|---|---|---|---|---|---|---|---|---|
| | | | obj | time (s) | mean | best | std | time | mean | best | std | time (s) |
| Model with constant power | | | | | | | | | | | | |
| 1 | 20 | 12 | 0.52 | 0.47 | 0.53 | 0.26 | 0.15 | 0.29 | 0.25 | **0.13** | 0.09 | 56.15 |
| 2 | 29 | 18 | 3.74 | 0.42 | 2.92 | 2.41 | 0.22 | 0.35 | 1.26 | **0.80** | 0.29 | 82.26 |
| 3 | 24 | 16 | 2.05 | 0.35 | 2.20 | 1.68 | 0.39 | 0.28 | 0.55 | **0.25** | 0.17 | 68.69 |
| 4 | 25 | 17 | 1.64 | 0.43 | 1.80 | 1.11 | 0.51 | 0.31 | 0.57 | **0.31** | 0.15 | 72.62 |
| 5 | 30 | 16 | 4.79 | 0.44 | 2.57 | 2.34 | 0.36 | 0.35 | 2.01 | **1.34** | 0.47 | 93.43 |
| 6 | 27 | 16 | 2.80 | 0.35 | 2.57 | 0.79 | 0.19 | 0.31 | 1.48 | **0.66** | 0.46 | 104.38 |
| 7 | 29 | 14 | 2.10 | 0.42 | 2.01 | 1.26 | 0.23 | 0.35 | 0.96 | **0.66** | 0.21 | 80.72 |
| 8 | 26 | 16 | 2.26 | 0.34 | 1.39 | 1.08 | 0.22 | 0.31 | 0.85 | **0.42** | 0.36 | 78.28 |
| 9 | 22 | 14 | 2.41 | 0.30 | 1.78 | 1.15 | 0.31 | 0.25 | 0.48 | **0.10** | 0.35 | 91.35 |
| 10 | 22 | 15 | 0.73 | 0.28 | 0.56 | 0.37 | 0.19 | 0.26 | 0.32 | **0.19** | 0.10 | 90.17 |
| 11 | 27 | 9 | 2.21 | 0.50 | 1.32 | 0.79 | 0.24 | 0.37 | 1.02 | **0.53** | 0.35 | 77.03 |
| 12 | 25 | 10 | 1.87 | 0.33 | 0.60 | 0.43 | 0.10 | 0.34 | 0.53 | **0.29** | 0.19 | 75.24 |
| 13 | 20 | 8 | 2.03 | 0.27 | 0.91 | 0.59 | 0.07 | 0.23 | 0.46 | **0.19** | 0.22 | 59.76 |
| 14 | 23 | 9 | 2.21 | 0.31 | 0.95 | 0.83 | 0.06 | 0.28 | 0.86 | **0.45** | 0.28 | 69.82 |
| 15 | 22 | 9 | 1.47 | 0.29 | 0.64 | 0.44 | 0.12 | 0.26 | 0.52 | **0.32** | 0.17 | 67.86 |
| Model with variable power | | | | | | | | | | | | |
| 1 | 20 | 12 | 0.27 | 0.30 | 0.25 | 0.02 | 0.14 | 0.23 | 0.03 | **0.00** | 0.05 | 61.50 |
| 2 | 29 | 18 | 3.34 | 0.39 | 2.53 | 2.09 | 0.23 | 0.33 | 0.80 | **0.20** | 0.28 | 90.74 |
| 3 | 24 | 16 | 1.79 | 0.32 | 1.73 | 1.03 | 0.28 | 0.26 | 0.23 | **0.00** | 0.16 | 76.03 |
| 4 | 25 | 17 | 1.26 | 0.35 | 1.40 | 0.62 | 0.56 | 0.28 | 0.13 | **0.00** | 0.09 | 85.89 |
| 5 | 30 | 16 | 4.51 | 0.38 | 2.52 | 1.98 | 0.56 | 0.33 | 1.48 | **0.66** | 0.46 | 104.38 |
| 6 | 27 | 16 | 2.46 | 0.34 | 0.81 | 0.46 | 0.24 | 0.30 | 0.61 | **0.13** | 0.25 | 105.85 |
| 7 | 29 | 14 | 1.90 | 0.36 | 1.59 | 0.61 | 0.42 | 0.32 | 0.68 | **0.27** | 0.23 | 101.39 |
| 8 | 26 | 16 | 2.00 | 0.33 | 1.07 | 0.28 | 0.35 | 0.29 | 0.48 | **0.10** | 0.35 | 91.35 |
| 9 | 22 | 14 | 2.06 | 0.28 | 1.24 | 0.33 | 0.39 | 0.24 | 0.48 | **0.09** | 0.25 | 97.57 |
| 10 | 22 | 15 | 0.48 | 0.27 | 0.38 | 0.13 | 0.27 | 0.24 | 0.05 | **0.00** | 0.05 | 88.00 |
| 11 | 27 | 9 | 1.89 | 0.34 | 0.76 | 0.32 | 0.28 | 0.34 | 0.67 | **0.09** | 0.30 | 90.27 |
| 12 | 25 | 10 | 1.64 | 0.32 | 0.37 | 0.16 | 0.27 | 0.28 | 0.35 | **0.04** | 0.25 | 87.20 |
| 13 | 20 | 8 | 1.73 | 0.25 | 0.59 | 0.50 | 0.02 | 0.23 | 0.17 | **0.00** | 0.16 | 70.88 |
| 14 | 23 | 9 | 2.04 | 0.29 | 0.59 | 0.57 | 0.05 | 0.26 | 0.60 | **0.10** | 0.27 | 77.36 |
| 15 | 22 | 9 | 1.14 | 0.28 | 0.30 | **0.00** | 0.20 | 0.24 | 0.20 | **0.00** | 0.10 | 74.38 |

**Table 3.** Comparison of results with $m = 20$.

| scenario | n | k | FCFS | | IGCH | | | | SA | | | |
|---|---|---|---|---|---|---|---|---|---|---|---|---|
| | | | obj | time (s) | mean | best | std | time (s) | mean | best | std | time (s) |
| Model with constant power | | | | | | | | | | | | |
| 16 | 48 | 22 | 6.04 | 1.86 | 3.33 | **2.49** | 0.50 | 1.41 | 3.87 | 2.52 | 0.61 | 320.80 |
| 17 | 50 | 22 | 6.22 | 1.50 | 3.59 | **2.09** | 0.64 | 1.40 | 4.25 | 3.45 | 0.42 | 340.73 |
| 18 | 59 | 30 | 4.95 | 1.67 | 3.68 | 2.55 | 0.63 | 1.68 | 3.01 | **2.26** | 0.42 | 391.15 |
| 19 | 66 | 28 | 7.39 | 1.98 | 5.90 | 4.78 | 0.71 | 1.91 | 5.39 | **3.84** | 0.74 | 428.59 |
| 20 | 51 | 23 | 4.73 | 1.52 | 3.51 | **2.08** | 0.61 | 1.38 | 3.33 | 2.58 | 0.52 | 323.44 |
| 21 | 58 | 33 | 6.60 | 1.70 | 5.61 | 3.88 | 0.86 | 1.65 | 4.10 | **2.80** | 0.71 | 381.73 |
| 22 | 53 | 26 | 3.76 | 1.59 | 3.49 | 2.40 | 0.69 | 1.41 | 2.66 | **1.30** | 0.60 | 322.97 |
| 23 | 68 | 37 | 9.24 | 1.98 | 7.80 | 5.41 | 0.85 | 1.90 | 6.08 | **4.60** | 0.67 | 448.97 |
| 24 | 52 | 27 | 4.21 | 1.49 | 3.02 | **1.76** | 0.78 | 1.43 | 2.57 | 1.96 | 0.37 | 340.59 |
| 25 | 54 | 28 | 5.89 | 1.59 | 5.17 | 3.82 | 0.81 | 1.46 | 4.02 | **3.09** | 0.58 | 370.09 |
| 26 | 40 | 20 | 4.11 | 1.17 | 2.18 | 1.55 | 0.36 | 2.01 | 2.10 | **1.29** | 0.42 | 283.48 |
| 27 | 40 | 18 | 4.05 | 1.13 | 2.05 | 1.33 | 0.28 | 2.11 | 2.28 | 1.42 | 0.51 | 259.32 |
| 28 | 37 | 20 | 3.52 | 1.12 | 1.53 | 1.03 | 0.31 | 1.97 | 1.37 | **0.96** | 0.30 | 246.28 |
| 29 | 33 | 15 | 2.88 | 0.91 | 0.90 | 0.62 | 0.14 | 1.58 | 0.85 | **0.44** | 0.30 | 206.95 |
| 30 | 39 | 18 | 3.48 | 1.13 | 2.04 | 1.47 | 0.27 | 2.21 | 1.58 | **0.88** | 0.34 | 240.35 |
| Model with variable power | | | | | | | | | | | | |
| 16 | 48 | 22 | 5.39 | 1.33 | 2.58 | **1.77** | 0.53 | 1.25 | 3.87 | 2.52 | 0.61 | 320.80 |
| 17 | 50 | 22 | 5.79 | 1.37 | 3.02 | **1.84** | 0.80 | 1.28 | 3.64 | 2.50 | 0.61 | 368.63 |
| 18 | 59 | 30 | 4.20 | 1.63 | 2.75 | 1.80 | 0.54 | 1.54 | 2.16 | **1.25** | 0.42 | 427.03 |
| 19 | 66 | 28 | 6.72 | 1.82 | 5.22 | 3.40 | 0.65 | 1.72 | 4.44 | **3.24** | 0.64 | 480.56 |
| 20 | 51 | 23 | 4.24 | 1.41 | 2.52 | **1.60** | 0.48 | 1.30 | 2.49 | 1.64 | 0.56 | 371.11 |
| 21 | 58 | 33 | 6.03 | 1.58 | 5.22 | 3.18 | 0.87 | 1.49 | 3.54 | **2.20** | 0.47 | 427.77 |
| 22 | 53 | 26 | 3.18 | 1.44 | 2.69 | 1.66 | 0.64 | 1.34 | 2.04 | **1.16** | 0.42 | 368.22 |
| 23 | 68 | 37 | 8.75 | 1.86 | 6.66 | 5.12 | 0.73 | 1.77 | 5.46 | **4.29** | 0.53 | 507.92 |
| 24 | 52 | 27 | 3.75 | 1.44 | 2.36 | **0.99** | 0.62 | 1.32 | 2.25 | 1.23 | 0.50 | 373.87 |
| 25 | 54 | 28 | 5.44 | 1.48 | 4.46 | 2.67 | 0.78 | 1.37 | 3.64 | **2.34** | 0.57 | 390.13 |
| 26 | 40 | 20 | 3.61 | 1.06 | 1.29 | **0.82** | 0.23 | 1.88 | 1.62 | 0.91 | 0.46 | 299.69 |
| 27 | 40 | 18 | 3.75 | 1.09 | 1.27 | **0.38** | 0.29 | 1.97 | 1.68 | 0.97 | 0.57 | 280.38 |
| 28 | 37 | 20 | 3.22 | 1.11 | 0.84 | 0.45 | 0.26 | 1.80 | 0.98 | **0.20** | 0.32 | 267.22 |
| 29 | 33 | 15 | 2.60 | 0.86 | 0.41 | **0.09** | 0.19 | 1.49 | 0.58 | 0.11 | 0.20 | 241.82 |
| 30 | 39 | 18 | 3.16 | 1.10 | 1.29 | 0.67 | 0.22 | 2.11 | 1.10 | **0.62** | 0.31 | 297.82 |

**Table 4.** Comparison of results with $m = 40$.

| scenario | $n$ | k | FCFS | | IGCH | | | | SA | | | |
|---|---|---|---|---|---|---|---|---|---|---|---|---|
| | | | obj | time (s) | mean | best | std | time (s) | mean | best | std | time (s) |
| Model with constant power | | | | | | | | | | | | |
| 31 | 93 | 45 | 6.27 | 6.30 | 3.92 | **2.43** | 0.67 | 5.34 | 4.76 | 3.91 | 0.58 | 1621.11 |
| 32 | 99 | 53 | 7.27 | 6.38 | 6.70 | 5.24 | 0.62 | 5.86 | 5.56 | **4.57** | 0.55 | 1433.02 |
| 33 | 79 | 41 | 6.64 | 5.74 | 3.59 | **2.19** | 0.55 | 4.45 | 4.81 | 3.50 | 0.69 | 1122.35 |
| 34 | 102 | 52 | 9.59 | 6.69 | 6.76 | **4.84** | 1.03 | 6.39 | 6.90 | 5.98 | 0.54 | 1442.63 |
| 35 | 93 | 52 | 7.77 | 5.97 | 5.61 | **4.10** | 0.70 | 5.51 | 6.11 | 4.83 | 0.48 | 1382.38 |
| 36 | 96 | 50 | 4.30 | 6.20 | 4.78 | 3.52 | 0.64 | 5.64 | 3.45 | **2.84** | 0.33 | 1462.46 |
| 37 | 96 | 52 | 9.47 | 6.47 | 5.09 | **4.02** | 0.69 | 5.61 | 6.77 | 5.89 | 0.71 | 1480.53 |
| 38 | 112 | 58 | 9.82 | 7.45 | 7.71 | **5.96** | 1.06 | 6.95 | 7.96 | 6.68 | 0.64 | 1648.97 |
| 39 | 95 | 43 | 7.04 | 6.26 | 4.70 | **3.53** | 0.80 | 6.04 | 5.34 | 4.40 | 0.55 | 1379.94 |
| 40 | 78 | 38 | 6.56 | 5.83 | 3.40 | **2.71** | 0.32 | 4.90 | 3.92 | 3.05 | 0.52 | 1127.74 |
| 41 | 85 | 37 | 6.40 | 5.22 | 4.09 | **3.13** | 0.68 | 9.35 | 4.44 | 3.20 | 0.55 | 1188.21 |
| 42 | 82 | 39 | 5.47 | 5.51 | 3.37 | **2.10** | 0.52 | 8.83 | 3.40 | 2.29 | 0.58 | 1245.61 |
| 43 | 88 | 40 | 6.36 | 5.47 | 4.27 | **2.96** | 0.68 | 9.86 | 4.56 | 3.86 | 0.41 | 1241.81 |
| 44 | 91 | 40 | 6.13 | 5.24 | 3.68 | **2.72** | 0.52 | 9.56 | 4.46 | 3.33 | 0.70 | 1239.36 |
| 45 | 79 | 39 | 6.81 | 4.70 | 4.36 | **2.98** | 0.56 | 8.13 | 4.69 | 3.41 | 0.51 | 1131.08 |
| Model with variable power | | | | | | | | | | | | |
| 31 | 93 | 45 | 5.32 | 5.75 | 2.59 | **1.21** | 0.77 | 5.07 | 3.59 | 2.34 | 0.63 | 1663.55 |
| 32 | 99 | 53 | 6.30 | 7.47 | 5.27 | **3.32** | 1.03 | 5.46 | 4.87 | 3.36 | 0.62 | 1629.41 |
| 33 | 79 | 41 | 5.79 | 5.15 | 2.47 | **1.44** | 0.62 | 4.37 | 3.88 | 2.97 | 0.49 | 1258.61 |
| 34 | 102 | 52 | 8.16 | 6.60 | 5.32 | **2.61** | 1.02 | 5.79 | 5.62 | 4.43 | 0.64 | 1645.17 |
| 35 | 93 | 52 | 6.69 | 5.86 | 4.02 | **1.71** | 0.84 | 5.18 | 4.75 | 4.01 | 0.39 | 1613.15 |
| 36 | 96 | 50 | 3.18 | 6.30 | 3.40 | **1.27** | 0.75 | 5.36 | 2.44 | 1.52 | 0.43 | 1640.23 |
| 37 | 96 | 52 | 8.31 | 6.27 | 3.39 | **1.72** | 0.60 | 5.37 | 6.24 | 4.67 | 0.70 | 1569.15 |
| 38 | 112 | 58 | 8.57 | 7.26 | 5.88 | **3.84** | 0.84 | 6.43 | 6.93 | 5.41 | 0.65 | 1784.67 |
| 39 | 95 | 43 | 5.79 | 6.24 | 3.02 | **1.21** | 0.80 | 5.65 | 4.17 | 3.27 | 0.62 | 1579.64 |
| 40 | 78 | 38 | 5.61 | 4.71 | 2.12 | **1.20** | 0.52 | 4.55 | 3.18 | 1.94 | 0.63 | 1307.81 |
| 41 | 85 | 37 | 5.56 | 5.20 | 2.78 | **1.23** | 0.76 | 8.74 | 3.47 | 2.49 | 0.52 | 1430.08 |
| 42 | 82 | 39 | 4.70 | 4.95 | 1.93 | **1.01** | 0.41 | 8.82 | 2.65 | 1.71 | 0.58 | 1259.53 |
| 43 | 88 | 40 | 5.33 | 5.19 | 2.93 | **1.71** | 0.76 | 8.08 | 3.43 | 2.25 | 0.54 | 1376.60 |
| 44 | 91 | 40 | 5.11 | 5.19 | 2.38 | **1.32** | 0.59 | 9.29 | 3.49 | 2.33 | 0.63 | 1357.30 |
| 45 | 79 | 39 | 5.98 | 4.38 | 2.86 | **1.60** | 0.63 | 8.67 | 3.95 | 2.51 | 0.60 | 1253.41 |

Therefore, the arrival time is modeled using a non-homogeneous Poisson Process with an arrival rate that varies $\lambda(h)$ at each hour $h = \{1, ..., 24\}$. The arrivals are likely high in the morning and low in the afternoon. The parking times $pr_j$ follow an exponential distribution with a mean parking duration that also varies over time. There is no correlation between the arrival time and the parking time so the two variables can be generated independently [15]. The departure time $d_j$ of each EV can be directly obtained $d_j = r_j + pr_j$. The initial state-of-charge at the arrival $(e_j^0)$ is considered uniformly distributed in the range of [20,70]. The desired state-of-charge of each EV $j$ $(e_j^d)$ is uniformly chosen from $[e_j^0, 100]$. The battery capacities are chosen randomly from the list of current real-world EVs battery capacities [2].

We generate 15 scenarios for each case. In the first 10 scenarios, the chromatic number (k) of the interval graph of the scenario is greater than the number of CPs while it is less or equal in the last five scenarios.

## 5.3   Simulation Results

Due to the stochastic nature of the IGCH and the SA algorithm, 30 independent executions were done for each scenario to obtain statistically significant results. The objective value is calculated for each scenario and we collect the mean, best, the standard deviation (std), and average execution time in seconds of each algorithm. Table 2, Table 3 and Table 4 shows the comparison of results obtained with $m = 10$, $m = 20$ and $m = 40$ respectively. We highlight in bold the best solution found.

About comparison between the model with constant power and the model with variable power, the objective value in the second one was averagely lower by 26.6% by the IGCH and by 20% by the SA than the first model. Thus, charging demands can be satisfied more using the variable power model.

For all scenarios, the best solutions found by the SA and IGCH always outperform the FCFS heuristic. For the scenarios with $m = 10$, the SA algorithm outperforms the IGCH in all scenarios whereas it outperforms the IGCH in 18 scenarios out of 30 with $m = 20$ and in 2 scenarios out of 30 when $m = 40$.

We perform the Mann-Whitney U test [13] to compare the results between the IGCH and the SA for each case. The Mann-Whitney U test is a non-parametric statistical test for determining whether two independent samples were drawn from a population with the same distribution. We compare the p-value to a significance level of 0.05. The p-value found was 0.0001, 0.3661 and 0.0038 for results with $m = 10$, $m = 20$ and $m = 40$, respectively. It suggests that there is no significant difference between the results with $m = 20$ of SA and IGCH algorithms. However, we can conclude that the difference between the SA and IGCH results with $m = 10$ and $m = 40$ are statistically significant. The average time taken by the three methods is always greater in scenarios with $m = 40$ when there are more EV charging demands. Also, the IGCH uses an average of 99.5% less computational time than SA. It is also noted that increment of computation time in the SA algorithm is due to solving multiple LP with CPLEX at each neighbor generation while there are one LP solved for each IGCH execution.

In conclusion, the proposed IGCH is significantly better than the FCFS mechanism and it is preferable to the SA regarding the execution time. Moreover, it can handle large scenario better than the SA.

## 6 Conclusion

In this paper, we addressed the electric vehicle charging scheduling problem (EVCSP) in a charging station with different charging modes to maximize the final state-of-charge of each EV by the departure time. To solve the optimization problem, we designed a heuristic based on interval scheduling and Simulated Annealing (SA) combined with linear programming. Variable power and constant power models were both studied and compared. Different scenarios were presented to evaluate the performance of the proposed algorithms. The results show that the variable power model is better for allocating power. The results also show that the proposed heuristic and SA can achieve an optimal solution and outperform the SA algorithm, and the performance is significantly better than the First Come First Serve (FCFS) mechanism. In this paper, we have assumed that the data on vehicle recharging (arrival time, departure time, state of charge, etc.) are known in advance. This assumption is realistic since many recharging service operators require a reservation of the recharging in advance to avoid queues. However, it is interesting to study the dynamic case of vehicle arrival. In future work, we will consider multi-objective optimization to reduce the charging cost.

## References

1. Connolly, D.T.: An improved annealing scheme for the QAP. Eur. J. Oper. Res. **46**(1), 93–100 (1990)
2. EVDB: Ev database (2020). https://ev-database.org
3. Franco, J.F., Rider, M.J., Romero, R.: An MILP model for the plug-in electric vehicle charging coordination problem in electrical distribution systems. In: 2014 IEEE PES General Meeting—Conference and Exposition, National Harbor, MD, USA, pp. 1–5. IEEE (2014)
4. García-Álvarez, J., González, M.A., Vela, C.R.: Metaheuristics for solving a real-world electric vehicle charging scheduling problem. Appl. Soft Comput. **65**, 292–306 (2018)
5. Gilmore, P.C., Hoffman, A.J.: A characterization of comparability graphs and of interval graphs. Can. J. Math. **16**, 539–548 (1964)
6. IEA: Global EV outlook (2020). https://www.iea.org/reports/global-ev-outlook-2020
7. Kang, Q., Wang, J., Zhou, M., Ammari, A.C.: Centralized charging strategy and scheduling algorithm for electric vehicles under a battery swapping scenario. IEEE Trans. Intell. Transp. Syst. **17**(3), 659–669 (2016)
8. Kirkpatrick, S., Gelatt, C.D., Vecchi, M.P.: Optimization by simulated annealing. Science **220**(4598), 671–680 (1983)
9. Kleinberg, J., Tardos, E.: Algorithm Design. Pearson Education, India (2006)

10. López-Ibáñez, M., Dubois-Lacoste, J., Cáceres, L.P., Birattari, M., Stützle, T.: The irace package: iterated racing for automatic algorithm configuration. Oper. Res. Perspect. **3**, 43–58 (2016)
11. Lundy, M., Mees, A.: Convergence of an annealing algorithm. Math. Program. **34**(1), 111–124 (1986)
12. Luo, L., et al.: Optimal planning of electric vehicle charging stations comprising multi-types of charging facilities. Appl. Energy **226**, 1087–1099 (2018)
13. Mann, H.B., Whitney, D.R.: On a test of whether one of two random variables is stochastically larger than the other. Ann. Math. Stat. 50–60 (1947)
14. Niu, L., Zhang, P., Wang, X.: Hierarchical power control strategy on small-scale electric vehicle fast charging station. J. Cleaner Prod. **199**, 1043–1049 (2018)
15. Pflaum, P., Alamir, M., Lamoudi, M.Y.: Probabilistic energy management strategy for EV charging stations using randomized algorithms. IEEE Trans. Control Syst. Technol. **26**(3), 1099–1106 (2018)
16. Rahman, I., Vasant, P.M., Singh, B.S.M., Abdullah-Al-Wadud, M.: On the performance of accelerated particle swarm optimization for charging plug-in hybrid electric vehicles. Alexandria Eng. J. **55**(1), 419–426 (2016)
17. Rose, D.J., Tarjan, R.E., Lueker, G.S.: Algorithmic aspects of vertex elimination on graphs. SIAM J. Comput. **5**(2), 266–283 (1976)
18. Sassi, O., Oulamara, A.: Electric vehicle scheduling and optimal charging problem: complexity, exact and heuristic approaches. Int. J. Prod. Res. **55**(2), 519–535 (2017)
19. IEC 61851–1: 2017 Standard: Electric vehicle conductive charging system-part 1: general requirements. The International Electrotechnical Commission, Geneva, Switzerland, 292, 7 February 2017
20. Tang, W., Zhang, Y.J.A.: A model predictive control approach for low-complexity electric vehicle charging scheduling: optimality and scalability. IEEE Trans. Power Syst. **32**(2), 1050–1063 (2016)
21. Wu, H., Pang, G.K.H., Choy, K.L., Lam, H.Y.: Dynamic resource allocation for parking lot electric vehicle recharging using heuristic fuzzy particle swarm optimization algorithm. Appl. Soft Comput. **71**, 538–552 (2018)
22. Wu, W., Lin, Y., Liu, R., Li, Y., Zhang, Y., Ma, C.: Online EV charge scheduling based on time-of-use pricing and peak load minimization: properties and efficient algorithms. IEEE Trans. Intell. Transp. Syst.(2020)
23. Yang, S.: Price-responsive early charging control based on data mining for electric vehicle online scheduling. Electric Power Syst. Res. **167**, 113–121 (2019)
24. Yao, L., Lim, W.H., Tsai, T.S.: A real-time charging scheme for demand response in electric vehicle parking station. IEEE Trans. Smart Grid **8**(1), 52–62 (2016)
25. Zhang, L., Li, Y.: Optimal management for parking-lot electric vehicle charging by two-stage approximate dynamic programming. IEEE Trans. Smart Grid **8**(4), 1722–1730 (2015)

# Author Index

Printed in the United States
by Baker & Taylor Publisher Services